The GlobalArctic Handbook

Matthias Finger · Lassi Heininen
Editors

The GlobalArctic Handbook

Springer

Editors
Matthias Finger
Institute of Technology and Public Policy
École Polytechnique Fédérale de Lausanne
Lausanne
Switzerland

Lassi Heininen
Faculty of Social Sciences
University of Lapland
Rovaniemi
Finland

ISBN 978-3-030-06345-0 ISBN 978-3-319-91995-9 (eBook)
https://doi.org/10.1007/978-3-319-91995-9

Softcover re-print of the Hardcover 1st edition 2019

Printed on acid-free paper

This Springer imprint is published by the registered company Springer International Publishing AG part of Springer Nature
The registered company address is: Gewerbestrasse 11, 6330 Cham, Switzerland

Contents

Chapter 1
Introduction

Matthias Finger and Lassi Heininen

Connections between the Arctic and the Earth System, such as the Arctic Haze and the Indian monsoon link, are long established. We have also known for quite a while that the Arctic is warming more than twice as rapidly as the rest of the planet, and that this already does and increasingly will affect the Earth System. However, the Arctic has, at least since the end of the Cold War, also become economically globalized, as a warming Arctic with less ice cover attracts the interest of corporations and other nations, both for its resources and its transport routes. Such globalization of the Arctic can also be identified in its governance, as the Arctic Council—the major regional council and high-level forum for inter-governmental cooperation in the Arctic—has become an important meeting place for both Arctic and non-Arctic states, as well as International and Non-Governmental Organizations, the latter three having the status of observers. In other words, the Arctic has now become global, ecologically, economically, politically and culturally.

With this book, we wish to offer a comprehensive and inter-disciplinary approach to this globalized Arctic, something we have labeled the *GlobalArctic*.[1] By that we mean the Arctic as a geographical region, which has now become part and parcel of the globalized world in all its dimensions: ecological, economic, geo-political, and cultural. Having entered this globalized world, the Arctic is actively contributing to further globalization. However, with this book we also want to offer a structured

[1]The term was coined in a brainstorming meeting of the Thematic Network on Geopolitics and Security in January 2014 in Copenhagen, and officially launched at the 2014 Arctic Circle Assembly. See also our Massive Open Online Course (MOOC) entitled GlobalArctic on the Coursera platform.

M. Finger (✉)
EPFL, Lausanne, Switzerland
e-mail: matthias.finger@epfl.ch

L. Heininen
University of Lapland, Rovaniemi, Finland
e-mail: lassi.heininen@ulapland.fi

M. Finger and L. Heininen (eds.), *The GlobalArctic Handbook*,
https://doi.org/10.1007/978-3-319-91995-9_1

methodology for studying the Arctic in its global context. In particular, we want to analyze the two-way interconnections between global and Arctic processes: global impacts within, and worldwide implications of the Arctic. As such, our book pursues the following three aims:

- First and foremost, we want to help conceptualize the Arctic as a multifaceted region within a changing global context, which is both affected by it and affecting it.
- Secondly, we aim to describe the major drivers of these GlobalArctic dynamics; namely, ecological changes, changes in resources extraction practices and corresponding infrastructure development, including urbanization, as well as changes in geopolitical configurations, and changes in Arctic economies, societies and cultures.
- Thirdly, we aim to define, analyze, and discuss concrete ways to address these changes in the GlobalArctic, including mitigation, adaptation, and resilience-building. The purpose is to offer the relevant GlobalArctic stakeholders innovative approaches, methods, best practices, and solutions to address these unprecedented dynamics. Here the GlobalArctic is a (new) geopolitical context.

While our GlobalArctic approach and method of analyzing the multi-dimensional and multi-functional state of the Arctic and its resources geopolitics is new and unprecedented, there already exists substantial literature on various aspects of the Arctic. This literature can be roughly divided into two categories. On one hand are social science approaches pertaining to geopolitics, security, governance, and institutions, mostly inspired by national policies and intergovernmental cooperation. On the other hand, there is even more literature on the Arctic ecology, especially climate change and ensuing implications on resources accessibility and indigenous peoples' lives and cultures. However, the link between these two types of literature, and especially the link between a warming Arctic and the global, Post-Cold War economy, has hardly been made yet. The scarce literature in this area is often quite normative, arguing either for or against a more or less sustainable development in and of the Arctic. Similarly, there exist quite different views as to what extent Arctic states, companies, and people have a responsibility to mitigate global climate change, and other changes for that matter, or should take advantage of the opportunities offered by a warming and resource-rich Arctic.

Such considerations have led us to gather leading senior experts and young researchers on Arctic matters in order to offer a comprehensive and interdisciplinary approach to and view of the GlobalArctic. Consequently, we have structured our book along the four main dimensions that we believe best capture the dynamics of the GlobalArctic: economics and resources, extraction, Earth system dynamics, institutions and governance, and culture. In each dimension, we use the various contributions to highlight how the global is driving the changes in the Arctic and, on the other hand, how the changes in the Arctic are affecting, or may affect, the globe. Table 1.1 summarizes our conceptualization of and approach to the GlobalArctic.

Table 1.1 The GlobalArctic matrix (see: www.globalarctic.org)

	The global affecting the Arctic	The Arctic affecting the globe
Resources and economics	Pressure to extract oil, gas, minerals, fisheries, and forest resources; development of corresponding infrastructures, including urbanization of the Arctic; opening of the Arctic to shipping and tourism	Significant oil, gas, and mineral reserves for further global industrial development; shortening of sea routes further facilitating global trade
Earth system dynamics	The Arctic as a sink of global pollutants; climate change impacts on ice cover, ecosystems, biota, and livelihood of indigenous peoples	Global climate forcing due to loss of sea ice and permafrost thawing (e.g., methane release); resulting changes in ocean currents and ensuing changing global weather patterns; contribution to sea level rise
Governance and institutions	Arctic regionalism; geo-strategic interests and attempts to cooperate; regional (e.g., Arctic Council) and global (e.g., UNCLOS) governance of the Arctic	The Arctic of the global Cold War; attempts to use Arctic climate and resources governance (e.g., Polar Code, the Arctic Ocean as 'global commons') as a model for global governance; potentially global geo-political conflicts around Arctic resources and trade routes
Peoples and culture	McDonaldization of Arctic indigenous peoples and corresponding loss of culture; threats to their food security and health; migration to and urbanization of the Arctic	Global recognition of Arctic peoples and that of indigenous peoples' rights, resources governance practices, knowledge and expertise

Source Authors

Below is a summary of the chapters in this book, following the above structure. The first section of the book, which contains five chapters, addresses the *economic and resource dynamics* in, and of, the Arctic.

Nikolas Sellheim (Kobe University), *Perceiving dignity, needs and rights: seal hunting communities and international human rights law.* This chapter critically examines the main elements of international human rights law in a seal hunting context. It argues that there is a cultural divide between geographical, seal hunting communities and functional, hunt-opposing communities that is almost impossible to reconcile. Drawing from the presentation of the seal hunt in popular media outlets, this chapter presents different approaches towards the (dis-)approval of certain rights based on specific needs, their definition, and scope.

Markus Kröger (University of Helsinki), *The global land rush and the Arctic.* This chapter discusses the recent Arctic land rush from the viewpoint of the larger literature on land grabbing, little of which has focused on the Arctic so far. The chapter argues that this literature can offer valuable knowledge on two key global/Arctic dimensions. First, looking at the Arctic from a global viewpoint can make a major contribution to our understanding of world politics and political economy of resources. Second, cases undergoing rapid climatic changes elsewhere in the world can offer insights on similar developments in the Arctic.

Andrey Krivorotov (Shtokman Development AG) and Matthias Finger (EPFL), *State-Owned Enterprises in the Arctic.* This chapter looks at the relationship between some of the corporations that extract oil and gas in the Arctic and their respective governments. In particular, it focuses on the two countries (Russia and Norway) whose economies predominantly depend upon the extraction of oil and gas resources in the Arctic, and the relationship that state-owned enterprises Gazprom, Rosneft and Statoil have with those countries. The authors argue that such relationships are key to resource extraction in the Arctic.

Soile Veijola and Hannah Strauss-Mazzullo (University of Lapland), *Arctic tourism at the crossroads of contesting paradigms of development.* This chapter addresses the future Arctic as a set of parallel realities, inhabited and formed by a variety of human and nonhuman factors. Tourism is used in this chapter as a prism that incorporates the key aspects of Arctic development, such as environmental and climate concerns, industrial rationalities, extractable economic resources, indigenous knowledge, gender, and the resulting and competing paradigms of land-use and energy solutions. By showing how these discourses and their underpinning ethics have materialized in tourism development, the chapter opens the way for more inclusive and sustainable development of the Arctic in its local and global context.

Frédéric Lasserre (Université Laval), *Arctic shipping: a contrasted expansion of a largely destinational market.* This chapter analyzes the evolution of commercial Arctic shipping (cargo and cruise) through the Northwest Passage through the Canadian Arctic archipelago, along Greenlandic coasts and the Northern Sea Route along the Russian Northern coast from the Kara Gate to the Bering Strait. It underlines the major differences between Arctic regions: for example, on the Northern Sea Route it is largely destinational (cargo) traffic, whereas cruise tourism markets are expending along Greenland and Svalbard. The chapter concludes that the receding ice may act as an enabler and facilitating factor, but is, of itself, neither a sufficient driver of the development of massive traffic along Arctic seaways nor of single-voyage cost-effectiveness.

The second section of the book contains three chapters and discusses the nature of *climate and other environmental changes* in the Arctic and their global implications.

Hayley Hesseln (University of Saskatchewan), *Boreal forests of the circumpolar world.* This chapter begins by putting the boreal forests, also known as taiga, into a global perspective (such forests account for approximately one-third of the world's total forested area) and by describing the physical resource across the eight Arctic countries. The author then discusses variations in ownership regimes and management and analyzes in detail forest management and governance, country by country. Finally, the chapter examines global forces that have increased the pressure for resource use. It also shows how society has responded to ensure future forest sustainability.

Sanna Kopra (University of Helsinki), *Climate change and China's rise to great power status: implications for the Global Arctic.* This chapter investigates the interplay of two significant changes that have fueled globalization of the Arctic region by examining the emergence of China's great power status and its implications for the Global Arctic. It focuses on China because of its crucial role in international climate politics, given that China argues that it has special interests in the Arctic due to adverse effects of climate change, and that it must have the opportunity to be involved in governance of the region. The chapter argues that because China is the largest emitter of greenhouse gases in the world, the fate of the Arctic—and therefore of the whole world—hinges substantially on China's climate change policies.

Teemu Palosaari (University of Tampere), *The Arctic Paradox (and how to solve it): oil, gas and climate ethics in the Arctic.* This chapter looks at Arctic oil and gas development from the viewpoint of global climate ethics. Indeed, new oil and gas resources become accessible as the sea-ice melts (and as technology advances), which raises new concerns about the so-called 'Arctic Paradox', the development of oil and gas resources in the Arctic arguably being incompatible with the efforts to limit average global warming to two degrees. The chapter then analyzes how climate justice and responsibility are covered in the current Arctic discourse. In doing so, the chapter finally provides an outline of a future research agenda on Arctic climate ethics.

In the third section of our book, which contains four chapters, the authors address issues of Arctic *geopolitics, security, and governance*.

Juha Käpylä and Harri Mikkola (The Finnish Institute of International Affairs), *Contemporary Arctic meets world politics: rethinking Arctic exceptionalism in the age of uncertainty.* This chapter begins by discussing on the Arctic as an exceptional "zone of peace" and a "territory of dialogue". However, the authors also argue that, as the contemporary Arctic is becoming increasingly global, it may face similar geo-economic and politico-strategic dynamics as other regions. This point has been recently highlighted by various spill-over effects of world politics into the region, such as the conflict in Ukraine or the United States' withdrawal from the Paris climate agreement. Finally, the chapter argues that while this analysis is supported by an increase of general uncertainty in the Arctic due to changed foreign policy behavior of Russia, and recent changes in US domestic and foreign policy, exceptionalism based on regional cooperation has shown continuing resilience, even in a difficult international situation.

Valery Konyshev and Alexander Sergunin (St. Petersburg State University), *The Changing role of military power in the Arctic.* This chapter argues that the nature and roles of military power in the Arctic have changed radically since the Cold War era. Instead of being a coercive instrument in a global military confrontation between the two superpowers and systems, military power now has new functions, such as ascertaining coastal states' sovereignty over their exclusive economic zones; protecting the Arctic countries' economic interests in the North; preventing illegal migration and terrorist attacks against critical industrial and infrastructural objects; fulfilling dual-use functions (such as SAR operations, monitoring air and maritime spaces, providing navigation safety, mitigating catastrophes); helping the academic community develop Arctic research; and performing some symbolic functions. The authors argue that, in order to avoid negative security trends, confidence- and security-building measures should be developed in and for the region.

Li Xing and Peng Bo (Alborg University), *The Rise of China in the emergence of a new Arctic order.* This chapter proposes a novel approach to understanding the emergence of an "Arctic order" influenced by Chinese interest, strategy, and policy. It argues that China is struggling to become an active player in the process of forming an Arctic governance structure where the rules of the game have not yet been fully defined. The chapter concludes that, as a major global economic power, China will inevitably become an indispensable and crucial actor in shaping the emergence of an Arctic order that will accommodate some "Chinese characteristics".

Lassi Heininen (University of Lapland), *Arctic Geopolitics with special features—a potential asset to (re)formulate world politics.* This chapter draws a holistic picture of Arctic geopolitics and international relations/IR in the post-Cold War period and analyzes how the high geopolitical stability of the Arctic was achieved and been maintained. It argues that the globalized Arctic based on intensive international, functional cooperation and high stability is an exceptional political space in world politics and international relations. It then discusses the recent transformation of Arctic geopolitics from classical to critical, and examines common interests between the Arctic states with an ultimate aim to transform from confrontation to cooperation, but also relevant features of Arctic geopolitics and security seen as prerequisites for high stability. The chapter concludes by asking if there are special features, based on the common interests and prerequisites that could become new themes of the Global Arctic geopolitics, and whether all this has potential to influence current turbulent world politics.

Finally, section four contains four chapters and focuses on Arctic *peoples, cultures* and their respective approaches to the governance of the GlobalArctic.

Maria Ackren (University of Greenland), *Diplomacy and paradiplomacy in the North Atlantic and the Arctic—a comparative approach.* This chapter focuses on the subnational jurisdictions, all of which are part of the Arctic territory and participate in Arctic affairs at local, national, regional, and international levels. The chapter argues that 'paradiplomacy' has become an increasingly important concept within international relations regarding non-state stakeholders, such as sub-state actors, non-governmental organizations, and other agents forming their own international policies and agendas. It subsequently analyzes international collaboration and (para)diplomacy in the Arctic through four case studies: the Faroe Islands, Greenland, Nunavut, and Svalbard. The chapter concludes as learned lessons that the four subnational units are international actors/stakeholders promoting their own international affairs.

Rasmus Gjedssø Bertelsen (UiT The Arctic University of Norway), *The Arctic as a laboratory of global governance: the case of knowledge-based cooperation and science diplomacy.* This chapter focuses on the role of Arctic triple-helix knowledge-based cooperation and science diplomacy in global governance under power transition. It argues that the Arctic is being deeply affected by post-Cold war struggles over both Russia's international position and power transition with the return of China. Knowledge-based cooperation between public, private, and civil society actors, as well as science diplomacy, is used in the Arctic to handle power transition from Western to Eastern states. The chapter concludes that found general lessons from Arctic science diplomacy can be used for global governance challenges under power transition.

Alexander Sergunin (St. Petersburg State University), *Subnational tier of Arctic governance.* This chapter examines the role that subnational actors play in the Arctic policy-making and governance. Indeed, the subnational actors have been gradually transformed from passive policy-takers to policy-shapers and become active players, not only in their traditional realms but also on all levels of Arctic governance. It then analyzes the role of the subnational actors in the desovereignization and deborderization dynamics in the High North, as well as in the process of institutionalization of subnational actors' international activities. The chapter concludes that paradiplomacy has become part and parcel of globalization process in the Arctic, while the subnational level has become an important and integral part of the regional governance system.

Laura Olsen (University of Lapland), *Sámi people at different levels of decision-making processes in the global Arctic.* This chapter discusses the Sámi people's position at different levels of societal decision-making structures in Finland, Sweden, Norway, and more widely in the Arctic region. The chapter elaborates on the current situation and evaluates the actual abilities of the Sámi people to affect decision-making processes. It concludes that after starting to claim their rights as indigenous peoples, being supported by international indigenous movements and actively developing their movement, the Sámi have built their own political bodies at the national level, been involved in international processes to strengthen indigenous peoples' position in world politics, and developed new forms to include their voice in decision-making processes at the international level.

Chapter 2
Perceiving Dignity, Needs and Rights: Seal Hunting Communities and International Human Rights Standards

Nikolas Sellheim

2.1 Introduction

Most readers of this paper are presumably familiar with the Canadian seal hunt, in one way or another. Furthermore, it seems fair to assume that the images that come to mind are not of a thriving cultural good that has for centuries been a part of eastern Canadian, particularly Newfoundland's, society, but rather that of an horrific, blood-drenched activity. This is not surprising given that the seal hunt, the seal hunters, and the entire industry have been vilified for decades while being linked with barbarism, the lack of necessity, and irresponsibility (Lamson 1980). This image has been elementary in shaping international legal responses to the hunt, for example, by instituting trade bans on seal products, such as in the European Union in 2009 (European Union 2009; Sellheim 2013). Also, the seal hunt has been, albeit for environmental reasons, part of the narrative that led to trade barriers such as the Marine Mammal Protection Act (MMPA) in the United States in 1972 (United States 1972) or the then-European Community's (1983) ban on trade in products stemming from harp seal and hooded seal pups (European Community 1983).

In combination with the successful campaigns by anti-sealing groups, with the adoption of the first trade bans as well as with the adoption of more recent bans, the markets for seal products effectively collapsed due to a lack of demand and associated declining value (Malouf 1986, p. 373; Sellheim 2015a, p. 284–285). However, apart from cause-and-effect relationships between market closures, market collapses and loss of employment (Sellheim 2015b, 486—495), sealers and industry workers feel misrepresented in the international discourse, while many Newfoundlanders consider the international anti-sealing campaigns as an assault on their history, culture, and dignity. Indeed, many consider the global attack on the seal hunt and the industry as a violation of human rights. For example, John Gillett, long-time fisherman and sealer

N. Sellheim (✉)
Kobe University, Kobe, Japan
e-mail: nikolas.sellheim@gmail.com

M. Finger and L. Heininen (eds.), *The GlobalArctic Handbook*,
https://doi.org/10.1007/978-3-319-91995-9_2

in Newfoundland, wrote: "The international human rights law states, 'All human beings are born free and equal in dignity and rights,' and it also states, 'people are entitled to the right to the free and full use and enjoyment of their natural wealth and resources'" (Gillett 2015, p. 149). He further added, "I sometimes think that if I had the money, I'd initiate a class action suit against these people [anti-sealing groups] for the indignity they have done to me, my fellow sealers, and our culture" (Ibid., p. 156). While, at first glance, the argumentation appears emotionally charged, it nevertheless holds significant indicators as to which elements of international (human rights) law are considered crucial within the self-understanding of Newfoundland sealers: the right to self-determination, the right to property, and the right to culture. Moreover, the infringement of human dignity by anti-sealing groups is a core part of self-perception of sealers.

This chapter looks at the seal hunt as a crucial and historical activity in the (sub-)Arctic through a human rights lens. The discussion revolves around how the applicability of human rights is perceived and what the notion of a "need" for the seal hunt entails. While I focus on the Newfoundland seal hunt, I also make excursions further north to the Inuit seal hunts and take a broad swing to discuss normative elements of international human rights exemplified by the contentious seal hunting issue. Therefore, the points raised in this paper are not only applicable in the context of the seal hunt, but can also be applied in the study of other human rights contexts. Furthermore, the significance of this paper lies in the human rights dimension relevant for a functional community, as explained below, that falls through any kind of roster of applicable rights. In other words, non-indigenous seal hunters are a community of people that have been subject to decades-long harassment, marginalization and ridicule. Unlike other groups that take a somewhat subaltern position in the national and international legal system—such as minorities, refugees, women, children or indigenous peoples—non-indigenous seal hunters, like similarly Iceland, Faroese or Japanese whalers, do not have clearly justiciable legal instruments at hand that protect their rights. Instead, rather nonjusticiable elements of international human rights law, such as the right to dignity, can be applicable (Ife 2009).

2.2 Geographical and Functional Communities

The title of this chapter contains the word "communities", which appears to be self-explanatory in the current context. Indeed, the main focus of the present analysis deals with communities that hunt seals. However, in order to further understand what this means, it is necessary to take a closer look at the term "community" since the debate on and controversy surrounding the hunt for seals, whether for commercial or subsistence purposes, makes frequent use of the term. The best example for this is the aforementioned EU ban on trade in seal products, which in its original implementing regulation (European Union 2010) refers to "other indigenous communities" in Article 2, as well as to "community" or "communities" on several occasions in Article 3; both outline the exemption to a trade ban in relation to those products being produced by indigenous peoples.

Without a clearer definition of what the term "community" refers to, one can assume that, in this context, it describes small assembly of residential facilities of predominantly indigenous inhabitants. This means that the community in question has specific physical and environmental conditions and that its main socio-economic characteristics are shaped by indigeneity and associated livelihoods, such as seal hunting. In other words, the community in question is closely linked to a specific locality and has an assumably long history of socio-cultural and economic development. It is inherently "local" and does not imply the voluntary conglomeration of community members based on political views, special interests, or other non-physical features.

The latter groups of "communities" are commonly referred to as "functional communities" and encompass non-local and often virtual gatherings of people that follow a specific goal or are part of a specific interest group. In order to differentiate between geographical and functional communities, the term "local" has often been inserted to define the geographical and locality-bound community, particularly in environmental and human rights contexts [e.g. CBD 1992, art. 8 (j)].

The debate surrounding the seal hunt brings together both types of "communities". First and foremost, "seal hunting communities" refer to small geographical communities in Newfoundland and in the Arctic, inhabited by non-indigenous and indigenous people whose livelihoods and community practices involve close interaction with the natural environment, including hunting for the endemic fauna as well as fisheries, which has spawned a deeply-rooted feeling of belonging and is part of the communal identity (Sellheim 2015c; Pelly 2001). It is, in particular, the members of these local communities who stand at the center of the sealing debate and whether or not the seal hunt is part of their needs. The question of what constitutes a "need" for a local community cannot be answered easily and I have argued elsewhere that the definition of "need" is blurry and non-linear (Sellheim 2015d, p. 34). I will pick this topic up again later in this paper.

This notwithstanding, functional communities constitute the other, if not larger, player in the sealing controversy. While I referred to the "non-indigenous seal hunting community" above, a geographical community denotes a *group of people* engaged in the same activity—that is, hunting seals as part of their livelihood, which is limited to specific localities. Given the small size of both the functional and geographical sealing communities, their impact on the global discourse surrounding seal hunting is limited. Much more impact is exerted by the opposing functional communities, which can easily take aim at the claim of local communities that seal hunting is part of their communal needs. These functional communities in this context are defined by their opposition to the seal hunt based on their perceived moral superiority due to ascribing fundamental rights to animals in general, but in particular to seals.

2.3 The Role of Knowledge: Determining Needs, Defining "Human" and (Dis)Approving Rights

The elevation of animals to a sacrosanct level that provides them with a right to life is deeply rooted in Western thought (Robertson 2015) and widely considered to be based on the 1975 book *Animal Liberation* by Australian philosopher Peter Singer (Singer 1975). Organizations and functional communities that strongly oppose seal hunting have largely adopted the perspective of animal rights, inevitably leading to a clash of philosophical approaches in the relationship between humans and animals. In other words, the understanding of a specific way by which humans are to treat animals is considered universal and any violation of this understanding is disapproved of. As is the case with human rights, the claim of universality is highly contested (Angle 2002; Pereira 1997). Especially in Inuit contexts, the animal rights movement has been perceived as a neo-colonial power that aims to force specific worldviews onto indigenous peoples (Wenzel 1991; Lynge 1991).

A problematic aspect in linking the role of human beings as utilizers of animals and their respective human rights with an animal rights discourse is the way in which human rights are philosophically constructed and animal rights justified. Landman argued that human rights lack a clear philosophical foundation and that "[a]ppeals to God, nature and reason have all in their own ways failed to provide the definitive source for human rights" (Landsman 2009, p. 21). On the other hand, Ife viewed the philosophical foundation of human rights in the tradition of the Enlightenment period and secular humanism. Thus, human rights are an "attempt to derive moral and ethical behaviour not from a position of divine revelation, but from human reason" (Ife 2009, p. 81). The latter approach is adopted in the present chapter and, in light of this view, animal rights and in particular the "rights of the seal" are considered.

The British philosopher Jeremy Bentham (1748—1832) is often described as the father of the idea of "Speciesism", a concept that rejects the idea of different rights for different species; specifically, more rights for humans than animals (Robertson 2015, p. 21). This view implies that no animal (vertebrate or invertebrate) is to be killed for the benefit of humans, irrespective of whether or not there was a "need" for it to be killed, and that all animals, including the human animal, are equals. Therefore, the adoption of this philosophical view rejects all animal utilization by humans and, in light of its radicalism, shall not be further discussed here.

Of greater relevance for the purposes of the present paper is the less philosophical and more emotional perception of the seal and the seal hunt. Here, the creation and dissemination of knowledge on the hunt, the hunters and the geographical communities is crucial. Of course, "knowledge" is a relative term and the emotionality of knowledge on the seal hunt that shapes judgment and inevitably (dis)approves of the rights of seal hunters is elementary. As argued elsewhere, groups opposed to the seal hunt have created key imagery and proclaimed knowledge on the subject. Cynthia Lamson has shown how anti-sealing protests strategically disseminate information on the seal hunt in order for public opinion to support their cause (Lamson 1980). In order to further discuss the implications on the sealing communities' rights, it is necessary to take a brief look at the presentation of the seal hunt in popular media outlets.

2.3.1 The Seal Hunt in European Newspaper Sources

Popular social media outlets such as *Facebook* or *Twitter* provide important pathways for seal hunting opponents to express their frustration over the seal hunt by issuing threats and insults. In the interest of scope, I have generally disregarded these. Instead, it is worthwhile screening the online versions of well-established and well-perceived newspapers with regard to the way the seal hunt is covered, while leaving aside the respective political bias. It appears especially important to consider European newspaper sources given the recently adopted ban on trade in seal products in the EU. Continuous screening of the publicly available information pertaining to the seal hunt and the sealing industry has occurred by subscribing to Google Alerts with the keywords "seal hunt", "seal ban" and "EU Arctic" since 2011. YouTube videos by anti-sealing organizations were also taken into consideration. Moreover, I conducted strategic searches for seal hunt-related articles with the keyword "seal hunt"[1] in the search engines of the following newspapers from EU Member States:

- Austria: *Wiener Zeitung* (http://www.wienerzeitung.at/)
- Belgium: *De Standaard* (http://www.standaard.be/)
- Denmark: *The Copenhagen Post* (http://cphpost.dk/), *Dagblad Information* (http://www.information.dk/)
- Finland: *Helsingin Sanomat* (http://www.hs.fi/)
- France: *Le Monde* (http://www.lemonde.fr/)
- Germany: *Frankfurter Allgemeine Zeitung* (http://www.faz.net), *Süddeutsche Zeitung* (http://www.sueddeutsche.de/), *Der Spiegel/Spiegel Online* (http://www.spiegel.de/)
- Italy: *La Repubblica* (http://www.repubblica.it/)
- Netherlands: *De Telegraaf* (http://www.telegraaf.nl/)
- Spain: *El País* (http://elpais.com/)
- Sweden: *Svenska Dagbladet* (http://www.svd.se/)
- United Kingdom: *The Guardian* (http://www.theguardian.com/).

Except for the *Wiener Zeitung*, which did not hold any articles on the seal hunt, the seal hunt is a recurring theme in the above newspapers, although they all began more frequent coverage of the issue in 2006. This may be because 2006 marked the beginning of the legislative process in the European Union (Sellheim 2013)—although very few articles directly mentioned this event— and the visit to the ice floes of the Gulf of St. Lawrence by Paul and Heather McCartney, which *The Guardian* in particular covered. In all of the abovementioned newspapers, the coverage of the seal hunt peaked during the last stages of the adoption process of the basic regulation and around May 2009, which was the date of the final vote in the European Parliament.

Out of approximately 100 articles, only one (Calonego 2011) deals with the socio-economic impacts of the dwindling markets for seal products on sealing communities in Newfoundland. Two recurring themes within the coverage of the seal hunt shape

[1] German: Robbenjagd; Finnish: Hylkeenpyynti; French: Chasse aux phoques; Dutch: Zeehondenjacht; Danish: Sæljagt; Swedish: Säljakt; Italian: Caccia alle foche; Spanish: Caza de focas;

the narrative of a fragile seal population: the increase of the TAC by the Canadian government, and the impacts of climate change on sea ice on which the seals whelp. This narrative is picked up by anti-seal hunt organizations such as the *Humane Society of the United States* (HSUS), which in a YouTube video suggests: "This is their home and it's one of the most beautiful places on the planet. But climate change is causing their habitat to literally melt from under them. And the seals that are lucky enough to survive the impacts of climate change are going to face the sealers" (HSUS 2013). As shown by the *International Union for Conservation of Nature* (IUCN), climate change has had no measurable effect on seal populations. While there is undoubtedly the potential for adverse effects in the future, no such effect on the seals subject to the commercial seal hunt—the harp seal—can currently be measured (Kovacs and IUCN SSC Pinniped Specialist Group 2008).

At the same time, the perceived cruelty of the seal hunt is especially focused upon in the German and English newspapers, with words such as "slaughter", "bloody" and "ravage" commonly used. Supported by imagery that suggests a high degree of cruelty—red blood on white ice—and interpretations by anti-seal-hunt organizations regarding the hunting methods leave no doubt about the perceived inhumanness of the Canadian seal hunt. For example, Ralf Sonntag of the German branch of the *IFAW* in an interview with the German private channel *VOX* said: "In dieser Schnelle können sie nicht sicherstellen, dass die Tiere tot sind und darum werden sehr häufig die Tiere lebendig abgehäutet ["Due to this speed they [the hunters] cannot make sure that the animals are dead and that's why very often the animals are skinned alive"; author's translation] (VOX 2014). Reference to scientific studies that deal with the humaneness of the seal hunt is absent in the media coverage, although the absence of a uniform interpretation of the applied killing methods makes it possible to reach definite conclusions about humaneness or inhumanness (EFSA 2007, p. 94). This aside, opposition to the seal hunt and therefore moral concerns over the killing methods are rooted in the belief in scientific fact, as the seal hunt is presented in media sources: "expert" voices speaking of cruelty and climate change, underlined by graphic imagery, inevitably raise doubts over the "necessity" of the hunt for the people involved. To the contrary, it seems even more unnecessary to pursue a hunt that is both harmful to the environment and the animal, while being globally opposed and economically seemingly unfeasible. This makes opposition significantly different to religiously motivated calls to end trade in certain products, which do not call for scientific proof or ask for a degree of necessity. However, public knowledge on the seal hunt, on animal welfare factors, and on the conditions in which the seal hunt occurs is sparse; Perišin noted that "EU citizens do think that their views in this regard are based on facts and scientific evidence, which is arguably untrue" (Perišin 2013, p. 396).

The Danish newspaper *Dagblad Information* is the only newspaper that deals regularly with the cultural dimension of the seal hunt and seal products in general, albeit in a Greenlandic context, given the island's self-rule status in the Danish realm. In other words, the seal hunt is culturally located within an Inuit/indigenous context. On the other hand, *Svenska Dagbladet* makes continuous reference to the seal hunt conducted in the Baltic Sea where hunts occur for the protection of fisheries.

2.3.2 *The Seal Hunt in Canadian and US Newspaper Sources*

Unsurprisingly, Canadian news sources cover the seal hunt differently. For example, one of the major Canadian news channels, *CBC News* (http://www.cbc.ca/), approaches the seal hunt from different angles. While taking into account the socio-economic contexts in which the hunt occurs, CBC also makes reference to the environmental impacts that a rising seal population could have on fish stocks and the overall ecosystem. The voices of the sealers, the sealing industry and other stakeholders find common representation at the same time as those opposing the hunt. In general, *CBC News* appears to approach the seal hunt as an open debate on whether or not the hunt should cease; its fairly balanced coverage provides a broad and informed context of different opinions and facets.

In the *New York Times* (http://www.nytimes.com/), the Canadian seal hunt is covered as an obsolete industry, while claims of cruelty and lack of necessity are substantiated by representatives of seal hunt protesters, primarily the public voices of *HSUS* and *IFAW*: Rebecca Aldworth and Sheryl Fink, respectively. This being said, an article dated April 5, 2004, constitutes an open and diversified representation of the hunt, its socio-economic dimension, and the position of hunters on the issue (Krauss 2004). Notwithstanding, while not overly present in the newspaper, the seal hunt is primarily presented under a negative pretext.

2.3.3 *Human Dignity*

The above examples illustrate that specific imagery of the seal hunt inevitably leads to the creation of imagery of the sealers themselves. While this is not a violable right, as such, it can be argued that the right to dignity constitutes a foundational element of international human rights law. For example, the preamble to the Universal Declaration of Human Rights (UDHR) reads: "Whereas recognition of the inherent dignity and of the equal and inalienable rights of all members of the human family is the foundation of freedom, justice and peace in the world." Therefore, especially since the Second World War, the concept of human dignity has become central to the major human rights treaties, such as the 1966 Covenants, that in their preambles echo the dignity clause of the UDHR.

Internationally, however, there are differences in how the concept of human dignity is applied, based on cultural differences and understandings of dignity. For instance, Ife highlighted that the way in which women are shown dignity and respect differs significantly between cultures, making it difficult to identify a generalized, universal way of doing this. In other words, the right to dignity does not imply "the right to be offered a seat on a bus by a younger person" (Ife 2009, p. 100) but rather constitutes a bridge between the relativist and monist stances concerning moral diversity, which Parekh refers to as "minimal universalism" (Parekh 1999). Contrary to the opposing views of relativism and monism—one celebrating moral diversity based on culture, and the other celebrating universal morality with little regard for cultural elements—"minimal universalism" seeks to find the lowest common denominator. Consequently, the rather broad right to human dignity does not exclude cultural differences (taking the relativist approach into account), but instead creates a universal moral basis for human interaction (the monist approach). In much more general terms, Hart defined the *teleos* (the specific end) of natural law as survival, which creates different moral avenues of how to achieve this goal (Hart 1961). Relating more specifically to human dignity, McCrudden noted that there is indeed a lowest common denominator for the understanding of what the right to human dignity implies, based on three criteria:

> The first is that every human being possesses an intrinsic worth, merely by being human. The second is that this intrinsic worth should be recognised and respected by others, and some forms of treatment by others are inconsistent with, or required by, respect for this intrinsic worth. [...]. The human rights texts have gone further and supplemented the relational element of the minimum core by supplying a third element regarding the relationship between the state and the individual. This is the claim that recognising the intrinsic worth of the individual requires that the state should be seen to exist for the sake of the individual human being, and not vice versa (the limited-state claim). (McCrudden 2008, p. 679)

Addressing how the right to human dignity and the breach thereof can be determined would be an arduous task that would require significant social and cultural research. However, for the purposes of this paper, it appears adequate to approach the issue through the depiction of the seal hunt and sealers in online media outlets as they constitute an important source of information for the public and decision-makers. However, this merely provides a snapshot on larger issues regarding online presence and human dignity (see also Balleste 2011). It is interesting that, in a recent judgment, the European Court of Human Rights (ECtHR) has ruled that websites that do not limit the use of insulting and threatening comments to individuals by their users can face legal proceedings and are not protected under the freedom of expression clause, as manifested in article 10 of the European Convention on Human Rights (*Delfi AS v Estonia* 2015).

Coverage of the seal hunt, with the exception of *CBC News*, does not appear to be balanced in terms of taking different perspectives into account. Instead, the seals are the focus of attention and this frames the applied narratives in screened sources. Consequently, the killing of seals for personal profit, in combination with declining markets, appears to be cruel and unwarranted. Therefore, it seems that media outlets primarily center the seal hunting debate around the charismatic seal pup—often referred to as "baby seal". It is unsurprising that a more balanced coverage is absent, given that "the voices of those opposed to the vested interests of media corporations and their clients are likely not to be heard" (DeLuca 1999, p. 88). The clients in this context seem to be functional communities; that is, non-governmental organizations (NGOs) that make use of anti-seal hunting rhetoric.

NGOs have developed as an important means of uncovering, monitoring, and alleviating human rights violations, as well as environmental degradation caused by states and companies. To this end, the Canadian seal hunt also came under close scrutiny when overhunting led to drastic population decline and concerns over animal welfare rose to the surface. Barry showed that *IFAW* was in fact founded to counter the Canadian seal hunt (Barry 2005). Through concerted and targeted efforts, information campaigns and other lobbying efforts, especially in Europe, the Canadian seal hunt has come under discursive, political, and ultimately legal attack. Also, during the legislative process leading to the EU's seal products trade ban, the notion of "barbarism", underlined by footage and information prepared by anti-seal-hunting groups, took hold (Wegge 2013). It can even be argued that no efforts have been made to separate the hunters as human beings from the hunt and to eradicate the stigma of being a "barbarian" when hunting seals.

The breach of the right to human dignity becomes obvious in this regard. Creating and maintaining the stigma of "barbarian" for a specific group of people that, after all, engage in a legal, well-regulated and strictly monitored activity, must be seen in the context of an infringement of human dignity. While NGOs are not bound to international human rights instruments—after all, it is states that ratify and implement them—some do follow codes of ethics or conduct. For example, the World Wildlife Fund for Nature (WWF) has developed a Code of Ethics that, for example, aims to "involve local communities and indigenous peoples in the planning and execution of our field programmes, and we will respect their cultural and economic needs" (WWF 2009). While the WWF does not oppose the Canadian seal hunt due to the stable seal population, the respect for human dignity is an inherent part of the organization's activities.

At the time of writing, no code of ethics can be found for three of the most prominent organizations opposing the seal hunt: *HSI*, *IFAW*, and *PETA* (People for the Ethical Treatment of Animals). This is not to say that that they do not exist internally, but they are not publicly accessible if they do exist. A conglomerate of NGOs, the World Association of non-Governmental Organizations (WANGO), developed a *Code of Ethics & Conduct for NGO*s in 2004, whose "standards are applicable regardless of an NGO'S focus" (WANGO 2004, p. 7). Section C of the Code makes exclusive reference to "Human rights and dignity" and cites the UDHR dignity clauses. To this end, "[a]n NGO should recognise that all people are born free and equal in dignity; An NGO should be sensitive to the moral values, religion, customs, traditions, and culture of the communities they serve" and "[a]n NGO should respect the integrity of families and support family-based life" (Ibid., p. 9—10). As shown elsewhere, family-based life is a fundamental element in the commercial seal hunt (Gillett 2015; Sellheim 2015c). Vilifying sealers jeopardizes the integrity of the sealing family and many consider the presentation of Newfoundlanders, and especially sealers, in public discourse to be an attack on their dignity. By following a path that does not counter trends of vilification of sealers and their communities, hunt-opposing NGOs consequently violate important elements of ethical conduct as stipulated in WANGO'S Code. Thereby, since the human rights clause on protecting human dignity is reflected in the Code, a violation of this principle occurs.

Similarly, but further developed, are the ethical codes applied in journalism. However, problematic in this regard is that ethical codes for journalists are predominantly confined to their work within a nation state and no global guidelines exist for applied journalistic ethics (Ward 2013, p. 2). Therefore, it is necessary to make use of one national example: the German *Pressekodex*, the ethical guidelines for journalists, makes several references to the protection of human dignity, the protection of personality and the protection of honor (available at: http://www.presserat.de/pressekodex/pressekodex/). The protection of an international human rights norm is consequently an integral part of German journalistic ethics. In the Preamble, the *Pressekodex* further states that journalists are to exercise their profession "fair, following the best of knowledge and judgement, without influence of their personal interests and motives outside the subject area" (Original: "Sie nehmen ihre publizistische Aufgabe fair, nach bestem Wissen und Gewissen, unbeeinflusst von persönlichen Interessen und sachfremden Beweggründen wahr."). Furthermore, research is considered an elementary part of German journalism. Therefore, it would seem fair to expect, based on the *Pressekodex*, that German news sources take a more balanced view on the issue of seal hunting. However, as shown above, the media outlets that were screened for this paper by and large followed the narratives created and disseminated by groups opposed to the seal hunt. A lack of investigation and an ultimately colored approach weakens the incorporation of the respect of the right to human dignity and, consequently, the integrity of the German journalistic ethos. After all, it can be argued, by *not* de-vilifying sealers and by *not* opposing discursive attacks on human dignity, that German media sources make themselves complicit in violating international human rights standards.

2.3.4 Defining "Needs"

A common narrative in local seal hunting communities is that the seal hunt is part of the history and identity of the population. The case of Newfoundland in particular shows that the seal hunt has been an integral part of the island's socio-economic development since humans first set foot on it. In current times, the seal hunt still plays an integral role in the life and maintenance of a local community (Sellheim 2015c).

Linking international human rights standards with specific geographical communities does bear difficulties with regard to the definition of "need" for these communities. In particular, involving functional communities that have different understandings of what these needs are challenges specific rights and calls for a more normative discussion on what comprises these needs, who defines them, and whose needs they are. By and large, as shown in the *Introduction*, one might argue that the right to self-determination is enshrined in common article 1 of the International Covenant for Economic, Social and Cultural Rights (ICESCR) and the International Covenant for Civil and Political Rights (ICCPR) 1966. The article reads:

1. All peoples have the right of self-determination. By virtue of that right they freely determine their political status and freely pursue their economic, social and cultural development.
2. All peoples may, for their own ends, freely dispose of their natural wealth and resources without prejudice to any obligations arising out of international economic co-operation,

based upon the principle of mutual benefit, and international law. In no case may a people be deprived of its own means of subsistence.

3. The States Parties to the present Covenant, including those having responsibility for the administration of Non-Self-Governing and Trust Territories, shall promote the realization of the right of self-determination, and shall respect that right, in conformity with the provisions of the Charter of the United Nations.

A vast body of literature exists that deals with the right to self-determination, how far this right stretches, and who the holder of this right is. Given the breadth of the literature, it shall not be revisited here, suffice it to say that common article 1 implies a self-definition of needs and the right of a people to use its natural resources in its own ways. However, does this also mean, as Gillett in the *Introduction* implies, that all geographical communities have the right to define their own needs? In other words, to what degree are geographical communities capable to actively engage in the process of their own socio-economic development?

As a general rule, and also constituting one of its biggest criticisms, human rights are implemented by the nation state, which is responsible for the way these rights are protected. This notwithstanding, public international law does not hold a clear definition of what constitutes a "people"—one of the most crucial question surrounding the rights of indigenous peoples under international law (e.g. Murray 2000). Consequently, the assertion that common article 1 also applies for seal hunting communities cannot be sufficiently supported.

Along with this difficulty is the definition of what a "need" actually implies from a human rights perspective or, to put it in Galtung's words, what "generation" of human rights the need belongs to: is it a civil or political need (first generation); an economic, social or cultural need (second generation); or a collective need (third generation) (Galtung 1994)? And this begs the question of whether the needs in question are collective or individual. A common dilemma is that specific needs of geographical or functional communities may impinge upon the needs of others. For example, the utilization of a river by a major oil company as a necessary element of oil production (for instance, in the Alberta tar sands and the waters of the Athabaskan River) may violate the need of local communities for clean fishing water. Or, more specifically in the seal hunting context, one could ask, in a somewhat provocative manner: does the right to culture (and therewith to hunt seals) of Inuit negatively affect the right to culture (and hunt seals) of non-indigenous Newfoundlanders? My own fieldwork has shown that seal hunters in Newfoundland wonder why, given the interwoven pathways of the seal trade and the similar means by which seals are hunted in Inuit and Newfoundland sealing communities, Inuit seal hunting is considered acceptable and Newfoundland seal hunting is not (Sellheim 2014). Indeed, also Inuit representatives argue that the Inuit and non-Inuit seal hunts cannot be separated and that it appears absurd to consider Inuit seal hunts as merely a subsistence activity. The most recent popular depiction of this interaction is the documentary *Angry Inuk* by Inuit filmmaker Arnaquq-Baril (2016).

At the same time, specific rights under international human rights law, such as the right to property, could potentially mean understanding the right as giving access to unlimited property at the expense of others. In other words, the existence of the right to property does not mean the right to property at all costs. This, it can be argued, is most crucial in the seal hunting debate and constitutes a major element in the opposition of many against the Newfoundland seal hunt: despite the negative economic effects of reduced seal hunts, this potential property deprivation is justified in light of the positive impacts it has on animal welfare. Or, to turn the argument around: seal hunts are not justified as a means for maintaining property in light of the adverse effects on animal welfare. The seal hunt, it is argued, is not justifiable given the cruelty it implies. Whether or not this is the case is subject to debate (e.g. Sellheim 2013, fn 14).

This notwithstanding, the question emerges of whether there is a principal right for Newfoundland local communities to hunt seals in light of the claim that there is a "need for seal hunting" in these communities? The simple answer to this question is that there is. This stems from the fact that the seal hunt is fully legal and highly regulated under Canadian law. From a human rights perspective, and taking into account common article 1 of the ICESCR and ICCPR, the Canadian government considers the seal, despite international opposition, as part of its "natural wealth and resources", the disposal of which falls under its rights as the Canadian people.

More challenging is the normative dimension this question holds, especially in combination with functional communities that oppose the seal hunt. This is a very specific question that focuses on the hunting of seals as a particular and obviously divisive element within the human species. Generally speaking, by denying sealing communities the ability to hunt seals, sealing opposing functional communities contribute to a discourse of denying local communities the right to define their own needs. Thus, they contribute to a colonialist narrative of judging from the outside and evaluating local activities without paying due regard to what local community members have to say. This approach stands in contrast to the postmodern view on human rights, an approach that Ife calls "human rights from below," but reiterates the top-down view on the rights of communities.

Furthermore, the normative denial of rights in light of the welfare of seals and the associated disadvantages for local communities can be seen in a context of utilitarianism or within the discourse of ends justifying the means. Here, Bentham and Immanuel Kant rise to the fore as two opposing philosophers on the rights of human beings. In the Benthamite tradition, specific rights can be sacrificed for the "greater good": if society overall can be improved, it is justifiable to violate the rights of a few who stand in the way of this improvement. More specifically, in this tradition and in line with many functional communities opposing the seal hunt, it is justifiable to contribute to the deprivation of property of sealers in order to improve the welfare of seals, thus contributing to the moral growth of a society—the greater good. At the same time, the image of specific rights as not applying to sealers is being created. As shown above, narratives of barbarism and cruelty accompany the common depiction of the Newfoundland seal hunt. I would even argue that the seal is elevated to an almost divine level, a species with a right to life, similar to whales (D'Amato and Chopra 1991), supported by the innocence of the harp seal pup and the pristine environment in which it dwells. The image of the sealer as the club-swinging enemy of the seals constitutes the antithesis of innocence and ultimately, and those who engage in the seal hunt are easily labeled barbarians or champions of cruelty. This label holds a dehumanizing stigma, allowing for the conclusion that specific rights do not apply to them. This depiction feeds into the much wider political and philosophical discussion on who counts as human (with associated rights) and who does not, a ‘debate’ which reached its sad peak during the Third Reich. Clearly, the discourse appears to imply that sealers do not, stripping them of their rights to define their own needs, all in the name of the “greater good.”

The Kantian tradition takes a different approach: the end does not justify the means, but the human being is an end in him/herself. The Kantian “categorical imperative” can be considered the lowest common denominator between different sets of ethical and moral conduct. Along with this argument goes Maslow’s hierarchy of needs (Maslow 1954), based on which it appears reasonable to derive specific rights. According to Maslow, the most fundamental needs are those without which the physical existence of a human being is not possible; food and water belong into this category. This corresponds to Hart’s *teleos* of survival. Ultimately, one might assume that there is a basic and therefore universal right to food and water. The fact that this is clearly not the case hints towards the inherently positivist notions of “needs” and “rights” (Ife 2009, p. 132). Also, the dilemma of need identification in the sense of the definition of how important a specific activity or cultural trait is for a specific geographical region is not remedied, leaving the dichotomous relationship between contextualized and universal needs unresolved.

So, even with different approaches towards what does and does not constitute a “need”, ideological, political, and cultural perceptions regarding the necessity of seal hunting continue to exist. If local seal hunting communities claim to have a need for the seal hunt, their perception of what this need implies is almost certainly different to functional communities’ stance on the same issue. I argue that the former locate their definition of “need” in a more holistic environment that takes the socio-economic and cultural life of a geographical community into consideration. These elements are not necessarily quantifiable, however, and are instead drawn from everyday experiences, shaped by the normative perception of the sea and the seal, and located within a wider discourse on life in remote coastal communities (Sellheim 2015c). Other communities, such as organizations that oppose the seal hunt, or political bodies, take a different approach and consider the hunt primarily through an economic lens. For example, the European Union in its 2006 Declaration on seal hunting noted that “sealers receive less than 5% of their income from sealing, which provides only a few days’ work each year” (European Union 2006, para. C), pointing to the lack of *economic* necessity. The IFAW went further, noting that “Canada’s commercial seal hunt costs more to support than it earns” (IFAW undated a). In both instances, therefore, a seemingly low economic yield does not justify the continuation of the hunt. Little consideration is given to the larger contexts of “needs” that take the above elements into consideration.

The situation is somewhat different when considering the “needs” of indigenous seal hunters, which are by and large discursively excluded from any opposition towards the seal hunt. In its trade ban, adopted in 2009, the European Union inserted an exemption for indigenous (Inuit) communities. As the preparatory process of the ban shows, at least politically and legally (although not empirically), the wellbeing of local Inuit sealing communities was considered holistically, meaning that the deliberations surrounding Inuit sealing incorporated cultural, social, and economic aspects. Whether or not these considerations corresponded to empirical facts or are just imagination is subject to debate (Sellheim 2016). Also, the IFAW "does not oppose the killing of seals for food, clothing and other products for local use by indigenous peoples" (IFAW undated b). The consideration of the wellbeing of Inuit, in terms of the degree to which they are able to hunt seals, indicates a difference in need-recognition between indigenous and non-indigenous local communities. In light of the modernized hunts in Arctic Canada and Greenland, which use essentially the same hunting methods, local as well as global distribution of indigenous and non-indigenous seal products, it seems outdated to make this distinction. It should be noted that, in the whaling context, indigenous communities are also exempted from the moratorium on commercial whaling that has been in force since the 1985/86 whaling season. However, in order to reach a quota set by the International Whaling Commission (IWC), indigenous coastal whaling communities must submit a "Needs statement" that "details the cultural, subsistence and nutritional aspects of the hunt, products and distribution" (IWC 2017, no pagination). While the terms sound indeed implementable and are furthermore considered and scientifically supported by the IWC's Scientific Committee, the realization of this exemption is difficult to achieve (Brand 2009).

2.4 Who "Needs" the "Right" to Hunt Seals?

The question that remains throughout our discussion—a question that is crucial from a local perspective—is whether there is a normative link between the existence of a "need" and a legal right: does a "need" imply an "ought"? To answer this question, we must decipher the particular element that gives the "need" its power: *why* is the seal hunt needed? For what end is it needed, and what justifies that end? Seal hunters themselves often refer to their tradition of seal hunting and to the notion of "identity." This implies that the seal hunt is a means to an end—the maintenance of history and identity. I would argue that this is a rather weak claim that is difficult to uphold in a normative context. This observation stems from the fact that identities can and do change and are comprised of a multitude of elements. The conclusion would be that coastal communities would disintegrate and lose their identity if the seal hunt was to end for good. In light of the normative role the sea plays in a community, the giving up of sealing would not be the cause of community dissolution, but would certainly contribute to it, as I have argued elsewhere (Sellheim 2015c).

This being said, the divide between perceptions again becomes a crucial part of the sealing controversy and the (non-)existence of specific rights. This is because, again, the level of relevance of sealing for a coastal community must be determined in order to derive the "ought" dimension from it. With it comes the definition of the relevance of the end the continuance of the seal hunt serves. Therefore, inside perceptions from the geographical community and outside perceptions from the functional community are likely to collide, with no way to bridge the differences. Indeed, there is hardly any *dialogue* between different understandings of "needs" and "rights" regarding seal hunting. Instead, those who argue *for* a needs-based right to hunt seals and those who argue *against* the needs-based right seek to dismantle and to weaken the other side's argument. By presenting either position as being emotional, greedy, or unscientific, the intention is to win a battle over the other and to substantiate the respective claim to a specific right (or lack thereof). Seal hunters claim to *need* the seal hunt, and thus claim specific human rights for themselves (there ought to be a right to hunt seals). At the same time, they consider and present opponents of seal hunting as being in breach of human rights by not granting seal hunters the right to make use of the abundant resources. Moreover, opponents of seal hunting are considered to oppose the seal hunt for financial reasons without any regard for scientific findings; that is, the overabundance of seals and the "humanness" of the hunting methods.

Similarly, seal hunters are presented as *not* needing the seal hunt and merely doing it for economic reasons without consideration for animal welfare or the environment. However, in light of the small economic yield the seal hunt generates, the need to hunt seals, and therefore the associated right to do so, are not justified, especially since animal welfare shortcomings appear to be a prevalent feature of the seal hunt. In other words, the end of historical continuity and maintenance of identity does not justify the means of maintaining the seal hunt.

These two views constitute a debate rather than a dialogue in which both sides could learn from each other. Particularly relevant in this context are, what Yankelovich called, "cultural fault lines" that span the adversarial seal hunting communities (geographical and functional). Particularly the "I–It" versus "I-Thou" dichotomy, which refers to the inability to communicate on equal footing, appears to be a common feature in the controversy surrounding seal hunting (Yankelovich 1999, pp. 149–152). Of course, the question remains as to how far dialogue is actually aspired towards. After all, sealing opponents wish that the hunt will end for good, while sealing proponents wish for it to continue. Therefore, an accompanying goal for both sides is the defeat of the other and the opposing opinion.

2.5 Conclusion

For decades, the Canadian seal hunt has been subject of heated debates and political and legal responses. This paper has shown that the perception of the relevance of the seal hunt differs significantly between geographical and functional communities. Taking selected aspects of international human rights law into account, I have argued that the divide between what constitutes a "need" for an activity and the "right" to pursue this activity—in this case, the seal hunt—is almost impossible to bridge. This is particularly the case when the champions of different opinions on an issue are not interested in a fruitful dialogue, but continuously engage in an inherently adversarial debate that aims to defeat the other.

By using examples from popular media outlets in the European Union, in Canada, and the United States, I have argued that the interests of geographical communities and the associated definition of "human rights from below" is largely absent in the discussion surrounding seal hunting. Instead, the created narratives contribute to a depiction of seal hunters as not being holders of specific rights. Even if they were, these rights appear justifiably violated in light of the achievement of the greater good: the welfare of seals.

Of course, the dimension of what constitutes a specific right and who the holders of this right are is a normative one and reflects upon the cultural charging of those defining it. A one-sided and somewhat simplistic approach towards geographical communities, their knowledge, and cultural development easily leads to their simplistic depiction based on which specific rights (or lack thereof) can be deduced. The aim of this paper is to contribute to a discourse that takes a more inclusive and self-aware approach when arguing for or against specific "needs" and "rights." However, in light of the powerful and loud voice of many functional communities calling for the end of specific means, the building of dialogue appears doubtful.

References

Angle SC (2002) Human rights and Chinese thought: a cross-cultural inquiry. Cambridge University Press, Cambridge

Arnaquq-Baril A (2016) Angry Inuk. National Film Board of Canada, Toronto

Balleste R (2011) Persuasions and exhortations: acknowledging internet governance and human dignity for all. Syracuse J Int Law 38(2):227–256

Barry D (2005) *Icy battleground. Canada, the international fund for animal welfare, and the seal hunt*. Breakwater Books Ltd, St. John's

Brand E (2009) The struggle to exercise a treaty right: an analysis of the Makah Tribe's path to whale. Environs: Environ Law & Policy 32:287–319. http://environs.law.ucdavis.edu/volumes/32/2/brand.pdf. Accessed 20 Mar 2017

Calonego B (2011) "Kein Interesse am Fell" ["No interest in the fur"; own translation], *Süddeutsche Zeitung*, 18 Apr 2011. http://www.sueddeutsche.de/panorama/robbenjagd-in-neufundland-kein-interesse-am-fell-1.1086472. Accessed 14 Mar 2017

CBD (1992) Convention on Biological Diversity (CBD), 5 June 1992. 1760 UNTS 79

D'Amato A, Chopra SK (1991) Whales: their emerging right to life. Am J Int Law 85:21–62

European Court of Human Rights (2015) *Delfi AS vs Estonia*. 64569/09 (Judgement), 16 June 2015

DeLuca KM (1999) Image politics. The new rhetoric of environmental activism. The Guilford Press, New York

EFSA (European Food Safety Authority) (2007) Animal welfare aspects of the killing and skinning of seals. EFSA, Parma

European Community (1983) Council Directive 83/129/EEC of 28 March 1983 concerning the importation into Member States of skins of certain seal pups and products derived therefrom (OJ L 91, 9.4.1983, pp 30–31)

European Union (2006) Banning seal products in the European Union. Declaration of the European Parliament on banning seal products in the European Union. P6_TA(2006)0369

European Union (2009) Regulation (EC) No 1007/2009 of the European Parliament and of the Council of 16 September 2009 on trade in seal products (OJ L 286, 31.10.2009, pp 36–39) (basic regulation)

European Union (2010) Commission Regulation (EU) No 737/2010 of 10 August 2010 laying down detailed rules for the implementation of Regulation (EC) No 1007/2009 of the European parliament and of the council on trade in seal products

Galtung J (1994) Human rights in another key. Polity Press, Cambridge

Gillett J (2015) Leaving for the seal hunt. The life of a swiler. Franker Press, St. John's

IFAW (undated a) Ending Canada's cruel commercial hunt. International fund for animal welfare. http://www.ifaw.org/canada/our-work/seals/ending-canadas-cruel-commercial-hunt. Accessed 20 Mar 2017

IFAW (undated b) Indigenous seal hunts. http://www.ifaw.org/united-states/our-work/seals/indigenous-seal-hunts. Accessed 20 Mar 2017

Ife J (2009) Human rights from below. Achieving rights through community development. Cambridge University Press, Cambridge

IWC (2017) Aboriginal subsistence Whaling. https://iwc.int/aboriginal. Accessed 20 Mar 2017

Hart HLA (1961) The concept of law. Oxford University Press, Oxford

HSUS (2013) Canada's 2013 baby seal slaughter begins, at 0:46–1:03. https://www.youtube.com/watch?v=foW_IiU2idQ (14 Mar 2017)

Kovacs K, IUCN SSC Pinniped Specialist Group (2008) Pagophilus groenlandicus. *IUCN 2014. IUCN red list of threatened species*. Version 2014.1. http://www.iucnredlist.org/details/41671/0. Accessed 14 Mar 2017

Krauss C (2004) New demand drives Canada's baby seal hunt. The New York Times, 5 Apr 2004. http://www.nytimes.com/2004/04/05/world/new-demand-drives-canada-s-baby-seal-hunt.html?pagewanted=1. Accessed 15 Mar 15 2017

Lamson C (1980) "Bloody decks and a bumper crop": the rhetoric of sealing counter-protest. Memorial University of Newfoundland, St. John's

Landman T (2009) Social science methods and human rights. In: Coonmans F, Grünfeld F, Kamminga MT (eds) Methods of human rights research (eds) Intersentia, Antwerp, pp 19–44, at 21

Lynge F (1991) Arctic wars, animal rights, endangered peoples. University Press of New England, Hanover and London

Malouf AH (1986) Royal commission on seals and the sealing industry in Canada report, vol 2. Supply and Services Canada, Ottawa

Maslow A (1954) Motivation and personality. Harper, New York

McCrudden C (2008) Human dignity and judicial interpretation of human rights. The Eur J Int Law 19(4):655–724

Murray R (2000) The African commission on human and peoples' rights and international law. Hart Publishing, Oxford

Parekh B (1999) Non-ethnocentric universalism. In: Dunne T, Wheeler N (eds) Human rights in global politics. Cambridge University Press, Cambridge, pp 128–159

Pelly DF (2001) Sacred hunt—a portrait of the relationship between seals and Inuit. Greystone Books, Vancouver

Pereira W (1997) Inhuman rights: the Western system and global human rights abuse. Rowman & Littlefield, Lanham

Perišin T (2013) Is the EU seal products regulation a sealed deal? EU and WTO challenges. Int Comp Law Q 62(2):373–405

Robertson IA (2015) Animals, welfare and the law. Fundamental principles for critical assessment. Routledge, London & New York

Sellheim N (2013) The neglected tradition? The genesis of the EU seal products trade ban. The yearbook of polar law V, pp 417–450

Sellheim N (2014) The right not to be indigenous—seal utilization in Newfoundland. In: Heininen L, Exner-Pirot H, Plouffe J (eds) Arctic yearbook 2014. Northern Research Forum, Akureyri, pp 546–552

Sellheim N (2015a) The goals of the EU seal products trade regulation: from effectiveness to consequence. Polar Record 51(3):274–289

Sellheim N (2015b) 'Direct and individual concern' for Newfoundland's sealing industry? When a legal concept and empirical data collide. The yearbook of polar law VI, pp 466–496
Sellheim N (2015c) Morality, practice, and economy in a seal hunting community. Arctic Anthropol 52(1):71–90
Sellheim N (2015d) Policies and influence—tracing and locating the EU seal products trade regulation. Int Community Law Rev 17:3–36
Sellheim N (2016) The narrated "other"—challenging Inuit sustainability through the European discourse on the seal hunt. In: Hossain K, Petrétei A (eds) Understanding the many faces of human security. Perspectives of northern indigenous peoples. Brill, Leiden, pp 56–73
Singer P (1975) Animal liberation. HarperCollins, New York
United States (1972) Marine mammal protection act (MMPA) of 21 Oct 1972 (16 USC Chapter 31)
VOX (2014) "Die Schlacht ist eröffnet" ["The slaughter has opened"; author's translation], *Vox Nachrichten*, 1 June 2014, 0:46–0:52. https://www.youtube.com/watch?v=A-ytrJdNoAI. Accessed 15 Mar 2017
WANGO (2004) Code of ethics & conduct for NGOs. http://www.wango.org/codeofethics/COEEnglish.pdf. Accessed 21 May 2015
Ward SJ (2013) Introduction: media ethics as global. In: Ward S (ed) Global media ethics: Problems and perspectives. Wiley, Hoboken, pp 1–9
Wegge N (2013) Politics between science, law and sentiments: explaining the European Union's ban on trade in seal products. Environ Polit 22(2):255–273
Wenzel G (1991) Animal rights, human rights. Ecology, economy and ideology in the Canadian Arctic. University of Toronto Press, Toronto
WWF (2009) Code of ethics. http://d2ouvy59p0dg6k.cloudfront.net/downloads/wwf_code_of_ethics.pdf. Accessed 17 Mar 2017
Yankelovich D (1999) The magic of dialogue. Transforming conflict into cooperation. Touchstone, New York

Chapter 3
The Global Land Rush and the Arctic

Markus Kröger

3.1 Introduction

This chapter discusses the recent Arctic land rush from the viewpoint of the larger literature on the global land rush and land grabbing, little of which has focused on the Arctic. This global view on the Arctic offers theoretical-methodological insights from a burgeoning literature on notable land control changes in different parts of the world (but not the Arctic) and can offer valuable knowledge on two key global/Arctic dimensions.

First, looking at the Arctic from the viewpoint of scholarship on global land grabs can contribute to our understanding of world politics and the political economy of natural resources, since the land-grabbing research has not yet focused on the Arctic. Bringing together scholarship focused either on the Arctic/global North countries or the Tropics/global South countries has tremendous potential for a fuller understanding of their similarities and differences in front of a globally felt new focus on investing in resource extraction.

Second, studying the Arctic does not mean simply adding one more case study region to complete the global picture in the sense of no longer having any unobserved cases, logical remainders, in a quantitative database. Instead, the Arctic case, and similar other cases undergoing major and rapid climatic changes from icy to less icy environments, are special because of their natural conditions and their change trajectories, and can be therefore used to specify how a global process—the land rush—operates in a different context and dynamics. Are there differences and, if so, are there more or fewer differences than we could hypothesize? Accordingly, the Arctic case is a qualitative one that may be used to offer insights into a number of theories about socio-environmental and political economic change processes; for

M. Kröger (✉)
Development Studies, Faculty of Social Sciences, Helsinki Institute of Sustainability Science, University of Helsinki, Helsinki, Finland
e-mail: markus.kroger@helsinki.fi

M. Finger and L. Heininen (eds.), *The GlobalArctic Handbook*,
https://doi.org/10.1007/978-3-319-91995-9_3

example, to see the extent to which they can or cannot be generalized. This is of interest to scholars of global issues, as well as to specialists focusing on the Arctic or, for example, on the Tropics, offering them understanding of their regions' specificities.

In the present chapter, I focus on carving out some potentially fruitful avenues for investigation, offering a panorama of different research focuses, questions, and hypotheses, that researchers should pay attention to in future empirically grounded research. The research for this article included a review of the research on the global land rush, to see how the Arctic has been treated or not treated therein, and a review of the literature on the Arctic resource rush of the past years, with a focus on land areas rather than maritime areas. Particular attention was paid to the literature on Arctic mining and forestry. Some empirical cases that are illustrative of the phenomenon are also mentioned in order to direct future research attention towards them. Empirical research for this contribution included field research and participant observation on mining governance and expansion in Finland and Canada during recent years.

3.2 The Global Land Rush

The world has seen a massive increase in land investment since 2007 (Borras et al. 2012), and considerable literature has been produced on the topic of land grabs to study the impacts of these land control changes in tropical countries (Edelman et al. 2013). The terms "land rush" and "land grabbing" refer here to "the large-scale acquisition of land or land-related rights and resources by a corporate, non-profit or public buyer for the purposes of resource extraction geared towards external consumers (whether external means simply off-site or foreign)" (White et al. 2012). To date, however, the literature has neglected the Arctic land rush, where the focus is on opening up new extractive operations. Furthermore, the reasons for the global land rush have been studied less than its impacts, which have been found to be typically negative.

The number of projects in the pipeline that aim to exploit Arctic resources indicates that the rush is gathering impetus. A huge iron ore mine was opened in August 2015 at the Mary River in Canada's Baffin Island; powerful forces in Greenland are seeking to open up uranium and other rare-earth elements mines; and the Nordic countries are also witnessing a boom (Kröger 2015). Russia has placed Arctic resource extraction as its number-one developmental project, and US corporations are considering the possibility of opening up wilderness areas in Alaska for resource extraction. Many new mines have already been opened and new timber extraction concessions have resulted in increased wood harvesting: these changes in Arctic land use and control do not just present a discursively constructed land rush—they are actually occurring.

With this in mind, it is striking that none of the 98 papers in the seven special issues published in leading journals since 2011 regarding the global land grab deal with the Arctic (for a summary, see Edelman et al. 2013). The same is true for books on the global land rush. A rare exception is Hall's (2013) *Land*, which includes notes on the Canadian Arctic. The existing, separate studies by Arctic specialists focus on climate change and geopolitics (Heininen and Southcott 2010), and consequently do not pay sufficiently rigorous attention to the new resource rushes penetrating the Arctic. More research is needed to fill this gap, and there is a particular need to produce a systematic study of the dynamics of the actually occurring land rush in the Arctic. To delimit such a broad research, focus could be given, for example, to the land areas above the Arctic Circle.

Approaching the issue from the perspective of the global political economy of the recent land rush and extractivism makes it possible to pay particular attention to theorizing the reasons for, and causes of, new resource rushes. The rise of commodity prices is often quoted as the principal driving reason behind land grabbing, but it alone cannot explain land investment booms. Therefore, enabling causes such as institutional contexts must also be studied.

Rather than a broad study of all economic sectors, a focus on new mining or forestry in some parts of the Arctic (such as Finland) would also allow deeper investigation of industry-specific factors than prior studies by restricting analysis to the sectoral specificities of mining and forestry. This is because a general study of all sectors involved in land grabbing in a single area cannot go beyond a limited view of new land and resource-controlling companies and states—the focus of prior study of the subject. The role of technology firms and other supporting industry-system actors that benefit from new resource-extracting investments has not been considered, while the new land-controlling companies and states have been framed as the major actors. I do not believe that this is the case, as capitalism goes much deeper than the soil, or even the subterranean level, cutting across varying geographical scales from the point of production to the world system, and is also based on political economic changes in social and symbolic spaces, not only the physical. Utilizing the global land rush-framework makes it possible to examine these scales and spaces and track down the causes of the Arctic land rush through a theoretically encompassing methodology, thereby bridging a large number of remaining gaps in the land-grabbing literature and also in broader theorization on the role of new resource frontiers in political economy.

The study of the reasons for land control changes offers particular methodologies and data choices, which could also be used in the Arctic. The causalities of post-2005 investment cases could be tracked and their parallels and specificities combined via in-depth interviews with key power holders. This line of inquiry begins by probing the path of the causal conditions that led to the investments, and continues downstream to the investment project areas (conducting field research at selected points of, for example, new timber and mineral extraction), thereby illustrating both the meaning of land control in the current global historical conjuncture and the changed political dynamics with regard to the new deals. Combining these studies through systematic comparative analysis would lead us closer to a minimal formula that explains the causal configurations under which the Arctic resource rushes have occurred, and also to examining whether there are multiple causal pathways. Once such an Arctic analysis and database on causalities have been created, it would then be possible to compare the findings with the bulk of theorizing elsewhere.

3.3 Examples of the Already-Happened Arctic Land Rush, and Speculation About the Future

There has been a notable increase in resource extraction since 2005 in the Arctic. Accordingly, while there is no need to speculate whether the Arctic land rush will take place or not, it has not occurred everywhere in the region, and not always in the same manner. There has been an increase in the volume of extracted natural resources and in the extent of land and landscapes being converted, existing nature and physical space being transformed into extractive operations, as well as a notable qualitative change in the style of extraction. Considering the change in the pace of extraction, "the volume of metallic ore and waste rock mining in Finland increased from fewer than 5 million tons to 46 million tons, mostly through the inauguration of four large mines in the east and north of the country" (Kröger 2015: 543). Causes of the qualitative change include mining becoming less focused on underground mining, and more on massive open-pit mines meant to excavate the lower-grade ores from a much larger land area. This qualitative change can be considered even more dramatic than the quantitative one, particularly for its larger use of land and wasteful transformation of landscapes and nature. In Finland, there was a 26-fold increase in waste-rock mining: "Illustrating the peak of the mining boom when new mines, particularly open-pit mines, were being opened, waste-rock mining in metal mines reached over 26 Mt in 2010, rising from only 1 Mt in 2005" (Kröger 2015: 547). Examples of new large mines that were opened in Lapland include the Canadian Agnico Eagle's hugely profitable gold mine in Kittilä, which has expanded rapidly; and FQM Kevitsa Mining's rich mixed-metal mine in Sodankylä, both of which have massive tailing ponds with at least some reported environmental problems (Kröger 2015). There has also been a massive increase in mining exploration, with one-third of Finnish Lapland being reserved for mining in one or another way. Resistance to the prospectors' increased nature-changing activity has been common and visible in overhauls such as drilling in nature reserve areas for uranium deposits by the Mawson company in Rompas, which has already destroyed important nature according to environmental officials; AngloAmerican's major Sakatti mining project on top of a Natura 2000 protected area and many other projects in their exploration or other pre-mining phase (but already with veritable changes in physical space, and land control—both crucial aspects in land grabbing).

Besides these greenfield projects, the Arctic has also witnessed a major expansion thrust in existing extractive operations, with a prime example being the Swedish state's Kiiruna iron ore mine expansion requiring the removal of the old city center. Both the greenfield and brownfield expansion need to be studied, and possible differences in their political and political economic dynamics highlighted. I would hypothesize that there is a major difference between the two, also considering the socio-environmental impacts, old and underground mining sites being potentially so distant in their dynamics from the new open-pit greenfield sites that their study as one "mining politics" would distort the reality more than bring light to it. Thus, I suggest starting the exploration of any land rush study by looking at the investment style, and differentiating the compared projects to different segments, and then studying quantitative factors (extraction volumes) along these comparable lines.

Besides mining, there has already been a veritable increase in deforestation of taiga forests of different types in the Arctic. For example, there was a downturn in deforestation in Finnish Lapland in the late 2000s due to the peace agreement between Greenpeace and other resistance actors such as the reindeer herders and the forestry companies, which ended the long-lasting "forest wars". However, deforestation has increased again in the present decade, and massive new production facilities are being planned in Kemijärvi (both for pulp and lumber) and close to the Arctic circle, and the continued high felling prices and high global pulp prices have led to expanded harvests that are no longer sustainable (Kröger 2016; Kröger and Raitio 2017). The case of pulp and forestry tells of a larger and important dynamic related to global commodity prices that needs to be assessed as a potential driving explainer of the Arctic land rush.

There was a global commodity super-cycle between 2007 and 2014, with the bulk of price increase in commodity indices coming from the high-volume and high-value oil price peak; falling oil prices led to a seeming end of this super-cycle by around 2014. The super-cycle gave massive windfall gains to many resource-exporting companies, and led several governments in the global South to increase export tariffs on commodities, which led to "resource nationalism" becoming the major fear for global mining bosses when moving from 2008 to 2012 (Kröger 2015). Some investment analysts even talked of a possible change in the global economic paradigm, where higher value-adding manufactured goods have increased their value over raw materials' valuation over the past century, towards a new "commodity paradigm", where this would again return closer to situation at the turn of 20th century, where resource extraction gave notable global advantages and inflows of capital and was a real strategy of wealth creation (the agricultural export giants of the time, Argentina and Uruguay, being the world's fourth and fifth wealthiest economies at that time). Commodity prices have generally come down and the commodity paradigm-change hypothesis seems to be an exaggeration, caused by the one-time buying of massive resources to build the budding multipolar world order of China and other rising powers, which required large amounts of base metals and resources for building infrastructure and military. Despite this, the prices of many commodities have not dropped and have in fact continued to rise. Nature and natural resources are limited and consumption is increasing rapidly. For example, pulp prices have not come down and energy wood prices have increased, which means for the Arctic that the demand for particularly pine-wood plantations will increase, leading to deforestation and conversion of forests into pine-fiber tree plantations. Making cardboard for e-commerce packaging and other high-demand paper products requires both the global South's eucalyptus or acacia-based fiber, as well as the pine-based fiber, to make the right mix. The Nordic pine-pulp has some of the best global qualities in this production value web. Thus, the land grabbing taking place to increase the size of eucalyptus plantations in the global South has also led to a sizeable impact in the global North and the Arctic, where there is now a market possibility, and a need, to increase the production of pine-pulp.

Due to these tendencies, I would expect a marked and rapid expansion of deforestation and tree plantations, for example in the Finnish Arctic, but also in other countries where there are governments and companies seeking to onboard the ongoing commodity super-cycle and unification of global value webs in the forestry sector's bioeconomy. In the mining sector, I expect a major rise to continue in the extraction volume of rare earth minerals, gold, oil, and gas in some places, as well as in many other minerals and metals, but not in all. For example, although iron ore has seen a major extraction volume increase via the Kiiruna mine expansion, and possibly via the Mary River mine expansion in the Baffin Island, it is likely not to continue at the same pace. This is because the massive iron ore boom between 2008–2014 led to wave of huge investments in new extraction capacities, which are taking effect now, new operations starting and pushing the price downwards, cutting out smaller and less profitable mines with higher costs (such as those in the Arctic). There is no reason why nickel production would not remain high and even increase, given the rising military spending in the rising economies and even among Arctic countries: other strategic minerals such as uranium are also likely to see price increases due to the wave of building new nuclear plants and nuclear weapons.

However, these are just speculative prognoses based on the current tendencies, and might change quite quickly through world and local politics, such as decisions in international negotiations and successful resistance by local populations, as well as new regulatory frameworks curbing extractivism at national and regional levels by states. The study of these enabling conditions is essential for understanding how the drivers of high prices and demand are affecting the land rush and making or not making it a reality.

Besides the forestry and mining sectors' increased land use, there has also been the reservation of land for new tourism, increased reindeer herding, military, and strategic or diplomatic international relations, which have already occurred. An example is the rise in Chinese and global South's interests in joining the circle of decision-making and land control in the Arctic; one such case is Chinese actors buying land in the Arctic for unclear purposes, and being the major capital investors in the new Finnish forestry investments particularly in the Arctic. These land use changes have already led to either a delimitation of land use possibilities and land control by some local populations that traditionally enjoy usufruct or other communal use rights over the targeted lands, or have led through a successful resistance process to a heightened sense of the already high consideration for indigenous and other local land control need among the locals. If and when the Arctic land rush continues, I expect there will be many clashes between local communities, environmental groups, and states and companies in their diverging views on how to use resources and land.

In the (unlikely) scenario of a dramatic downfall in global demand for resources, I would still expect some notable and even major land rushes to take place in the Arctic, both because its rapidly changing climatic conditions making some valuable deposits accessible in a manner that has never been possible, and because there are so few people and far fewer resource nationalists there than in the other parts of the globe. In this sense, what happens in the Arctic is, from a global perspective, defined even more by what has already happened in terms of resource governance in the global South and elsewhere than by what happens in some "isolated" Arctic politics. Commodity frontiers are becoming or are already closed in much of the global South, such as the Amazon, where massive destruction of biodiversity and nature and human rights abuses—virtual ethnocides—are now occurring in order to expand the commodity frontiers (to existing indigenous reserves, for example, as is occurring right now in Ecuador and Brazil; see Kröger and Lalander 2016), which means that while the Arctic may be more expensive, it might be considered by some global capitalists as a more ethical or accessible resource extraction point.

3.4 The Wider Scientific Importance of the Arctic Case

While Arctic scholars agree that most of the drivers behind the Arctic resource rush come from outside the region (Smith 2011; Arbo et al. 2013), they have not linked this claim to the increasingly sophisticated methodologies of land grab researchers (see, e.g., Scoones et al. 2013) explicitly studying this new global land rush (e.g., Edelman et al. 2013). In recent years, the translocal flow of land control has cut far more deeply than existing theorizing on past commodity price booms would suggest (Klare 2012). Comprising a new phenomenon, the contemporary global land rush has placed resource geopolitics in the limelight of world politics (Borras et al. 2012). Land is no longer an arcane subject with minimal relevance to theories of economic growth, governance, or political economy (Hall 2013), which means that land rushes are a timely and important research topic for the social sciences, the humanities, and environmental studies. Furthermore, all the Arctic states are powerful members of the global North, which adds a vital and fascinating element to the bulk of theorizing: how is it that land grabs can also occur in these countries and not just in the global South? There is land grabbing in all the Arctic countries, but not in all their territories and not in exactly the same way.

The present contribution draws on my own land-rush field research in Finland (Kröger 2015) and Canada for future research. It would be important to analyze the Arctic countries separately as parts of the Arctic region because they have similar characteristics in terms of climate, resources, and supra-national governance through the Arctic Council. However, there is also a need to compare different levels of democracy, welfare, civil society development, land tenure, class hierarchy, and transparency, among other political economic conditions—and differing biophysical and political ecological relations—on the assumption that the way these factors influence the land rush in the North is likely to differ from the role they play in the global South.

The main impetus for and argument in studying the Arctic as a case of global land grabbing is that the presumed global North/global South difference in terms of institutional contexts (such as democracy, welfare, and transparency), where the North is seen as more "developed", misses a key element in recent political and economic changes that can be made visible through an empirical investigation of the Arctic land rush. The alleged superiority of Northern political systems over their Southern counterparts has become highly problematic given that land grabs can occur even in supposedly democratic and transparent countries such as Finland and Sweden, or in Canadian First Nations' land tenure arrangements, which are considered comparatively advanced in terms of indigenous land rights. The appearance of the new resource frontier is a symptom of changed world-political and intra-national power relations and the rise of new powers. For example, the Russian government recently rented 1 million ha of forest to China (although this happened somewhat South of the region typically considered to part of the Arctic); in Finland, publicly owned mineral deposits worth billions of Euros have been given practically free to multinational companies or private entrepreneurs for export. There are currently several major forestry investment projects in Finland, focusing on Lapland and Northern Finland, funded by Chinese capital and aiming to dramatically increase the use of trees growing (presumably faster and faster) in the Arctic to produce fuels, energy, fibers, feed, and other wood products from them (for example, the Kaidi project in Kemi and the biorefinery project in Kemijärvi; see Kröger and Raitio 2017). The empirical study of the causal paths leading to the new Arctic resource frontier is a tool with which to analyze how and why the world political and national power structures and political games have changed, as well as the currently unfolding political economy whose reality has not yet been grasped by decision-makers, the public, or even social scientists.

Therefore, changes in land use and land control in particular—which are no longer arcane subjects but centerpieces of politics and power relations—can reveal larger, unperceived globalizing changes that have made the world a flatter place in terms of institutional frameworks that could regulate capitalism's negative socio-environmental impacts. I suggest several hypotheses through which to explore this overarching argument.

3.5 Key Research Questions and Hypotheses About the Arctic Land Rush

The existing land-grabbing literature has missed some political causalities due to an overly strict focus on land and resources. Little attention has been given to the roles played by companies and countries that provide machinery, technology, logistics, and diplomatic support for actors visibly engaged in the land rush. The result of this neglect has been that these players—central in the sense that, without their technologies, large-scale mines or wood-energy tree plantations, for example, would be much harder to establish—often fall outside the parameters of discussion. Therefore, there is a strong need to investigate the role of supporting companies, an analysis that complements the current, horizontally-focused literature, in order to produce not only a geographically more encompassing, but also a deeper, examination of the global land grab in vertical terms. Different industries have varying chains of operation, corporate agencies, and relations to contingency—for example, the availability of choices and the scale of independence from natural endowments inherent to the requirements of mineral-specific and general forestry-resource using companies—resulting in different industry systems.

How are the frontiers of resource extraction expanded? This key question can be divided into three key research tasks: investigating why, how, and when land rushes occur across different contexts, scales, and industries. The Arctic focus provides a particularly fruitful approach to answering these questions, one greater than the simple addition of another geographical case to the study of global land rushes. This is because the Arctic constitutes an example that diverges in critical dimensions from the cases on which the bulk of existing theories are based, thereby offering a theoretically insightful field of research. The radically different landscape, extremities, and cold of the Arctic can breathe new air into the theories of resource frontiers, and the role of land use and control in political economy, political dynamics, and social power relations.

An initial factor to be considered is that profitable extraction of resources in Arctic climatic conditions is no easy feat. The failure of pioneering projects can quickly change sentiments about the receptivity of the Arctic, and thus have a remarkable causal effect on investment in potential future projects. Secondly, the Northern governments and their populations, which are seen as receptive to new large-scale investments, might ultimately be a much more contentious force than investors anticipate when predictable problems and conflicts inherent to land control changes start to surface. While the tropics have previously been subject to large-scale projects of the current magnitude, the Arctic has not: for example, huge open-pit mines and so billion-euro, wood-based bio-refineries are novelties. People are just starting to grasp what these enterprises entail, and it is beginning to affect prior, enthusiastic attitudes towards such investment.

As a consequence of the initial boom years (2005–2013), the politics of the new projects are currently unfolding as a much more conflictive and complex mesh than was expected. In *Contentious agency and natural resource politics* (Kröger 2013a, b), I demonstrated how social-movement strategies can influence investment outcomes, even in unlikely contextual settings where there is strong corporate and government support for land investment. The toolbox of contentious politics research would be useful for grasping the dynamics of controversy in the Arctic. Furthermore, by using previous expertise on the South as a backdrop, scholars could bring fresh viewpoints to Arctic politics.

In terms of current Arctic expertise, the innovative aspect of looking at the Arctic based on what has been learned by the global land rush is to highlight the importance in studying similarities and dissonance between Arctic countries, and sector-specific land rush phenomena, from the perspective that the Arctic is possibly not as discrete as has been portrayed (e.g., Heininen and Southcott 2010). A novel approach is to link the forces of land use transformation in the area to global tendencies rather than considering them as unique (there is a lot of Arctic exceptionalism ongoing in the current literature, as well as Arctic hype). Therefore, one thread that is important to maintain is the parallel between Arctic and global dynamics and if, where, and why they diverge.

A new resource frontier can be studied in many ways. If I were to focus on its impacts, I would start by scoping local divergences through in-depth case studies; but when attention is directed towards tracking down causes and their dynamics, it makes more sense to start from examination of similarities across broad case units, and then proceed to specificities.

Next, I suggest several hypotheses that should be assessed in doctoral and master's theses, and broader collaborative research projects. The hypotheses include the major geographical scales from the point of production to the world system, considering both world-political and local drivers. This is a rarely seen research design, and a research area that needs methodological innovation.

Hypotheses 1 and 2 below are drawn from the substantial literature on land rush in the global South; there is a need to test whether these causes also apply to the Arctic situation:

1. Land and resources have been perceived as a good investment alternative since the 2008 global financial crisis and rise in land prices. However, the rise in land investment is linked with, and dependent upon:

   (1a) A long history of land grabbing in the investment area;
   (1b) Legal contexts giving the extractive operation stronger land use rights than competing land claims;
   (1c) The support of labor;
   (1d) Government policies driving and/or impeding the development of capitalism; that is, the gradual separation of workers from their means of production, especially land (Hypotheses 1a–1d are drawn from Edelman et al. 2013);
   (1e) A free international investment regime whereby the mutual interest between recipient governments and investors is not constrained by third parties' key interests or by international investment banks; and
   (1f) A clear majority of locals who do not oppose increased resource exploitation.

2. Resource frontiers, in that they comprise physical space changes, can be expanded more quickly by making correlating changes in power relations in social and symbolic spaces, such as in class relations and by legitimating discourses (Kröger 2013a, b, 2015). More specifically:

   (2a) Extractive operations must be preceded by changes in power relations in physical, social *or* symbolic spaces;
   (2b) The boom has been based in all settings on a simultaneous social space hierarchization and symbolic capital enjoyed by capitalists; and
   (2c) The trust and honor that the (new) capitalists enjoy as wealth- and job-creators in the pre-boom symbolic space cements the transformation of physical space.

The study of the roles of ideology and symbolic power exemplified in this hypothesis, building on Bourdieu's (1991, 1998) conceptualization, has been absent in prior studies on resource frontiers, particularly on the Arctic. Hypothesis 2 would make it possible to complement spatial analysis with a deeper theorization and, thus, fuller analysis of the role and changes in symbolic power. Hypothesis 3 is suggested by my research on global (including Arctic) tree-use increase (with new, multiple and flexible deployment; for example, for biodiesel) (Kröger 2014a, b, 2016), and mining booms and governance in the Nordic countries (Kröger 2015). The validity of Hypothesis 3 with regard to the rest of the Arctic needs to be assessed.

3. The Arctic countries are not only witnessing a material/territorial expansion of capitalism in the form of land rushes, but are also embarking on national industry development based on increased extraction-technology sales. In decision-making, the existence of this industrial-production-promoting dynamic trumps the weight of 1f (grievances concerning the land rush), even if the conditions identified in hypotheses 1a–1d are also present in the causal configuration (Tracking down technology chains to investigate possible motives by domestic machinery and consulting companies to push extraction is a helpful methodology for pursuing this hypothesis.).

Based on a review of the existing literature explaining the historical drivers of Arctic resource rushes, I have also formulated the below claims and hypotheses that require further testing. I suggest the following hypotheses for the study of the global/Arctic land rush, with the goal of testing the extent to which they apply across the Arctic:

4. Costs of extraction in Arctic resource frontier regions have been found to be especially high and therefore particularly sensitive to price fluctuations (Haley et al. 2011). Thus, (4a) long-term price expectation has been the key driver in all cases. Alternatively, as evidence from the Finnish mining boom suggests (Kröger 2015), other reasons, such as (4b) (expected) changes in legislation, (4c) job creation discourses, (4d) sudden increase in access to geological information, and (4e) the nature of the exploration system, are more important and precise explainers of booms than commodity prices.
5. An increased number of shipping lanes in the Arctic areas have encouraged investments.
6. The greatest challenges to expansion are still the local social conditions, such as the absence of a skillful/willing local labor force and regional social backing (these not enabling investments), rather than Arctic climate or geopolitics (Avango et al. 2014 argued that this has been the case historically).
7. The breaking of existing Arctic "historic natures"—for example, the human ecologies of indigenous peoples that rely on hunting or herding—creates a new labor force and support for large-scale projects. The new "lived environments" of post-2005 Arctic mining and forestry expand comparatively faster in places that lack socially, symbolically, and physically rooted traditional human ecologies (See Taylor (2015) for the helpful concept of "lived environments", which he originally coined as an alternative of political ecology to the problematic adaptation/resilience concepts, which assume a Cartesian nature/society dualism).

8. The absence of existing infrastructure or its high cost are no longer a bottleneck, as there is (8a) global surplus financial capital, and (8b) very strong industry lobbying that has secured subsidies.
9. The investment-area's (9a) political autonomy, (9b) fiscal capacity, (9c) institutional independence, and (9d) state's industrial policy (for example, the state performing the role of custodian, demiurge, midwife, and/or husbandry for industrial transformation; see Evans 1995) have a weighty causal impact.
10. Political instruments for managing conflicts are considered by the industry to be more developed in the Arctic than in the global South. Such instruments encourage investment (Haley et al. 2011).
11. The evolving situation of multiple global crises, such as resource scarcity and climate change, have a strong causal relation, but are invoked in a radically different way depending on the stakeholder.

These hypotheses could be studied separately, but most benefit would be gained when they are studied in conjunction, as possibly forming causal condition configurations that illuminate specific situations, combinations of drivers and enablers in which land rushes occur. For example, the function of enabler 1e (free international investment regime) may be rendered ineffective if 1d (capitalism-supporting government policies) becomes inactive (as happened, for example, in Greenland in the 2013 elections, when a government critical of China was elected).

The research agenda I have sketched and recommended here, through elaborate research questions and specific hypotheses that could be tested and thus further specified, would offer us a broad and deep understanding of how and why a global land rush is (or is not) taking place in the Arctic.

3.6 The Importance of Studying the Global/Arctic Land Rush

Studying the global/arctic land rush (and its potentiality) is both vital and timely for a number of reasons. Firstly, considering that the area north of the Arctic Circle comprises 15% of the earth's land, and is currently undergoing a resource boom (Borgerson 2013) the absence of rigorous and systematic study of the phenomenon is a striking gap in the literature on global land rushes. Secondly, there is increasing concern for the future of the Arctic among Arctic specialists, visible in the number of publications on the subject (see Arbo et al. 2013; Bruun and Medby 2014), although most of these largely lack the critical attention, methodological rigour, and global theorizing that the research questions and hypotheses identified above would make it possible to employ. Historical studies of the Arctic's long trajectory of land grabs and colonization (e.g., Sale and Potatov 2010; Stuhl 2013) and resource extraction (e.g., Massa 1999) are useful for contextualizing current events, but cannot explain their dynamics.

There is much to be learned for those studying the Arctic in the work of land-rush scholars in agrarian political economy and critical agrarian studies. Peluso and Lund (2011) argued that the global land rush (the large-scale acquisition of land or land-related rights and resources leading to notable landscape changes) comes in many forms and consists of a range of differing drivers and enablers that facilitate shifts in de jure and de facto land control. Most of the study of extractivism has focused on Latin America, where it has been identified that de jure and de facto ethno-territorial rights have undergone major changes during the past decade (often, but not only or everywhere for the worse), and that these two do not often overlap (local populations often have either de facto or de jure land rights, and the role of governments play out differently based on what the constitutional rights are) (Kröger and Lalander 2016). Those taking a broader perspective have argued that, taken together, contemporary land-rush moves constitute a new enclosure of commons (Borras et al. 2012), while Klare (2012: 15–16), in a rare study of both the Arctic and the tropics, argued that the global land rush is "the race for what's left". In Klare's view, the "invasion of the last frontiers" represents a drive without true precedents as "virtually all accessible resource zones are now in production", there being nowhere else to go except for extreme areas such as the Arctic.

In their review of the literature on Arctic futures, Arbo et al. (2013) argued that trans-arctic shipping routes and offshore oil and gas exploration have already received substantial scholarly attention; therefore, there is a need to study industries that have not been at the core of existing theorizing, such as mining and forestry. This would accord with current scholarly thinking that calls for expansion of the parameters of empirical and theoretical inquiry into land rushes (Borras et al. 2012).

When conducting research on this topic, it is also important to confront the literature critique that many publicized global land-rush deals have not materialized, although assumptions have been made on the basis that they have (Edelman et al. 2013); thus, in order to provide a valid study of causalities, those investments that have been realized, rather than merely projected, should be examined. This is particularly the case when considering how widespread the "Arctic hype" has been, and knowing how, in extractive investment projects, the announcement of resource reserves in itself generally constitutes a source of wealth. For example, Talvivaara mining company's Pekka Perä managed to obtain over 2 billion euros by pronouncements of ore deposits in Finland for global investors, which proved to be exaggerations; mining experts such as Mauno Vilminko have told me that the Talvivaara mine will never be profitable since there is no ore (due to low-grade minerals). As both the Arctic and the mining sector thrive on spectacular rumours of wealth—similar to the myths of Eldorado that drove the Spanish conquistadors in their loot of the Americas, of which those after the Aztec and Inca exploits proved to be catastrophic "investments" as no such blunders were to be found—research should use actual profit as a baseline to separate cases of real and imagined land rushes. When profit/loss is used as a baseline, it is possible to compare the extent to which the new Arctic extractive projects have been profitable or not. For example, I found that less than one-third of the Finnish mineral mines were profitable; most made massive losses and were unable to produce cheap resources (due to the Arctic conditions of hard operating environments and not-so-rich deposits) (Kröger 2015). The profitable and unprofitable new investments should also be studied and compared separately to see whether their causal dynamics differ (the drivers and enablers making them possible in the first place, and their subsequent life and possible expansion), and also to determine how their political economic and ecological impacts differ. Besides the profitable/unprofitable making ventures, we should also study, as a separate category, the anticipated and imagined but as-of-yet not realized projects, for their investment dynamics. What politics do they entail? This category of anticipated project, which has not (yet) been realized, can be accompanied by studies on how and why were those projects discontinued (for the time being). This makes it possible to answer the many open questions on land grabs that did not occur. Usable cases could involve, for example, studying Yara's Sokli phosphate-uranium-niobium mining project in Finland's Sokli (which was discontinued), comparing this to a similar project in Canada, and many more. The Arctic provides ample opportunities to study how the global dynamics of creating expectations and capturing investors' attention based on rumours is a source of wealth creation in today's world.

3.7 Concluding Remarks

Currently, what we know about global land rush causalities is based on empirical evidence from limited geographical and sectoral studies: large-scale plantations in sub-Saharan Africa ensuring food security in East Asian and Middle-Eastern states; animal feed production to service increased consumption of meat (Cotula 2012); the expansion mechanisms of tropical mining capitalism (e.g., Kirsch 2014) and its conflicts (e.g., Bebbington and Bury 2013); and conflicts connected to tropical industrial forestry expansion (Kröger 2013a, b, 2014a). New research into land grabbing in some parts of Europe has commenced (but not in the Arctic countries) (Franco and Borras 2013), and there is excellent new research on land grabbing in Ukraine and Russia (Visser et al. 2012). However, there is an absence of interdisciplinary studies on the drivers of new exploitation projects in the Arctic (however, see Haley et al. 2011 for an analysis of pre-2010 mining drivers from the perspective of economics), and no experienced Arctic-focused scholars among the land-grab research networks. These lacunae should be filled.

I would suggest that mining and forestry are key sectors to be studied (of course, oil and gas is also of crucial importance). Mining is a resource interest that capital has found geographically extant and politically enabled by Arctic governments. Massive reserves of minerals in the Arctic are expected (Borgerson 2013) and exploration investment in the region has peaked in recent years. Historically, this has been a predecessor to extraction booms—that is, to the establishment of resource capitalism (Moore 2014)—and studies should examine whether this also applies to the Arctic, mapping the new projects and the paths leading to them.

Meanwhile, Arctic forests have not only gained profile in the bio- and green economy strategies of companies and governments (Kröger 2013a, 2014b; Kröger and Raitio 2017), but are also facing increased harvesting pressure to serve wood demands (for example, that of China on the Siberian Taiga), to make space for mineral, oil, or gas extraction projects (for example, in Alaska, Yukon, and Lapland). Environmentalists have started to talk about the current deforestation in the Taiga as being as brutal as the devastation of the Amazon rainforest, and Greenpeace has been mobilizing forces to push for a new boycott of Finnish wood products, for example. These are understudied trends, along with the recent Russian–Chinese timber flow, the retreat of the boreal tree line northward caused by climate warming, and new wood-using investment aspirations. This explains why I suggest industrial forestry expansion as a sector to be studied as well as mining (naturally, the tundra and permanent ice-cover regions such as Greenland and Baffin Island cannot be studied for forestry expansion). There are important benefits in comparing the two sectors. For example, the identification of key causal conditions shared by the projects across the differing sectors of mining and forestry can show how institutional contexts, politics, and political economy impact the way commodity frontiers expand.

Prior Arctic studies have emphasized the region's specificity (Heininen and Southcott 2010), but I suggest the novel approach of studying the forces of Arctic land use transformation as being largely shared with global dynamics rather than unique: beginning with similarities and proceeding to specificities. Such an approach better links Arctic studies and scholars that have so far been isolated to wider global discussions and theoretical developments, escaping the somewhat skewed notion of Arctic exceptionalism (which I consider a less applicable category in today's world, given the global land rush).

The literature on new extractivism—which largely overlooks the Arctic—charts governmental recognition of the primary sector as a key growth element in the economy after decades of theoretical assertions that extractive economies lead to underdevelopment (Bunker and Ciccantell 2005).

There is a need to explain the spread of the phenomenon of extractivism, even to political economic contexts that could be assumed to resist the lure of pursuing capitalist growth via the creation of their own *domestic* "Cheap Natures"—a term that refers to new extractive projects wherein profits accrue primarily to foreign or limited actors, but costs are high and unevenly shared, with negative socio-environmental impacts being born locally (Moore 2014). Research that has tracked down causalities in the fast advancing capitalization of Arctic natures has made it possible to re-assess both the truly global expansion of the capitalist project of creating cheap natures, and the role of natural endowments in political economic transformations.

Pursuing all the hypotheses and research questions suggested herein simultaneously will result in multi-scalar and interactional analysis of spatial changes. This would make it possible to study *how* land rushes occur, in exact detail, which is of key importance and goes far beyond including an "Arctic case" to the global land rush studies. There is also a need to conduct this study by looking at the multiple geographical scales through which they take place, which include:

(1) The point of production: case studies of key new mining and forestry projects;
(2) Regional political economies: comparing the land areas above the Arctic Circle in all the macro-regions with differing political economies, such as the Nordic Countries, Canada, and Russia;
(3) The position of the larger production zone in the world system: how the Arctic is being rendered a new periphery, an ultimate resource frontier for global capitalism;
(4) The world economy's historic conjunction: studying how the Arctic land rush epitomizes (a) a global move to a new commodity paradigm where the time of Cheap Natures is ending (thus testing this theoretical proposition by Moore 2014), and (b) an extractives super cycle that augments the scales of production, raises costs, and consolidates production in the hands of the largest producers (testing the global applicability of the claim that the current era is in such a cycle made by Bebbington and Bury 2013).

A change in any of the above scales tends to lead to changes in the others, and these inter-scalar dynamics need to be taken into consideration. To answer the question of how the land rush has occurred, I suggest analyzing the political, ideological, and symbolic acts that are used and required "to map, code, survey, quantify and otherwise identify and facilitate *new* sources of cheap nature" (Moore 2014: 4) in the Arctic, in all of the four above geographical scales. Recently, Arctic governments have issued many strategies and mineral policies to attract investment, which downplay the economic, socio-environmental, and cultural value of existing land uses by, for example, indigenous people. It will be essential to compare the role of such discourses and framings, visible in official documents, identifying their causal role in land rushes by contrasting the official policies with new primary data, thus examining how the new lived environments of the Arctic are being created.

References

Arbo P et al (2013) Arctic futures: conceptualizations and images of a changing Arctic. Polar Geogr 36(3):163–182

Avango D, Hacquebord L, Wråkberg U (2014) Industrial extraction of Arctic natural resources since the sixteenth century. J Hist Geogr 44:15–30

Bebbington A, Bury J (eds) (2013) Subterranean struggles. University of Texas Press, Austin

Borgerson S (2013) The coming Arctic boom: as ice melts, the region heats up. Foreign Aff 76–89

Borras S et al (2012) Land grabbing in Latin America and the Caribbean. J Peasant Stud 39:845–872

Bourdieu P (1991) Language and symbolic power. Polity Press, Cambridge

Bourdieu P (1998) Practical reason: on the theory of action. Polity Press, Cambridge

Bruun J, Medby I (2014) Theorising the thaw. Geogr Compass 8(12):915–829

Bunker S, Ciccantell P (2005) Globalization and the race for resources. Johns Hopkins University Press, Baltimore

Cotula L (2012) The international political economy of the global land rush: a critical appraisal of trends, scale, geography and drivers. J Peasant Stud 39(3–4):649–80

Edelman M, Oya C, Borras S (2013) Global land grabs: historical processes, theoretical and methodological implications and current trajectories. Third World Q 34:1517–1531

Evans P (1995) Embedded autonomy: states and industrial transformation. Princeton University Press
Franco J, Borras S (eds) (2013) Land concentration, land grabbing and people's struggles in Europe. TNI, The Hague
Haley S et al (2011) Observing trends and assessing data for Arctic mining. Polar Geogr 34:37–61
Hall D (2013) Land. Polity Press, Cambridge
Heininen L, Southcott C (2010) Globalization and the circumpolar north. University of Alaska Press, Fairbanks
Kirsch S (2014) Mining capitalism: the relationship between corporations and their critics. University of California Press, Berkeley
Klare M (2012) The race for what's left. Metropolitan Books, New York
Kröger M (2013a) Contentious agency and natural resource politics. Routledge, London
Kröger M (2013b) Globalization as the 'pulping' of landscapes: forestry capitalism's north-south territorial accumulation. Globalizations 10(6):837–853
Kröger M (2014a) The political economy of global tree plantation expansion. J Peasant Stud 41:235–261
Kröger M (2014b) Flex trees: political and rural dimensions in new uses of tree-based commodities. Think piece series on flex crops & commodities, no 2. TNI, The Hague
Kröger M (2015) Spatial causalities in resource rushes: notes from the Finnish mining boom. J Agrar Change 16(4):543–570
Kröger M (2016) The political economy of 'flex trees': a preliminary analysis. J Peasant Stud 43(4):886–909
Kröger M, Lalander R (2016) Ethno-territorial rights and the resource extraction boom in Latin America: do constitutions matter? Third World Q 37(4):682–702
Kröger M, Raitio K (2017) Finnish forest policy in the era of bioeconomy: a pathway to sustainability? For Policy Econ 77:6–15
Massa I (1999) Risk economy in the circumpolar north. In: Goldfrank WL, Goodman D, Szasz A (eds) Ecology and the world system. Greenwood Press, London, pp 123–146
Moore J (2014) The Capitalocene II. Available at: https://jasonwmoore.wordpress.com/
Peluso NL, Lund C (2011) New frontiers of land control: introduction. J Peasant Stud 38:667–681
Sale R, Potatov E (2010) The scramble for the arctic: ownership, exploitation and conflict in the far north. Frances Lincoln, London
Scoones I et al (2013) The politics of evidence: methodologies for understanding the global land rush. J Peasant Stud 40(3):469–483
Smith L (2011) The new north: the world in 2050. Profile Books, London
Stuhl A (2013) The politics of the "New North": putting history and geography at stake in Arctic futures. The Polar J 3(1):94–119
Taylor M (2015) The political ecology of climate change adaptation. Routledge, London
Visser O, Mamonova N, Spoor M (2012) Oligarchs, megafarms and land reserves: understanding land grabbing in Russia. J Peasant Stud 39(3–4):899–931
White B et al (2012) The new enclosures: critical perspectives on corporate land deals. J Peasant Stud 39(3–4):619–647

Chapter 4
State-Owned Enterprises in the Arctic

Andrey Krivorotov and Matthias Finger

Economic activities in the Arctic play an increasingly important role in the globalizing economy, given that such economies are largely based on natural resources (oil, gas, and minerals), as well as the long-distance transportation of these resources and related goods (such as shipping). The Arctic appears to be one of the last opportunities for global resource exploitation that, ironically, is made possible by the negative effects of this very exploitation in the form of global warming and subsequent receding ice coverage of the circumpolar North. In other words, the Arctic already has, and increasingly will, become more, a part of economic globalization. However, this paper will not discuss if this evolution is either plausible or desirable. Instead, we assume that, given the currently unaltered dynamics of industrial development, this evolution will take place. However, its pace will significantly depend upon the combined effects of several factors: The speed at which Arctic ice is receding; the long-term development of global oil and gas prices; the speed at which fossil fuels will be substituted by alternative energy sources; technological progress in matters of resource; extraction; and (geo-)politics.

This chapter focuses on another factor that significantly determines if, and to what extent, Arctic oil and gas resources will be exploited; namely, the relationship between some of the corporations that will perform this extraction and their respective governments. In particular, we will look at the two countries whose economies predominantly depend on Arctic oil and gas resource extraction, Russia and Norway. We will examine two case studies: The relationship between Gazprom, Rosneft and the Russian government; and the relationship between Statoil and the Norwegian government. We want to better understand if these companies, as state-owned

A. Krivorotov (✉)
Shtokman Development AG, Zug, Switzerland
e-mail: krivorotov@starlink.ru

M. Finger
Ecole Polytechnique Fédérale Lausanne, Lausanne, Switzerland
e-mail: matthias.finger@epfl.ch

M. Finger and L. Heininen (eds.), *The GlobalArctic Handbook*,
https://doi.org/10.1007/978-3-319-91995-9_4

enterprises (SOEs), have a competitive advantage over non-SOEs in developing Arctic resources because of their privileged relationship with their respective states. If so, why and how do they make their privileged relationship with their respective states work to their advantage, as well as that of their respective governments? The cases of Gazprom, Rosneft (Russia) and Statoil/Equinor (Norway) will be followed by an analysis to identify the main characteristics of the relationships between these SOEs and their respective governments.

4.1 The Case of Gazprom, Rosneft and Russia

Russia has always been a northern country. More than half of its landmass contains permafrost and is characterized by negative average temperatures and very cold winters (although Siberia can have very hot summers). In Soviet times, 70% of Russian territory was regarded as high north, or areas equal thereto.

Meanwhile, the Arctic has only emerged as a political term within the past decade, following the global trend. The relevant policy guidelines were adopted in 2008, and the Strategy for Developing the Russian Federation Arctic Zone followed in 2013. The land areas of the Arctic Zone, as defined in a presidential decree of 2014, only included administrative units crossed by the Polar Circle. Therefore, the Russian Arctic Zone is essentially smaller onshore than the traditional high north, but covers vast economic zones and a continental shelf in the Polar Sea.

The USSR pioneered global Arctic oil and gas exploration and production. Approximately 90% of Russian gas and two-thirds of its oil, are produced in the northwestern Siberian swamps and tundra originally developed in the 1960–1970s. This included constructing enormous upstream facilities, cities, airports and roads, the world's biggest centralized pipeline network, and developing brand new technologies, industries, and creating R&D and education institutions.

In the 1980s, Soviet geologists also did extensive shelf exploration, especially in the Barents and Pechora Seas, discovering numerous oil and gas deposits such as the giant Shtokman and Ludlovskoe fields. However, these remained undeveloped, as the Siberian resources were enough to cover domestic demands, plus export to Europe.

This was all financed by the Soviet state through its many geological and production entities. The problem of inter-relations between SOEs and TNCs only emerged after the break-up of the Soviet Union in late 1991, when Russia started far-reaching economic reforms.

The oil and gas industries followed remarkably different reform paths. The old Soviet Ministry of Gas Industry was transformed into the state-owned conglomerate Gazprom in 1989 and merely reorganized as a joint stock company in 1993. Viktor Chernomyrdin, the founder of Gazprom, saved it from numerous dismantling attempts when he was the Russian Prime Minister from 1992–1998. Moreover, the company possesses legal monopolies to operate the national, 171,400-km network of gas trunk pipelines and export natural gas from Russia.

Gazprom has always been the world's biggest gas producer and a strong national champion. In the 1990s, when the Russian economy was in deep economic crisis, Gazprom reluctantly turned into a nationwide, social policy tool, supplying gas to countless domestic consumers at fixed retail prices, which were kept artificially low by governmental regulators. Exporting gas to Europe was the only way to survive. Therefore, the economics of Gazprom's operations is very sophisticated. The company was criticized in the 1990s for its obscure financial schemes, which allegedly covered up massive fraud by top managers. Gazprom also abused its dual role as a gas production and transportation company to discriminate against independent producers such as oil companies that had to flare much of their associated gas. These were the main reasons why there were few attempts by private investors to enter into the gas business. Changes started in 2001 when Alexey Miller, a close ally of President Vladimir Putin, was put in charge of Gazprom to transform it into a streamlined business-like group and a global energy leader.

In comparison, a highly competitive market emerged in the Russian oil industry. By President Yeltsin's Decree of October 1992, it was split into numerous production, refining, and sales enterprises. These were largely privatized by Russian tycoons (oligarchs) in the 1990s. These enterprises were torn piece-by-piece from Rosneft, the state oil company, which turned into a loose array of assets spread all over Russia. It nearly escaped a sell-off, only thanks to the Russian economic crisis in 1998. There were also several local companies owned by federal or regional authorities (such as Onako, Tatneft, and Bashneft), but most of these were later acquired by the major companies. However, the government retained control of Transneft and Transnefteprodukt, the natural monopolies operating the oil and product pipeline networks, respectively.

The very liberal *Law on Foreign Investments*, adopted in 1991, attracted petroleum transnational corporations (TNCs), which set up many joint ventures in the Russian oil industry, especially in Siberia and the northern republic of Komi. TNCs strongly lobbied for production sharing agreements (PSAs) with the Russian state to guarantee a stable, legal, and tax environment for a project, and get a portion of the oil or minerals in return. The respective, highly controversial law was passed in 1995. Many oil and gas fields, including those in the Arctic, were made eligible for PSAs, but only three agreements were actually signed and implemented in the mid-1990s: Sakhalin-1 and Sakhalin-2 in the Pacific shelf, and onshore Kharyaga in the Arctic part of European Russia. Later on, Russian SOEs (Rosneft, Gazprom and Zarubezhneft) farmed in all three areas.

Relations between Gazprom and Rosneft were close and complicated, especially after an unsuccessful merger attempt in 2004. In the early 2000s, they cooperated in the Arctic shelf. Now, they pursue different strategies, expanding their operations into each other's core businesses. In 2005, Gazprom acquired Sibneft (now Gazprom Neft), a large, integrated oil company possessing upstream assets in Arctic Siberia and Chukotka. Rosneft, headed since 2004 by Igor Sechin, a close friend of Putin, increased natural gas production and set up a separate gas division.

The Russian domestic oil and gas markets have lately moved in opposite directions. Gazprom increased competitive gas sales at the St. Petersburg Mercantile

Exchange and now faces growing challenges from Rosneft and Novatek, a private gas producer (Henderson and Moe 2017). From 2000–2017, Gazprom's share of Russian natural gas production decreased from 88.3 to 60.7%, while Rosneft increased its share of the oil production from 4 to 41.2%, largely through acquiring Yuganskneftegaz, the key upstream asset on the former Yukos, and TNK-BP (a JV between BP and its Russian partners).

In October 2016, Rosneft also bought Bashneft, a regional producer with excessive refining capacities, although key government officials supported a purchase by Lukoil, a private company. This acquisition was accompanied by the arrest of Alexey Ulyukaev, the Russian Minister of Economic Development, accused by Mr. Sechin of demanding bribes from Rosneft to issue a positive opinion on the deal. In December 2016, Rosneft, not the Russian state, orchestrated the sale of 19.5% of its own shares to a consortium of Qatar sovereign wealth-fund investors and Glencore for more than US$10 billion. Many observers regarded this as clear manifestations of Rosneft's heavy political influence.

The Russian government heavily relies on both SOEs in its economic and foreign policies. The state, which nearly lost control of Gazprom in the 1990s, was keen to increase its stake to 50.1%. Only then did it lift the limitation on foreign ownership in Gazprom (so-called *ring fence*). Rosneft, originally completely SOE, underwent several privatization rounds, but the state still has the controlling stake. Both companies have played key roles in the national goal to return to the Arctic over the past decade.

4.1.1 Rethinking the Arctic

In the 1990s, Russian oil and gas companies focused predominantly on utilizing their Soviet heritage. Exploration drilling fell dramatically, and new developments only involved minor fields adjacent to those already in operation. On the one hand, the domestic market collapse, and the resulting permanent cash shortage, made the industry severely cut costs and concentrate on existing facilities. On the other hand, the oligarchs, who bought the oil companies at tailor-made privatization auctions for only a small percentage of their real value, did not dare make any long-term plans. At the same time, they were accustomed to unusually high rates of return, which greenfield investments could never yield. The biggest groundbreaking projects at that time were implemented by foreign investors (led by ExxonMobil and Shell), under the two Sakhalin PSAs.

However, the prospects for new developments became more tangible in the early 2000s. First, most of the old fields were heavily depleted, which could not be compensated for by new minor developments. The Zapolyarnoe gas field in the Arctic region of Yamal was producing more than 100 bcm/y. This was put on-stream in 2001 and finally helped reverse Gazprom's falling output. It was the last, giant, untapped deposit adjacent to the existing facilities. The industry needed to enter into new provinces in the Arctic, eastern Siberia, and the Far East.

Second, oil and gas prices started a long rally, following the US invasion into Iraq in 2003. This helped improve both the economy of Arctic projects and the companies' financial status, giving them confidence to invest.

Third, US-Russian political relations dramatically improved for a while after 9/11. Russian oil and gas, which could be supplied at competitive prices and free of political risks, was welcome to the US market, justifying joint projects in new geographical areas. However, Transneft, the monopoly SOE, halted an attempt by four major oil companies to construct an independent export pipeline to Murmansk (the largest, Russian ice-free port in the north) for oil shipments to the US. Since then, developments have been concentrated in the more central and eastern areas of the Russian Arctic. The government reinforced these trends by introducing regional tax breaks for petroleum activities in the Arctic and the Far East, which is Russia's gateway to the Pacific Rim.

4.1.2 Russian Arctic Development: SOE and Private Mix

In the gas industry, Gazprom is implementing a megaproject in Yamal Peninsula. Its proven gas resources are estimated to be approximately 11.7 tcm, which can justify a long-term production of more than 300 bcm/y, or approximately two-thirds of Gazprom's present overall figure. The giant Bovanenkovo field, launched in 2012, was a groundbreaking effort involving construction of an airport, a unique railway with a bridge of record length, and living quarters. Gazprom has also entered into JVs with its traditional European partners in the well-developed Arctic area of Nadym-Pur-Taz, as part of broader asset swaps. A deal was signed with Austrian OMV, whereby Gazprom may also get 38.5% of OMV's subsidiary in Norway, thereby entering the Norwegian Arctic shelf.

The biggest challenge for Gazprom in the Russian gas market, specifically in the Arctic, comes from Novatek. Its shareholders include powerful businessman Gennady Timchenko and Total, a French-based TNC. Novatek's Yamal LNG project with Sabetta offloading port, commissioned in late 2017, is a key industrial development in the contemporary Russian Arctic. Each of its three trains is set to produce 5.5 mt of LNG annually, or more than all of the traffic across the Northern Sea Route (the Northeast Passage, east of the Novaya Zemlya islands) which hit a post-Soviet record in 2015 with 5.4 mt.

The project is strongly backed by the Russian government for regional, foreign, and energy policy reasons. The Russian state granted Yamal LNG unprecedented tax relief and donations from the National Wealth Fund, as well as constructing the Sabetta port facilities, an airport and other relevant infrastructure, which was all financed from the federal budget to make the project viable (Lunden and Fjaertoft 2014). Yamal LNG also obtained an individual exemption from the *Federal Law on Export of Gas*, breaching Gazprom's export monopoly.

Yamal LNG manifested a heavy entry of Total and Chinese SOEs into the Russian Arctic gas industry. Novatek holds 50.1% of the company. The rest belongs to Total

(20%), the Chinese National Petroleum Corporation (20%), and the Silk Road Foundation (9.9%). Additionally, Chinese corporations and banks have covered more than 40% of the debt financing, supplied more than 70% of the equipment, and purchased no less than 18% of the LNG produced.

In the oil business, the two Russian SOEs work on several breakthrough projects in the Arctic that do not involve private investments. Rosneft has developed Vankor oilfield, opening a new province in the Arctic part of eastern Siberia. Its annual production is 22 mt and may be increased (involving adjacent fields) to 55 mt, or 30% of Rosneft's present total output. Oil from Vankor is transported through a new pipeline to the south, to ultimately feed into the East Siberia-Pacific Ocean (ESPO) trunk pipeline to China.

Foreign investors were only invited to the project at a later stage. In 2016, Rosneft inked agreements with four Indian oil companies, which would obtain up to 49.9% of shares and bring the investments required to maintain the production level. The Indians also purchased 29.9% shares of Rosneft's project in Central Yakutia, which is not actually in the Arctic area, but also has a severe climate (Kozlov 2016). In return, Rosneft has farmed into Vadinar refinery in India, creating what it calls a *global integrated chain*: "This laid the foundation for the Energy Bridge between the countries" (Sechin 2016).

Gazprom Neft delivered two landmark oil projects in the Arctic in 2016. In May 2016, it started shipments from Novy Port, the first oilfield developed in the Yamal Peninsula, with plans to produce 6.3 mt of oil in 2018 (Gazprom 2016a). In September 2016, Gazprom Neft and Rosneft jointly launched commercial production in the Vostochno-Messoyakhskoye field in the Yamal region. It became the northernmost oilfield in operation in Russia, approximately 150 km from the nearest settlement (Gazprom 2016b).

In contrast, the biggest Russian private oil company, Lukoil, focused on the Arctic in its traditional Timan-Pechora area, which Vagit Alekperov, the corporate president, defined as a priority. The plans are to increase output from 15.3 to 21 mt/y.

SOE domination is even more visible in the Russian Arctic shelf. The *Federal Law on Subsoil*, as amended in 2008, provides that licenses for Arctic offshore blocks may only be granted to legal entities that are more than 50% controlled by the Russian state. This clause confines the potential operators to Gazprom (therein Gazprom Neft) and Rosneft, Sintez Group, the only private actor that held several exploration licenses in the Barents Sea, sold this business to Rosneft in 2012. Lukoil, which successfully put several fields in the Caspian and Baltic Seas on-stream and indicated a clear interest for offshore Arctic sites, had no other way to do this but to enter the Norwegian shelf.

The increased political interest in the Arctic brought about a major shift in the government's licensing policy. Within a few years, the Russian Subsoil Agency granted numerous offshore blocks to the two SEOs, covering some 80% of the nation's Arctic shelf. Recent amendments to the *Law on Continental Shelf* made that possible without competitive auctions. While Gazprom gives priority to near-shore blocks close to its existing infrastructure, Rosneft tends to obtain larger acreage in high seas, which are more challenging to develop (Buzovsky 2016).

The SOEs acknowledge the need to cooperate with foreign partners. Due to its traditional focus on onshore fields, Russia lacks offshore technologies, which are yet to be developed for many Arctic areas. "So far, we do not possess import independence as regards offshore equipment," Valery Golubev, Deputy CEO of Gazprom, admitted in October 2014 (*Shelf Urges New Technologies* 2014). Prirazlomnoye, the only Russian Arctic offshore field in operation, is a clear example. This oilfield was put on-stream by Gazprom Neft in 2013 in the shallow waters of the Pechora Sea, which is often covered by ice, and the bulk of the production equipment was delivered by Norwegian suppliers. This oil field now produces approximately 2 mt/y (Savina 2016).

To attract technologies and investments, Rosneft formed partnerships with international leaders such as ExxonMobil (USA), Eni (Italy), and Statoil (Norway) regarding its Arctic offshore blocks. Gazprom had a common JV, Shtokman Development AG, with Total and Statoil from 2008–2015, and has discussed cooperation opportunities with Shell. The two SOEs have also invited Chinese, Vietnamese, and Indian partners. Foreign investors forge these partnerships in anticipation of Russia easing up on legal constraints on access to the shelf resource base.

However, cooperation with Western TNCs stalled in 2014, once the US, EU, and Norway imposed sanctions on supplying deep-sea and Arctic drilling equipment to Russia. The Rosneft-ExxonMobil alliance suffered most. After a successful drilling in the Kara Sea, where they discovered the giant Pobeda (Victory) oil and gas field, ExxonMobil had to withdraw from the partnership to comply with the sanctions. Any further cooperation in the Arctic shelf remains subject to serious changes in the global policies.

Summing up, SOEs play a key role in the Russian Arctic. There are also opportunities for private investments, but they require goodwill from the authorities and the companies themselves.

There is much criticism of Gazprom and Rosneft, both outside and inside Russia, especially on behalf of liberal politicians who dislike governmental intervention in the economy. One set of arguments points at some obvious blunders and failures of the SOEs and concludes that they are inherently inefficient, because they merely follow policy-motivated orders from the Kremlin. Another criticism depicts the SOEs as almighty industrial monsters that impose their narrow domestic and international interests on the Kremlin. The paradox is that both sets of arguments are often expressed by the same analysts, depending upon the circumstances, and reflect a clear political bias.

Their opponents, who are often affiliated with the left-wing and patriotic political forces, point out the unfair nature and negative economic effects of oil-industry privatization in the 1990s. They support big state enterprises as the tool for Russia to keep its natural resources in the public domain and benefit all people (although the high salaries of their top managers and the regular corruption charges undermine this argument).

In reality, the two SOEs show elements of a monopolist way of thinking. This is the case with most Russian big companies, whether state-owned or private, due to Soviet traditions and lack of experience in competitive markets. However, the SOEs are willing to cooperate in the Arctic to share the high technological and economic risks, as well as to forge broader partnerships as part of their global business strategies.

International TNCs are not scared by the state control of Russian SOEs. Well aware of the high increment risks of Arctic operations and the frequent changes in Russian legislation, international TNCs prefer to have strong political backing. After two decades of operations in Russia, ConocoPhillips sold out its 20% in Lukoil and its stakes in all of Lukoil's subsidiaries in 2010–2015, once it saw that it wouldn't get either strategic assets or access to the Russian shelf (Focht 2015). In the same period, Rosneft entered into strategic deals with several TNCs, and British Petroleum (BP) acquired 19.75% of its shares in 2013. However, Western sanctions have created major obstacles for such cooperation in the Arctic, which resulted in a major shift toward joint projects between Russian and Asian SOEs.

4.2 The Case of Statoil/Equinor and Norway

Norway's oil history began in the early 1960s, but the industry was gradually moving to the north. Prior to 1979, any exploration and development in the Norwegian continental shelf (NCS) was only allowed south of 62°N, or in the southern part of the North Sea. Drilling in the Norwegian Sea started in 1981 and in the Barents Sea in 1987. By that time, Norway had already established a robust system of relationships between its government, petroleum TNCs and SOEs, which was merely projected to the north.

4.2.1 SOEs and the Government

Norway has always welcomed foreign investments, subject to tight public control to ensure budgetary revenues, technology transfer, development of national capabilities both off- and on-shore, and high HSE standards. The government stake was extremely high, up to 85% (now 78%) of net corporate income. However, TNCs have accepted these terms, thanks to high political stability, transparent predictable legislation, strong law enforcement, and short distances to European markets. Unlike in Russia or the UK, investors in Norway did not voice tax reliefs even after the dramatic oil price falls in 2008 and 2014.

Developing national petroleum competence has been a high political priority for a long time, resulting in traditionally close relationships between Statoil (the principal SOE) and the Norwegian state. The company and the key government agency, the Norwegian Petroleum Directorate, were established in 1972 when Norway first charted its oil and gas policy. Statoil was ordered to consult with the Ministry of Industry (now the Ministry of Petroleum and Energy) and submit annual reports to the Storting (Norwegian parliament).

Statoil became a favorite of the Social Democrats from the Norwegian Labor Party, which had ruled the country throughout most of the post-war period and wanted oil to benefit all Norwegians by helping to create a "qualitatively better society." For many years, only prominent Labor Party figures were appointed as Statoil CEOs.

The authorities ensured Statoil would have 50% or more of stakes in all offshore licenses granted. However, the company acted as a so-called *carried partner* until the mid-1980s. Its portion of investments was covered by its license partners, while it received its share of any revenues. Exxon, a minority stakeholder in the Gullfaks field, also did most of the actual development when Statoil was first formally appointed as field operator in 1981. In this respect, Statoil enjoyed preferential treatment compared to Norsk Hydro, a semi-SOE with nearly 50% government stake. It originally was a large fertilizer and aluminum producer, which entered the oil business in the 1960s, but had to do everything at its own cost and risk (Skjeldal and Berge 2009).

The idea of a privileged, completely state-owned oil company has traditionally been opposed by right-wing politicians who disliked strong government involvement in the economy. In 1985, they managed to split Statoil's license shares, whereby the company only retained approximately 20% in all blocks. The rest was handled directly by the Norwegian state as a regular financial investor, paying its portion of costs from the budget and cashing in the dividends. The idea was to prevent Statoil from getting too much financial strength, commensurate to the budget itself. However, this new, unique, system did not bring about any change in operational terms, as the government entrusted Statoil to manage its interest.

Both Statoil and Norsk Hydro were quickly learning the business. By the mid-1980s they had acquired excellent expertise in petroleum geology, offshore field development, and harsh-climate operations. However, their decision-making remained heavy and bureaucratic, influenced by political considerations.

In the early 2000s, Norwegian petroleum policies underwent major changes. The government became concerned that upstream production in Norway was about to peak, so set the dual policy goal of extending the nation's oil age and expanding its worldwide operations. To this end, Statoil was partly privatized in 2001 and entered listings at the New York and Oslo Stock Exchanges. The direct, state financial interest in oilfields remained in Statoil's trust, but it was transferred to a new 100-percent SOE, Petoro, to prevent a partly private company from running public assets.

At present, the Norwegian state (represented by the Ministry of Petroleum and Energy) controls only 67% of Statoil's shares, while the others are split among private owners, largely from Norway, the UK, the rest of Europe and the US. The company has exploration and production assets in more than 30 countries around the globe. Within a decade, growing public exposure and overseas operations dramatically transformed Statoil's operations into a streamlined corporation with little bureaucracy. It is compatible to international majors in terms of corporate governance, project management, and financial engineering.

Under a center-right cabinet, Helge Lund, an active conservative, was appointed the CEO of Statoil in 2004 for the first time. He retained the office even after a left-centrist coalition, led by the Labor Party, gained power a year later. That was due to his strong track record in crisis management. For several years, Lund had been saving Aker, the biggest Norwegian offshore contractor, from bankruptcy. When he suddenly left Statoil in 2014, the state, as the majority shareholder, took another step forward. For the first time in its corporate history, the new CEO, Eldar Sætre, was not a political appointee, but a professional manager with 35 years of experience inside Statoil.

In 2007, Statoil merged with Norsk Hydro's petroleum business, which reflected the global industry trend at that time, and created an oil company that ranked among the top 20 worldwide. However, the deal provoked skepticism at home, as it produced a national champion that became not merely an unparalleled energy SOE,[1] but Norway's by far biggest manufacturer, employer, and taxpayer (besides the dividends paid to the Norwegian state as the key shareholder). Norwegian observers regard Statoil CEO as second only to the Norwegian prime minister in terms of power and influence.

In 2017, Norway's oil and gas output was 238.4 million tons of oil equivalent (mtoe), therein 163.7 mtoe operated by Statoil. Statoil's equity production was 76.3 mtoe, plus it handled Petoro's production of 64.5 mtoe, for a total of 140.8 mtoe. This constituted nearly 60% of the nationwide figure, or 3.5 times more than all four TNCs (ExxonMobil, Total, Shell and ConocoPhillips) put together. The remaining 54 mtoe were supplied by medium-sized national and independent oil companies from Norway, the EU, and Japan (Norsk Petroleum 2017).

Although the present right-wing cabinet favors private enterprises, it sees the merits of retaining a controlling stake in Statoil, though in commercial terms: "The objective of the state's ownership of Statoil ASA is to maintain a knowledge-based and high technology industrial group with head office functions in Norway. The company is to be run on a commercial basis and with the aim of delivering a competitive return. On the basis of guidelines specified in a sale and marketing instruction, Statoil is responsible for managing, transporting and selling the Norwegian state's oil and gas in conjunction with Statoil's own reserves. This arrangement presupposes that the state is the majority owner of Statoil" (Government of Norway 2014).

However, Statoil's domestic position is always challenged. As the North Sea province was entering maturity, Norwegian authorities adjusted their policies to promote competition and diversity of actors. Since 2000, more than 110 new domestic and foreign oil companies were prequalified by the Petroleum Directorate to work and operate offshore fields. Many of these were small or medium-size high-tech startups (so-called *mosquitoes*); ready to develop small reservoirs or mature fields in the late production phase, which are of less interest for the big players. The government introduced a simplified fast-track procedure to issue clearances for development of minor fields, often done by subsea tie-backs to the existing infrastructure.

Statoil disliked this development. Lund harshly publicly criticized the authorities in his speech at the *Offshore Northern Seas* conference in 2008, stating that the "mosquitoes" could not maintain production and geological exploration

[1]The Norwegian state is also the sole owner of several other companies in the energy sector: Petoro, Gassco (operator of the privately held Gassled gas pipe system), Statnett (power grid), plus Enova and Gassnova, which were established to introduce environmentally friendly technologies.

under crisis, as they did not possess necessary financial strength. Nevertheless, the smaller companies did play an important role in reversing the fall in petroleum production. They helped monetize or extend life cycles of several fields making NCS more competitive. Lundin Norway, a Swedish medium-size independent producer, even outclassed Statoil by making a major find in a North Sea block Statoil had abandoned. Meanwhile, Statoil's organic production in Norway was decreasing, compensated for only by new acquisitions. The company was criticized for expanding overseas operations, while scaling down domestic involvement. Developing Arctic resources helped counter these negative trends.

4.2.2 National Arctic Strategy and Approaches to Developing the Arctic Shelf

Norway was the first nation to launch an Arctic policy paper in December 2006, and its principal goals remained unchanged, despite the power shift in late 2013. In its first High North Strategy the Norwegian government aspired to take a leading role in global Arctic research and governance. At that time, Norwegian SOEs were delivering landmark offshore projects. In 2007, Statoil commissioned Snøhvit, the first field in the Barents Sea, and the respective Melkøya LNG plant, the first one in Europe and north of the Polar Circle. In the same year, Hydro successfully developed the mid-Norway, offshore Ormen Lange gas field. Both were megaprojects, each worth more than $10 billion, successfully applying seabed completion. The authorities were keen to use this opportunity, setting a specific policy objective to enhance development of petroleum activities in the Barents Sea (Government of Norway 2006).

Snøhvit, discovered in 1984, is an interesting case study of the interplay between commercial and policy considerations. As the field operator, Statoil lobbied its rapid development, strongly supported by central and regional authorities. However, economic and technological challenges were growing, in addition to the fact that the plant was testing a new liquefaction technology, the Statoil-Linde process. Several stakeholders, including Hydro, opted to leave the project. Statoil management was also aware of the problems, but could not abandon the project, due to its importance to the region. Development of Snøhvit was finally sanctioned in 2002. Only after the bulk of the LNG was booked by potential buyers, did Statoil establish a market presence in the US, and the project was granted unparalleled individual tax breaks by law.

Snøhvit development was prone to delays, incidents, and massive cost overruns. However, gas prices started to rise during this time, The Arctic entered the global politics agenda, and the Russian gas monopoly Gazprom invited foreign companies to the giant Shtokman offshore field in the Barents Sea. These trends economically improved the Snøhvit project and turned Statoil into a global Arctic pioneer and trendsetter. The project also produced major ripple effects for the region, although the central government, local producers, and politicians initially had to press Statoil to increase the local content.

Statoil invests heavily in petroleum-related R&D, focusing on seabed solutions, which is highly relevant for ice-covered waters. Seabed compression has been tested at Ormen Lange, and Statoil has promised to introduce full-scale seabed factories in 2019 (Knott 2013). The company has introduced a practical division of Arctic offshore areas into three categories: Workable (completely ice-free); stretch (seasonally ice free, requiring some new technological development, but well within the company's capabilities); and extreme (areas covered by ice throughout the year and regarded as a distant future option) (Statoil 2013). Most of Arctic Norway is workable. Statoil has also established an offshore presence in eastern Greenland and the Alaskan North Slope, as well as in offshore Newfoundland (not actual Arctic, but a harsh climate area with a high risk of iceberg impact) (Arctic Journal 2013).

Cooperation with Russia, which dates back to the early 1990s, is particularly important, as the two nations share the promising petroleum province in the Barents Sea. Projects are underway in both parts of the sea, with new major finds on the Norwegian side. In 2010, Norway and Russia agreed to divide the 175,000 km^2, previously disputed area, which may contain large petroleum resources (Overland and Krivorotov 2015).

Since 2004, Statoil and Norsk Hydro (and later the unified company) signed a series of memoranda of understanding with Gazprom and Rosneft, the two Russian SOEs operating in the Barents and Pechora Seas. Between 2008 and 2012, Statoil held 24% in Shtokman Development AG, a JV with Gazprom and French Total, was established to start Phase 1 of the Shtokman field. This project was postponed in 2012, due to the unfavorable market situation, but Statoil entered into cooperation with Rosneft that same year on several blocks in both the Russian and Norwegian parts of the Barents Sea, in the Sea of Okhotsk, and onshore in Russia. This cooperation had strong political backing, but suffered from the sanctions on the supply of Arctic offshore drilling equipment to Russia imposed by the Norwegian government in 2014 (Bourmistrov et al. 2015).

For Statoil, developing the Arctic shelf also manifested a greatly desired comeback to Norway, as the key player in the area enjoys high political priority. TNCs, such as ConocoPhillips and ExxonMobil, also clearly indicated a desire to get more acreage in the Norwegian and Barents Seas, highlighting Norway's advantages of good infrastructure, skilled labor, stable framework conditions, and easy access to the market (Helgesen 2012). The Norwegian authorities and Statoil welcomed an intensive international cooperation. In 2012, at a conference of oil ministers and TNCs from the US, Canada and Iceland, organized by the Norwegian Ministry of Oil and Energy, Lund said: "Realizing the whole potential of the Arctic requires a joint focus on technology development, No one of us can tackle the Arctic alone, we ought to cooperate" (Ree 2012). In the Norwegian part of the recently delimited area, 16 oil companies, including Statoil, established the Barents Sea Exploration Collaboration to cooperate in regional seismic shooting, ice management, environment, and oil-spill response.

However, Statoil remains the leader and trend-setter in the north, with Snøhvit finally in full-scale operation with a strong progress in several new projects, such as the Aasta Hansteen in the Norwegian Sea, north of the Polar Circle. This big gas field marks the first deepwater development (1300 m sea depth); the first SPAR type platform in Norway and the largest worldwide; and the first gas pipeline, Polarled, to connect Arctic Norway with European markets (Lien 2016).

In December 2017, Statoil also tabled a plan for development and production for the principal Johan Castberg oilfield in the Barents Sea. Applying new advanced technologies, the SOE cut development costs by half and the breakeven oil price from $80/barrel to $35/barrel within a few years. The average figure for Statoi's portfolio is now $27 (Statoil 2017). By comparison, foreign companies only completed one project in Arctic Norway; the Goliat oilfield, delivered by the Italian firm Eni in Fall 2015, after long delays, and project management and HSE issues.

Statoil interest in the Arctic peaked in 2012, when the company was planning to triple its R&D spending on Arctic technologies and drill nine exploratory wells there within one year (Stensvold 2012). Since then, investors over the entire Circumpolar Arctic have postponed or halted numerous projects due to the global market turmoil, especially after the 2014 fall in oil prices. Statoil halted its offshore Alaskan drilling plans in 2012 and closed down all Alaskan activities in 2016, following the path of Shell and other TNCs.

Statoil stated in its 2015 Sustainability Report: "Our approach to Arctic operation is not to move faster than technology allows and to ensure safe and responsible operations" (Statoil 2016a). This was well in line with the precautionary approach to Arctic oil and gas operations reflected in the 2016 U.S.-Nordic Leaders' Summit Joint Statement (U.S.-Nordic Leaders 2016). Over the past few years, the company has concentrated on the workable Arctic areas in offshore Norway and developing new, efficient, environmentally friendly technologies.

In December 2016, Statoil CEO Eldar Sætre stated: "In a low-carbon future, we need oil and gas producers who can deliver energy at low cost, with low emissions. Therefore, the NCS will become even more attractive and important" (Statoil 2016b). In 2018, the company even changed its name to Equinor to avoid any indications of affiliation to either the state or the oil industry, reflecting its increased focus on renewable energy. In November 2016 CDP, a UK-based advisory firm, recognized Statoil as the major petroleum company that is best preparing for a transition to a low-carbon economy (Soliman et al. 2016).

At home, Statoil's HSE record is often criticized especially the numerous incidents onboard offshore platforms, and the failure to introduce total carbon capture and storage at both the Snøhvit LNG plant and Mongstad refinery, despite massive government support. For years, the company failed to overcome the influential environmental and fishery lobbies and persuade the government into allowing petroleum exploration and production in the fish-rich Lofoten-Vesterålen area. As operator, Statoil had to postpone Snøhvit Phase 2 in the Barents Sea. Its proposal to construct another LNG train raised opposition from both its license partners and another Norwegian SOE, Gassco, which lobbied for construction of a gas pipeline to the south. Then-Foreign Minister Jonas Gahr Støre strongly supported this idea.

Nevertheless, the SOE is indispensable for ensuring continuity in the Norwegian offshore industry, especially when many foreign investors bail out from the Arctic.

Ola Borten Moe, an independent oil businessman, and former Petroleum and Energy Minister of Norway, wrote: "In my own opinion, this is a challenging issue, On the one hand, Statoil occupies a large and very dominant position in the Norwegian shelf. This impedes diversity and real competition of [technical] solutions and supplies. This presses prices up and efficiency down. On the other hand, Statoil is the locomotive which the Norwegian shelf needs to go further and to open up new areas. It is Statoil who has put Snøhvit and Aasta Hansteen on stream. It is Statoil who bears the burden in the north" (Borten Moe 2016).

The open, democratic decision-making in Norway has also greatly contributed to charting efficient and socially responsible oil policies, including SOEs' performance. The Storting, Petroleum Safety Authority, competitors, environmental NGOs, and business media all keep a close eye on Statoil/Equinor. This public scrutiny, combined with the company's growing international exposure, make it rely more on technological leadership and a strong environmental record than on political connections or individual tax privileges.

4.3 Analysis

Our two case studies—Russia and Norway—yield very interesting observations, as well as lessons (which we will discuss in the conclusion). Seven such observations are worth highlighting.

- It is obvious that developing Arctic oil and gas resources has a number of economic features, which all require strong government relations. In particular, developing such resources is very costly in terms of large-scale infrastructures such as ports, airports, housing facilities, access roads, and communication and navigation facilities. In addition, the incremental costs of exploiting these resources, such as, the need for icebreakers to support operations, are comparatively higher than in other parts of the world. These costs eventually justify subsidies (or tax breaks), even for operations. Typically, private companies do not undertake such risky investments, unless they are supported by government agencies, public money, and long-term government guarantees. SOEs have a significant comparative advantage in this respect, as government guarantees and support are generally easier to mobilize because of a more direct (virtually immediate) contact line to the government. Even more so, SOEs virtually have an implicit government guarantee, at least in terms of their operations, as government typically will not abandon their own enterprise in times of troubles.
- For the Russian and Norwegian governments, oil and gas exploitation is strategically important, particularly for national economic development. For example, more than 80% of Russia's exports are oil, natural gas, metals and timber, approximately half of which stem from the Arctic region. Similarly, Norway's export of oil and gas constitute 45% of its total exports and more than 20% of its GDP. As such, both Russia and Norway are resource-based economies, similar to Saudi

Arabia, Kuwait, and Qatar. Because oil and gas are strategically important for these countries' economy and development, their governments typically tend to rely on SOEs more than on private firms.

- Developing oil and gas resources in the Arctic also requires high environmental sensitivity, given both its ecological sensitivity (such as the possibly devastating effects of an oil spill) and great relevance for the Earth's system and its dynamics (such as the unpredictable effects of ice melts on ocean currents and the general global climate). This is the so-called "Arctic tipping point" argument (Stockholm Environment Institute and the Stockholm Resilience Centre 2016). Such political sensitivity makes any major Arctic resource extraction, and related infrastructure project, relevant for any Arctic nation's domestic and foreign policies. Governments typically trust their own SOEs more than foreign SEOS, and/or private companies, when it comes to being politically sensitive.
- In addition, the Arctic remains a politically sensitive region. It was largely demilitarized after the end of the Cold War, but its importance for global policies has increased dramatically over the past decade. Issues of Arctic governance, control over its territories and resources, and establishing and maintaining a national presence attract great attention of all countries involved (Pelaudeix and Basse 2018). This adds to the environmental sensitivity of oil and gas exploration, and actually puts it into a bigger geo-political context. Again, governments typically trust their own SOEs more when it comes to being militarily sensitive.
- Oil and gas TNCs have comparative advantages related to their global experience, strong technological skills, and economies of scale greater than SOEs operating just in one geographical area. However, for (Arctic) national governments, such TNCs also mean greater risks, since they are less politically sensitive, less controllable, and may divest from the Arctic and move their assets to other regions. Conversely, and for exactly the very same reasons, TNCs try to establish partnerships with local SOEs, rather than work alone. In some cases, TNCs may actually be forced to do so.
- There is an ongoing domestic debate, in both Russia and in Norway, on how efficient the SOEs are as public policy tools for Arctic oil and gas development. The discussion is highly politicized and often relies on sweeping ideological concepts, rather than factual analysis. However, most experts acknowledge the key role of the SOEs in the Arctic, no matter if they welcome such development or see it as an indispensable evil.
- Finally, one may also ask if, and to what extent, national governments are actually capable of actively steering their own SOEs in extractive industries, and, their SOEs in general. SOEs, especially large ones that are active in resource extraction, not only possess strong resources and are financially independent from government, but play an important role in their countries' economies and supply the national budget with tax revenues, jobs, regional development and other additional revenues and advantages. In addition, they have their own business-development strategies and need to be competitive in global markets, which all diminish the government's grip on their own SOEs.

4.4 Conclusion

In this chapter, we presented and analyzed the cases of three SOEs—Gazprom, Rosneft and Statoil—in Russia and Norway, respectively, in the Arctic. Because Russia and Norway are the two countries that are most dependent upon their Arctic oil and gas resources for their (national) economic development, they heavily rely on their SOEs to explore and exploit these resources. In turn, these SOEs also heavily depend upon their respective governments for both their operations and development, at least in the Arctic.

One can say that these three SOEs and these two countries constitute quite unique cases, which are interesting to discuss in their own right, but are not really generalizable. As such, this chapter primarily helps the reader understand the dynamics of oil and gas development in the Russian and Norwegian Arctic, which is, itself, a contribution to a poorly understood phenomenon.

However, in this conclusion, we would like to go a little further and explore the generalizability of the lessons learned from the above cases:

- There is first the question of the generalizability from oil and gas exploitation to other Arctic mineral resources, the most important ones being nickel, cobalt, copper, zinc, lead, tungsten, titanium, zirconium, gold, silver, platinum, palladium, diamonds and uranium (Bortnikov et al. 2015). Most of the economic features mentioned above also apply, as the exploitation of such mineral resources is equally costly and, probably in need of government support and guarantees. However, all other above-mentioned points probably apply less, as these minerals are generally regarded to be of less strategic importance for a country, and/or of less ecological and political sensitivity. Therefore, one cannot automatically generalize from oil and gas to other minerals in the Arctic.
- However, there is a certain amount generalizability that can be made to oil and gas exploitation in other parts of the world. Developing new oil and gas fields becomes more costly, requiring heavy investments and government support. In parallel, oil and gas resources will be found in increasingly remote and ecologically sensitive areas (such as rainforests and deep-sea ocean beds). This adds to the political sensitivity of such activities and, to an extent, will speak in favor of SOEs, or at least joint ventures with SOEs.
- Finally, there is the question of the generalization of our cases to oil and gas exploitation in other Arctic countries, namely the United States (Alaska), Canada, and Greenland. All the above considerations also apply in the cases of these three countries, although to a much lesser extent in Greenland, as there seem to be no known oil and gas resources, as well as in Canada, which mainly seems to dispose of tar sands and not oil and gas fields. However, generalizability is somewhat limited in Alaska, considering that the importance of Alaskan oil and gas as part of the US economy will always be marginal, so will probably not warrant development of SOEs for Alaskan oil and gas exploitation. Finally, SOEs are clearly not part of the American economic tradition.

References

Arctic Journal (2013) Statoil this year's leader in striking oil and gas. Huge Canadian discovery guarantees number one spot for Norway's oil giant. Retrieved from http://arcticjournal.com/oil-minerals/statoil-years-leader-striking-oil-and-gas

Borten Moe O (2016) Også et sterkt Statoil trenger konkurranse [A strong Statoil needs competition as well]. Retrieved from http://sysla.no/2016/05/04/meninger/ogsa-et-sterkt-statoil-trenger-konkurranse_99890

Bortnikov NS, Lobanov KV, Volkov AV et al (2015) Strategic metal deposits of the Arctic zone. Geol Ore Deposits 57(6):433–453

Bourmistrov A, Borge Doornich J, Krivorotov A (2015) Driving forces for Norwegian–Russian petroleum B2B cooperation: implications for the Barents Sea. In: Bourmistrov A, Mellemvik F et al (eds) International Arctic petroleum cooperation. Barents sea scenarios. Routledge, New York, pp 77–96

Buzovsky V (2016) Факторный анализ условий освоения арктического шельфа РФ. Различия стратегий ПАО «НК Роснефть» и ПАО «Газпром» [Factor analysis of development conditions in the Russian Arctic shelf. Differences between Rosneft and Gazprom strategies]. Pro-Arctic.ru. Retrieved from http://pro-arctic.ru/01/11/2016/resources/23925

Focht E (2015) ConocoPhillips покинула Россию после 25 лет работы [ConocoPhillips leave Russia after 15 years]. RBK Daily, 22 Dec

Gazprom (2016a) First cargo of Yamal oil shipped from Arctic gate offshore terminal [Press release]. Retrieved from http://www.gazprom.com/press/news/2016/may/article274906

Gazprom (2016b) Russia's northernmost active oil field enters commercial production [Press release]. Retrieved from http://www.gazprom.com/press/news/2016/september/article284341

Government of Norway (2006) The Norwegian Government's high north strategy

Government of Norway (2014) Report to the storting (white paper) Meld. St. 27 (2013–2014) Diverse and value-creating ownership

Helgesen OK (2012) Vil bore helt opp til Nordpolen [Wish to drill all the way to North Pole]. Retrieved from http://www.tu.no/artikler/vil-bore-helt-opp-til-nordpolen/244769

Henderson J, Moe A (2017) Russia's gas "Triopoly": implications of a changing gas sector structure. Eurasian Geogr Econ 58(4):442–468

Knott T (2013) Subsea gas feels the pressure. Upstream Technol Q2:14–25

Kozlov D (2016) Ванкор снова утекает в Индию [Vankor to leak to India, again]. Kommersant, 17 June

Lien T (2016) Gas from Norway's high north bringing energy security and opportunities to Europe. Speech at the Atlantic council, 25 Feb 2016. Retrieved from https://www.regjeringen.no/no/aktuelt/atlantic-council/id2477645

Lunden LP, Fjaertoft D (2014) Government support to upstream oil & gas in Russia. How subsidies influence the Yamal LNG and Prirazlomnoe Projects. International Institute for Sustainable Development/WWF, Geneva-Oslo-Moscow

Norsk Petroleum (2017). Produksjon [Production]. Retrieved from http://www.norskpetroleum.no/fakta/produksjon/#arlig

Overland I, Krivorotov A (2015) Norwegian-Russian political relations and Barents oil and gas developments. In: Bourmistrov A, Mellemvik F et al. (eds) International Arctic petroleum cooperation. Barents sea scenarios. Routledge, New York, pp 97–110

Pelaudeix C, Basse EM (eds) (2018) Governance of Arctic offshore oil and gas. Routledge, London

Ree M (2012) Ministermøte om Arktis i Trondheim: Arktisk utvinning krever samarbeid [Minister summit in Trondheim: Arctic upstream requires cooperation]. Retrieved from http://www.tu.no/artikler/arktisk-utvinning-krever-samarbeid/244278

Savina A (2016) Gazprom Neft Shelf: Prirazlomnoye Field Development. Rogtec 45:12–21

Sechin I (2016) Energy market balance formula. Retrieved from https://www.rosneft.com/press/today/item/184179

Shelf Urges New Technologies from Russia (2014) Шельф требует от России новых технологий. I-Mash.ru, Oct 9 2014. Retrieved from http://www.i-mash.ru

Skjeldal G, Berge U (2009) Feber. Historia om norsk olje og gass [The Rush. Norwegian Oil and Gas History]. Cappelen Damm, Oslo

Soliman T, Fletcher L, Fruitiere C (2016) In the pipeline: which oil and gas companies are preparing for the future? Executive summary. CDP

Statoil (2013) Statoil annual report 2012. Sustainability report

Statoil (2016a) Statoil. Sustainability report 2015

Statoil (2016b) Well positioned for a low-carbon future. Retrieved from http://www.statoil.com/en/NewsAndMedia/News/2016/Pages/06dec-autumn-conference.aspx

Statoil (2017) How we cut the break-even price from USD 100–USD 27 per barrel. Retrieved from https://www.statoil.com/en/magazine/achieving-lower-breakeven.html

Stensvold T (2012) Statoil tredobler arktisk forskning [Statoil to triple Arctic research]. Retrieved from http://www.tu.no/artikler/statoil-tredobler-arktisk-forskning/244782

Stockholm Environment Institute and the Stockholm Resilience Centre (2016). Arctic resilience report. Stockholm

U.S.-Nordic Leaders (2016) U.S.-Nordic leaders' summit joint statement, 13 May 2016. Retrieved from https://www.whitehouse.gov/the-press-office/2016/05/13/us-nordic-leaders-summit-joint-statement

Chapter 5
Tourism at the Crossroads of Contesting Paradigms of Arctic Development

Soile Veijola and Hannah Strauss-Mazzullo

5.1 Introduction

Tourism, like oil and gas development, has made the Arctic a hub of industrial activity. A lot of Arctic tourism is built around nostalgic ideas of frontier expeditions, creating curiosity for a sparsely or uninhabited and untamed environment, enriched by special treats such as Aurora Borealis, whale-watching, and other fruits of nature that professional tourism operators aspire to turn into safe and luxurious experiences. The perceived weakening of the presence of indigenous peoples and their cultures, as well as unspoiled nature has led to these becoming objects of tourist desire. Imaginaries, spread through media and social networks, create a rush due to a growing awareness of "last-chance-to-see" landscapes.

While the economic benefits of tourism are substantial for the region, they do not help to solve the wider problems in the Arctic. The list of issues that sustainable research and public policies need to address includes competing land uses (such as reindeer herding, extractive industries, forestry, landscapes of everyday living and leisure), unresolved political and cultural struggles between indigenous and local populations and global business agents, youth outmigration from rural areas, and of course the distinct gender divisions between tourism-related and extraction industry-derived employment and livelihoods, combined with the seasonality of tourism activity.

There is a wide range of topics within the multidisciplinary field known as tourism studies. However, due to different knowledge interests, research results remain largely unconnected in their practical applications. Here, the field resembles

S. Veijola (✉) · H. Strauss-Mazzullo
University of Lapland, Rovaniemi, Finland
e-mail: soile-veijola@ulapland.fi

H. Strauss-Mazzullo
e-mail: hannah.strauss-mazzullo@ulapland.fi

M. Finger and L. Heininen (eds.), *The GlobalArctic Handbook*,
https://doi.org/10.1007/978-3-319-91995-9_5

its topic: tourism is diverse and multiform, not just *one* discourse and reality. While public discussion tends to reduce it to news on rising or declining market segments or numbers of bed-nights sold, globalized hospitalities and mobilities mean that tourism is actually everywhere (see, e.g., Urry 2000).

In this chapter, tourism is on display as *a prism depicting a wide spectrum of Arctic realities across cultural, political and industrial spheres*. In terms of thematizing Arctic development, we find that the issues related to environment, extractable economic resources, cultures and communities of Arctic peoples, and geopolitical governance are solidly inbuilt in tourism development and analysis, and vice versa. Hence, we shall look into the points of connection between these key aspects of Arctic futures and tourism.

The tourism industries must not be seen as a separate industry, but as an assemblage of hotel industries, transport and logistics industries, catering industries, and firms providing tourism program services, all of which consist of small, medium-sized, and large operators; and these are only the ones directly involved in handling tourists and their aspirations. Indirectly, tourism occupies and benefits grocery stores and other shops, cultural institutions, and events related to art and handicraft, sports facilities and arenas, medical professions, alcohol sales, and providers of local food. Through services like Airbnb and Couchsurfing, it also enters the private lives of local people. Thus, the tourist landscapes are multiple and varied and impossible to set apart from the rest of the social life and environment.

We start with a short account of contemporary Arctic tourism across Russia, Finland, Sweden, Norway, Greenland, Canada, and Alaska. We also include the Faroe Islands and Iceland in our list, more because of their climatic conditions than their exact geographic locations (for the same reason, Northern Scotland is also sometimes included in the discussion).

We then define Arctic tourism with the help of Arctic geographies: as *seasons in nature* that fluctuate increasingly in the tempo set up by urbanized working life and the consuming society. In the Arctic, nature-related seasons have traditionally marked both work and leisure more significantly than in areas with less dramatic seasonal changes. Thus, our overview is premised in nature, land, and landscape, which managerial thinking and economic calculations have started to harness (see, e.g., Ingold 2000; Descola and Pálsson 1996).

When discussing northern regions that are rich with natural resources and indigenous landscapes, *land* can be understood as *amenity landscapes* for both tourists and locals, as well as *natural resources* such as timber, minerals, and oil and gas that interest multinational companies. In the first place, however, land is *home* to many of the First World's indigenous peoples. Therefore, tourism stands or falls between economic and political pressures of utilizing nature industrially, on the one hand, and the impacts that climate change has on the livable lives of local cultures, on the other.

For the sake of clarity, we will address two alternative perspectives on developing the Arctic through research, both of which carry practical, ontological, epistemological and ethical consequences. The first viewpoint is that of *industrial rationality*, which advises us to approach the region in terms of its economically exploitable natural and human resources; the second is that of *indigenous cultures* that have traditionally related to the land without making a major distinction between it and themselves. If Arctic tourism follows the example of industrial rationality of seeing nature and culture as extractable economic resources, it will struggle with environmental concerns, together with (other) industries of scale and, just like them, face the need to engage in common efforts for sustainable energy solutions to balance the books. Insights from technology studies promoting sustainability will hopefully be transferred and incorporated in future tourism development.

The other option is to learn from local cultures about ways of "developing" land and landscape for both human and nonhuman species in an entirely different scale and responsibility; that is, not in human-centric ways. Indigenous ontologies and epistemologies of living of and with the land (e.g., Smith 2012) question the narrow, modernist parameters of "development" that tend to deal with mere economic investments, employment, and competitive edge when, for instance, planning land use (see, e.g., Dredge 2010, pp. 105–106).

Thus, the present chapter aims to balance the interplay of natural (and naturalized) sciences of economics and technology, on the one hand, and fields such as indigenous studies, cultural studies, sociology, and political philosophy, on the other. Particularly, when indigenous cultures are recognized as an essential part of the Arctic imagery upon which tourism is sold, alongside images of heroic explorers from centuries past, and the increasing need for silence and slowness craved by modern urbanites, it makes no sense to talk about culture, nature, society, forests, or economy as separate and immobile categories of their own. The Arctic is being constantly mobilized on many levels: not only through tourism and commerce, but also via images, noise, pollution, waste, and meanings that travel and cross borders quickly and widely. Besides, throughout history, people have realized that they have a culture, a language and a place first when a stranger arrives for whom to show and explain it. Places are in many ways created by mobilities, visitors, and connections with other places.

5.2 Features and Figures of the Arctic as a Tourist Destination

The Arctic tourism market integrates mass tourism and more elitist venues. It includes sport fishing and hunting, ecotourism, adventure tourism, culture and heritage tourism, animal watching and photography, and northern lights tourism, and many tourists participate in activities across these divisions. Adventure tourism, for instance, builds on the history of pioneer expeditions (Snyder 2007), although all other market sectors benefit from the adventurous (that is, "harsh and dangerous") image of the High North to some degree, even when it is experienced in the most safe and comfortable way. With activities becoming more regularly conducted and highly controlled, with clothes and means of transport becoming better adapted to the environment, the risk for the tourist has become minimal compared to the adventure tourism of a century ago (ibid.). Indigenous cultures, from Aleut, Yupik and Inuit to Saami and Nenets, Khanty, Evenk and Chukchi cultures (from West to East, starting in Alaska), are drawn on forcefully as an imaginary or real link to experiencing life near nature and with the handicraft skills needed to survive in these milieus.

Tourism is increasing across branches in the Arctic—sometimes dramatically, as in Iceland; very few cases are displaying stagnation (Lasserre and Têtu 2015), and there are no examples of decrease in the long-term (see, e.g., Heimtun and Viken 2015; Johnston 2011; Maher et al. 2014). The following numbers have limited statistical value because they reflect great variations in chosen methods of data collection and the year accounted for. Rather, they provide readers with a tentative idea of the volumes we are actually dealing with and depict noticeable current trends. Further, this overview has to be read in the context of small population numbers (not given here), as well as our more specific discussion of infrastructure, institutions, and land-use plans.

The leaders in Arctic tourism are Alaska, Sweden, and Finland, with each receiving about two million visitors in their Arctic territories annually (Maher et al. 2014). Russia makes up almost half of the Arctic region and offers cruises to the North Pole from its ports, but hosts relatively few visitors; estimates range around 500,000 visitors (Tzekina 2014, cited in Maher et al. 2014), or about 2% of the cumulative number of tourists in the Arctic (Ilkevich 2015). Generally, travelling in the Russian Arctic is restricted by military zones and resource extraction, but also by the need for foreigners to apply for a visa and the region's general lack of infrastructure (specifically, roads). In order to gain greater control over its resources, the Arctic shoreline has been turned into a border zone, limiting access even for Russian citizens (Stammler and Ivanova 2016).

In the Canadian territories, Yukon tops the list, with about 300,000 visitors per year, followed by Nunavik (88,000), the Northwest Territories (64,000), Nunavut (31,000), Manitoba (21,000) and Labrador (10,000) (Maher et al. 2014, with numbers from 2012 or 2013, rounded). A large proportion of visitors to the Canadian Arctic are relatives of locals and business people.

Most of the approximately 68,000 overnight stays in hotels in Greenland by foreign tourists in 2015 were Danish (Visit Greenland 2016). A further 22,000 people visited Greenland on cruise ships, with large vessels landing on the main ports of Nuuk, Ilulissat, Kangerlussuaq, etc., but smaller vessels seek increasingly more remote locations (ibid.).

Iceland has experienced dramatic tourism growth over the last seven years, with an increase of 31% in foreign exchange earnings from 2014 to 2015 (Icelandic Tourist Board 2016). In 2016, an exceptional number of overnights stays was registered ("Overnight stays in 2016 just over 8.8 million", Statistics Iceland, 12 April 2017). This has led to concerns among Icelanders about adverse effects on the landscape, Iceland's main attraction (Icelandic Tourist Board 2016). Among the three northernmost counties of Norway, Troms county has seen the biggest growth, with 32% more foreign overnight stays in 2016 (compared to the previous year, The Local, 28 December 2016). This recent trend was kicked off by the film *Frozen* in 2013, the result of collaboration between Disney and the Norwegian Tourism Agency. With business travel numbers going down, leisure trips are becoming more important in the Norwegian Arctic. Svalbard has seen similar growth recently, with 130,000 visitors in 2015 (Statistics Norway 2016), attracting increasingly Russian tourists and thus marginalizing the community's income from mining (Staalesen, 18 April 2017). The Faroe Islands has seen about 270,000 visitors arriving by air in 2015, displaying a steady increase since the 1980s, while the number of cruise tourists has remained around 50,000 since the 1970s (Statistics Faroe Islands 2016).

The more southern destinations of many Arctic territories are more developed and frequented than the northern parts; because this fact is regularly not accounted for, average numbers can give a misleading impression as they convey little information on the relationship between tourists and host communities, with the latter often being very small. This mismatch has become especially problematic in cruise ship tourism, where great numbers of tourists regularly disembark and almost overrun small coastal towns, such as in Nunavut and Greenland. Hence, in small places, the benefit as well as problems caused by tourism are often significant.

In the framework of tourism, *places* become *tourist destinations* and their key characteristics have been their *accessibility* from crowded urbanized regions around the world. Hence, in the vast area of the Arctic, tourist destinations have developed rather unevenly over the last two centuries. One reason for the growth of tourism hubs lies in the availability of an affordable transport system and available services (Kaae and Råhede 2011) that have catered remote communities and extractive industries in the past. Therefore, while accessibility is important, it is not the only crucial issue.

Experiencing the vastness of the region by car, bus, and boat is a motivating factor for tourists to travel to the Arctic, making the trip itself a priority. Ice roads allow visitors to roam the Canadian North, and especially its roadless areas, in winter. In the Western Canadian Arctic, since the construction of the Dempster Highway in 1979, the mostly indigenous communities of Aklavik, Inuvik, and Tuktoyaktuk have seen an influx of "non-consumptive" or eco-tourists; that is visitors other than sports hunters (see Dressler et al. 2001, p. 38), whose trophies are photographs (Lemelin and Wiersma 2007) rather than animal corpses. Alaska, Nunavut, and Greenland have seen significant growth in touristic activity on cruise ships and leisure boats to visit ports, islands, and icebergs. Travel on land often requires helicopters or zodiacs, which remain the means of transport in the Russian Arctic, at least during summer months. Icebreakers along the Northern Sea Route are few and expensive, although a currently weak currency has positive effects here. Compared to other parts of the Arctic, Russia's remote northern areas invite pioneer tourists for a less professionalized experience. Naryan-Mar and Salekhard in particular are considered key destinations in future tourism development in the Russian Arctic (Ilkevich and Strömberg 2017).

Places that are less accessible provide an extra challenge, but certainly offer "an original, distant, and remote location visited by few" (Grenier 2011, p. 8), as in the case of Nunavik. As in the old days, when the Arctic had generally less developed routes for travel, this option is chosen by those seeking a "different" experience on a less beaten path. This elitist discourse dates back to the 18th century (Abram and Lund 2016, in reference to Butler 1990), and considering current growth rates, such prevailing conception will most probably push the tourism frontier into ever more remote areas of the Arctic.

Compared to the North American and Russian Arctic, as well as Greenland, the Nordic countries are both highly accessible and developed for the European masses and the growing group of visitors from Asian countries. They combine car or bus road trips (on the South-North axis, or East-West), cruises along the Norwegian coast, and charter flights to Rovaniemi, Finland and other destinations, altogether catering for different kinds of holidays and budgets. The attempt to develop destinations elsewhere ("Greenland wants tourism," Christensen 1992, meaning especially the Greenlandic government, see Nuttall 2008) is not always successful and not always wished for (Johnston 2011), despite the Arctic region's general potential in terms of exotic cultures, pristine landscape, and extreme climatic conditions.

Increasing attention in tourism and its studies has been given to animal ethics (e.g. Kristoffersen et al. 2016), but the situation is complex. On one hand, whale watching vessels and low-flying airplanes encroach on the animals' habitats by sea and air (Cunningham et al. 2012). Similarly, concerns have been raised about viewing cabins from which to look at bears and wolves in the wilderness. On the other hand, catch quota and campaigns of environmental NGOs aim to limit the consumption of local communities' traditional staple food. Taken together, this entails environmental, social, and legal pressure on the people inhabiting the Arctic, who are forced to adapt themselves to modern lifestyle and thereby not only to industrial landscapes but also to modern tourism. For communities that have not drawn a clear distinction between themselves and the nature, a pipeline drawn across a forest or alongside a river is not just a pipeline. Nor do they enjoy the presence of animal rights activists disguised as tourists taking pictures of the slaughtering of sea mammals (Dressler et al. 2001).

Finally, while Arctic regions can advertise their thick layers of snow needed for winter sports and pristine, relaxing views—unlike areas further down south with declining snow resources—climate change is going to make weather conditions more unpredictable and finally alter the conditions for various tourism activities. The lengthening of the warm season, increased precipitation, and unstable weather are producing irreversible changes in the landscape. Polar tourism at both poles will be affected by climate change in more drastic ways than in any other region on the face of the earth.

As diverse as the development of tourism has been in the Arctic, it will continue to grow at different speeds and in different directions, creating niches and opportunities as they afford themselves, while in the long run, communities will increasingly be confronted with the tourism frontier reaching the remotest places in attempts to find pristine nature and society.

5.3 Seasons in Nature Defining Work and Leisure

Our approach views Arctic tourism in terms of *seasons in nature* rather than a self-sufficient cluster of hospitality and transportation industries. By doing this, we can understand it as part of everyday life—work and leisure—in the Arctic. People in the Arctic areas have always lived with and of different seasons afforded to them by nature (e.g., Länsman 2004; Mazzullo and Ingold 2008). However, tourism, like other globalized drivers of economy, forces local life to follow a different logic. Instead of seasonal affordances in the cycle of eight different seasons adapted to climate and weather, which is typical of Saami culture, for example, Arctic living is also being pushed to follow modern, Western rhythms and divisions of work and leisure. The old way of drawing from different livelihoods in each season, thus having nature-induced breaks from certain kind of work, allowing various forms of free-time and multi-purpose travelling, has been largely replaced by delivering tourist services through high-season-driven tourism development in a constant competition over being the number-one tourist destination in the region or continent. Hotels are then vacated for some or most of the year, yet they need maintenance and consume energy year-round.

Accordingly, much of contemporary tourism is out of sync with traditional lifestyle in Arctic conditions. Indigenous people may habitually stay in one place and move to another according to the seasonal changes. Theirs is not the rhythm of a five-day working week in an office or factory, nor of a one- or two-week-long holiday. But working life in urbanized global world means mass-based mobilities that require hospitality services only seasonally, evacuating Arctic areas from visitors for several months a year. Even if there were tourists on the move during the off-season, it is not feasible to maintain services in big hotels and resorts only for a few tourists.

For tourism workers, jobs are available seasonally but not all year round. Work is mainly temporal, intensive, precarious, and low-paid (e.g. Valkonen and Veijola 2008; Koikkalainen et al. 2016). It is considered suitable for young people, students, workers from poor countries, and women. People in these categories tend to settle—as they usually have no choice—for lower pay than the average white, middle-class or working class, male person does. Even if working with tourists is often quite demanding, requiring multiple skills and consisting of responsibility over people's lives, it is not valued highly in society partly since the sphere of work in question is part of leisure, which is seen as marginal in society. On the positive side, even if it is a less respected job market, tourism provides opportunities to have additional (or only) income by working out niches, or by working for a bigger tourism company.

The fact that tourism jobs are framed as gendered is crucial. Tourism, along with other sectors of work involving care, catering, cleaning, services and social relations, is considered women's work and feminine skills in most societies (e.g. Veijola and Jokinen 2008). Since women's employment is still largely seen as filling up gaps in men's employment in regional development, their jobs easily give way to grand-scale promises of investments in traditionally male jobs, usually in the production, extraction, or transportation of material resources or outcomes. This makes tourism jobs even more precarious than other lines of work. Because the Arctic areas on both land and at sea are pressurized to impart with their natural resources, such as oil, gas, minerals, and hydro-power, should multinational companies request them, small or medium-sized tourism companies do not have a say in the matter.

In reality, the rigid thinking of tourism as less important women's extra job is costly in both economic and social terms. Tourism gives work and livelihoods to women,

men, families, and entire communities. Moreover, skills that have been traditionally considered as feminine or masculine are now equally demanded from practically everyone in working life (see Adkins 2005). Deeply rooted, gendered hierarchies that are still enforced in the north deem tourism as "light-weight" compared to "heavier" industries.

For young people living in the Arctic, indigenous or otherwise, the situation is particularly cumbersome. Tourism does not automatically offer a buoyant career and permanent livelihood with its seasonal, low-paid jobs and labor-intensive business models, but the alternatives in industrial jobs like mining are not necessarily attractive. The younger generations have increasingly grown up believing that their identity is based on and displayed through the social worlds of leisure and hobbies rather than on a life-long career in one profession. Fittingly, tourism offers many aspects of intensive and vibrant encounters with the global world, but it is also tied to drinking and promiscuous communalities—temporary perhaps for the tourist, but permanent for local youth (e.g. Örnberg and Room 2014). With declining populations in local communities, future prospects of living in thriving communities for the youth are not colossal. Finally, the lack of commitment goes both ways in today's working life: with the help of "zero-hour-contracts" in high-season jobs, companies can hire workers whose salaries depend entirely on factors such as whether the sea freezes or whether there is enough snow to build a snow castle or a ski slope.

5.4 Industrial Rationalities in Tourism Development

It is safe to say that industrial rationality is a guiding principle of tourism development globally, and also in the Arctic. Lands and landscapes that have been inhabited and "scaped" (Ingold 2000), or sculptured, by local and indigenous people are being turned into tourist destinations by way of industrial and managerial logics of making business. The results can eventually turn against both local life and subsequent business opportunities through "re-scaping" of nature and culture.

Quantifying logics equals *coding logics*, for which everything that is solid in terms of progress and success is countable and codable (MacKenzie and Vurdubakis 2011). Yet, in tourism as in all other spheres of life, quantity always affects quality. Expansion of tourist destinations leads, among other things, to individual tourists having less time and space to enjoy the diminishing natural areas of unique amenity values or to experience sincere and spontaneous encounters with the locals.

Quantifying rationality also affects tourism from the outside. In order to compete in terms of controlling land and landscape against other industries and interests, tourism businesses would be wise to promise and promote industrial jobs in large building projects. And, when commodification of the tourist experience is the goal and consequence of industrialized tourism that seeks quantitative rather than qualitative growth, this can lead to standardized tourist products that are easy to consume, but also to grave disappointments regarding the authenticity, hospitality, and credibility of the Arctic experience.

Of course, if the aim is to build tourist destinations that can compete with Las Vegas and all-inclusive holiday resorts planted in poor countries by multinational hotel chains, then the guests will not be disappointed by the lack of authenticity. The value of authenticity is floating in today's tourism; especially in the case of growing tourism from Asia. Across the Arctic, we witness the efforts of the industry to offer a full experience that combines features and activities that are otherwise spread across space and time, and that conveys tourism products that are "pseudo-real, imitative and imaginative [in] nature" (Saarinen 1999, 241). However, if the idea is to pull in tourists who want to experience Arctic wildlife, indigenous livelihood, and local habits and customs of people living in the north—or even to help protect Arctic nature from harm and danger—then the grip of commodification may embrace its marketing assets lifeless. However, the Nordic countries offer everything from Santa Claus to dog sledding. Who knows, one day we might even witness penguins as tourist attraction in the Arctic, if any combination of things you can do in the cold works for some tourists.

Industrial rationality has the potential to have an even more severe impact on the Arctic than hazarding the credibility of wilderness and locality; namely, environmental concerns caused by unsustainable energy solutions.

While the environmental impact of tourism might not be perceived as equally devastating as other land uses, such as mining and forestry, it is not negligible (Roura 2011). It is also not without irony: In the rush to witness a diminishing Arctic landscape, tourists reinforce the decline by adding to the pollution in a highly sensitive environment (Lemelin et al. 2010). The large "footprint" left by Arctic tourists stems from the long distances that have to be covered to reach the Arctic and move within the region, as well as all the resources used during the stay, most of which have to be transported to the destination in the first place. When the mining and fur trades dominated, the Arctic used to be known for its exported goods rather than for its dependence on southern products to feature a rather homogenous lifestyle and tourist experience. Creative innovations are being sought in order to maximize comfort as well as environmental sustainability in tourist dwelling in increasingly extreme locations (Qu et al. 2015). With new forms of tourist lodging, the experience of "sleeping in extraordinary places or spaces" (Salmela et al. 2017, p. 69; see also Valtonen and Veijola 2011) is shifting the focus towards a new market segment. When outdoor temperatures are −30 °C, a tourist can gaze at the sky from inside a glass igloo with heated ceiling, floors, and motorized beds for perfect vision of the Northern Lights. Just how sustainable the new forms of "glamping" (glamorous camping) will be remains to be seen.

In nature-based tourism of the Arctic, the presence of humans entails a threat to the habitat of different species, already diminished by resource extraction and climatic changes (see also Hall 2016 on tourism consumption and biodiversity loss in the Anthropocene). Growing awareness of species at risk—the effort to protect wildlife in national parks—can lead to friction with the local communities where traditional hunting and fishing are prohibited while tourists are welcomed to visit the place (Pashkevich et al. 2016). As has been pointed out elsewhere (e.g., Einarsson 2009), there is a need to involve locals in the planning of new forms of livelihood.

Hence, Arctic tourism cannot turn a blind eye to uncontrolled processes of growth that can gradually destroy its prerequisites: pure nature, woods, water and food, winters with snow, peace and silence. It needs to actively promote renewable and sustainable systems of energy production and waste management in order to prevent unwanted consequences. Tourists, especially those from polluted metropolises, appreciate the experience of pure air in ways that people who breath it on a daily basis can hardly understand.

5.5 Rights and Copyrights of Indigenous Peoples in Tourism

If we turn the tables to approach "the tourist seasons in nature" from the perspective of indigenous peoples and first nations in the Arctic, after looking at them through Western working life, the only ethically consistent tourism involving their lands and cultures would be run, controlled, and profited by these peoples, in their own ways and scales.

However, developing tourism in culturally sensitive and sustainable ways requires also infrastructure, natural scenery, and entrepreneurship. The latter sometimes lacks the social environment to thrive; local people do not necessarily want to engage in tourism. Conscious of the need to promote small enterprises, the Greenlandic Home Rule had set up a company in 1993 to help entrepreneurs start up handicraft-producing companies through financial support and advice. Dressler et al. (2001), for their part, described how interest among the Inuit of the Western Canadian Arctic in developing beluga whale watching varies significantly between communities. Initial funding and managerial capacity need a supportive attitude among community members. To better understand how businesses develop in these areas, Lemelin et al. (2015) listed a range of indicators (including economic, planning, leadership, community and legal aspects) that help evaluate indigenous entrepreneurship especially in their capacity to benefit the wider community.

Cultural copyright and usage in marketing of Indigenous imageries is one of the key issues in Arctic tourism. The use of Saami cultural symbols and the mass reproduction of artifacts by non-indigenous, but also by members of the indigenous group, for tourism purposes has provoked criticism and debate in the Nordic countries (Kramvig and Flemmen 2016). Building on a history of marginalization and suppression, elements like Saami dresses with their typical colorful patterns, symbols from shaman drums, reindeer herding gadgets, and so on have become popular in mainstream culture and part of Lapland's imaginary repertoire, both of which appropriate and reproduce elements as they seem to fit—and sell. One of the rotated activities is the dog sled ride from North American and Greenlandic traditions.

Since the 16th century, when museums in central Europe started to exhibit Arctic cultures, collections of exotic practices, artifacts and ideas have been translated for specific (white, Western) audiences. Today's representations of an imaginary culture are understood both as threats and as opportunities. They pose a threat where justice has not been reinstated, where rights to land use and self-determination are still looming, and where mainstream culture continues to perceive of indigenous livelihood as being naïve, inferior, or folkloric at best (Nuttall 2002). Moreover, members of the Arctic's ethnic minorities hardly resemble the "emblematic" indigenous that is being advertised in tourist brochures, and remain invisible in their contemporary livelihood (Olsen 2003). In addition, the use of cultural markers by non-indigenous entrepreneurs is rarely agreed upon beforehand. While guidelines have been established to prevent inappropriate use in the future, they are often easier to write than to put into practice. The best scenario would, of course, be that indigenous copyrights are respected and that profits benefit the indigenous community through the sale of items produced locally and by its members (Sejersen 2007). While this will not change souvenir business on a large scale, it benefits local craftspeople and caters to a specific group of tourists that crave authentic products.

On a positive note, representations of indigenous cultures in touristic events can also be understood as a way to revive and promote what would otherwise remain marginal, and eventually become history to be encountered only in museums and archives. Hence, tourism can shape and strengthen identity (Aikio 2015, p. 170) when indigenous peoples participate in tourism development. However, hundreds of years of marginalization and misappropriation will not disappear at once; disrespectful appropriation leaves a stale taste, as our minds tend to stick to those meanings established long ago (see, for instance, Denzin 2013, p. 27, on the construction of whiteness in Native American performances).

Considering the legacy of assimilation policies and as-yet-unresolved questions of sovereignty, the exploitation of cultural symbols and practices by non-indigenous peoples for place branding purposes (Saarinen 1999) often appear as another cruel twist of history. Where indigenous peoples initiate, manage, and benefit from touristic activities, tourism has shown significant potential to encourage cultural revival (Dressler et al. 2001), and to offer opportunities for employment and entrepreneurship where other policies have failed to prevent out-migration, especially of the young generation. The prospects and offerings of livelihoods, employment, cultural and social belonging, and self-realization for indigenous youth, and other young people, in the remote, sparsely populated Arctic regions are crucial for sustained habitation based on sense of belonging in the Arctic.

In interviews with the Inuvialuit elders of Inuvik, Aklavik and Tuktoyaktuk Dressler et al. (2001, p. 35) investigated the question, "Can local employment be created through tourism development, in ways that support the local mixed economy and minimize conflict with the traditional sector?" The results of their study highlight how it is crucial to involve the local population directly, not only in the production of guidelines and policies, but as guides, producers and tourism company owners in order to secure place-sensitive interaction between people, traditions, and environments. Interviewed elders in this study showed satisfaction over those encounters with tourists that were full of curiosity for the other one's livelihood, but that any engagement had limits that needed to be respected. Discrepancies over the hunting of beluga whales versus their sightseeing and protection was the most conflictual aspect in the meeting of different cultures (see also Kramvig et al. 2016, on whale watching/hunting in the European High North).

When Indigenous tourism is operated and designed by indigenous peoples, it can lead into creativity, cultural critique, and living well in ways that can even be shared with tourists. Numerous examples from across the Arctic tell successful stories of living and vital culture in areas such as music (throat-singers, yoikers, rappers), modern art, handicraft, hunting, fishing, and herding. New and artistic representations of indigenous livelihood encourage visitors to critically examine cultural images and forms of colonialism in museums and art galleries. For instance, Gauriloff's (2007) documentary *A Shout Into the Wind* vividly depicts the struggles of the young generation in a Skolt Saami village in Finnish Upper Lapland.

Examples from the Nordic countries show the development of *slow tourism* outside the mass market. Interested visitors can join the life of a reindeer herding family for a limited amount of time and experience migration with the animals. These very intimate encounters with people's livelihood are a rare opportunity, and the cost of a week with reindeer herders is rather high. The experience of stepping out of the modernist temporality is perhaps one of most sustainable, if not authentic, promises that can be given to tourists. Tourism can indeed provide a more democratic and sustainable method of community income, if community members are able to benefit from the visitors—which is not always guaranteed, as in the case of tourists disembarking cruise ships (Stewart et al. 2011), or where indigenous peoples are part of a folkloristic landscape rather than wilful entrepreneurs in their own right.

5.6 Responsible Land-Use Planning Forming Tourist Landscapes

Discussions of the Arctic often involve references to its vastness. However, the Arctic is not formed as one geometrical space or container, consisting of measured distances between locations. Life, land, and landscapes in the Arctic are site-specific realities; they are often places with cultural, social, and spiritual meanings and histories. This poses a challenge to any all-inclusive notion of land-use planning that promises to "serve public interest" yet fails to include a considerate analysis of multiple interests in different times and scales among a variety of stakeholders—also tourists (see e.g. Dredge 2010, p. 111; Tuulentie and Mettiäinen 2007). Reindeer herding, forestry and mining form the key parameters of the tourist landscape.

Biological accounts of Arctic landscapes usually emphasize the region's limited carrying capacity, meaning that any harvesting of its natural resources has stricter limits than elsewhere and that it takes relatively long for flora and fauna to recover (Kumpula et al. 2011). Habitats are easily destroyed, pastures quickly overgrazed and, with these longer time scales, any change entails a danger to the animals that feed on them. Similarly, the social and spiritual landscape of the Arctic's inhabitants is considered vulnerable to drastic changes. While traditional land uses are commonly understood to be well adapted to the Arctic's social and biological environment, rapid tourism growth needs to be embedded and negotiated carefully so that it does not exceed region's carrying capacity (Maher et al. 2014).

In this debate, the term "landscape quality" (e.g. Tyrväinen et al. 2014) assumes objective and countable features of natural settings, even when they are often assessed through subjective means. In contrast, the term "therapeutic landscapes" (Gesler 1993; Wilson 2003) gives equal importance to the *concept* of visiting a pristine, vast, sparsely populated area that offers both locals and visitors a strong sense of place. Centuries-old wisdom is sought through the encounter of reindeer herders, shamans and more fictive characters and activities associated with the Arctic.

Herding, hunting and fishing, berry-picking and other nature-bound livelihoods are still alive in the Arctic, also as a tourist activity. The traditional staple food continues to be a major source of nutrients and a healthy diet for Arctic communities; moreover, it offers a sustainable form of living off the land as it has been practiced over centuries. Clashes over bureaucratic requirements are frequent regarding hunting and fishing licenses for subsistence purposes (rather than "sports" hunting and fishing), as in the case of the Teno River in Finnish Lapland, where recently Saami have been deprived of their exclusive salmon fishing rights for the sake of leisure fishing open to everyone. Similarly, traditional livelihood struggles with environmental protection schemes where the maintenance of a reindeer herd must cope with increasing numbers of protected predators. Large numbers of foreign berry-pickers are usually perceived as an encroachment on local people's space, while berry-picking hikers who observe local customs are a welcome sight. Against this background of a conflict over environmental protection, leisure hunting, and traditional engagement with the Arctic landscape and its resources, certain activities seem more acceptable than others, clashing with the formal and informal rules of those still living off the land.

The establishment of national parks is one way to boost environmental protection and tourism at the same time. The world's largest national park is situated in Greenland and encompasses 1 million square km. The recently established national park Russian Arctic aims to both protect wildlife and attract tourists and has involved an attempt to clean the area of waste dumped during Soviet times.

Forestry, another field of stakeholders in land-use planning, has been the backbone of industrial activity in areas below the treeline. Harvesting methods have changed the landscape significantly; in particular, clear-cuts and monoculture plantation have turned old forest with its abundant biodiversity into highly homogenous patches across the forested Arctic (Brotons et al. 2003), making the forest even more vulnerable to climate change. To tourists, the forests continue to be marketed as pristine and deserted, whereas they are in many areas actually far from untouched and are full of activity. Each party benefiting from the forest sees it with different eyes: for its timber, as pasture, as berry-picking ground, or as a spot for skiing, hiking, or motor-sledding.

Extractive industries comes in third, accompanied by industrial tourism. Recent trends in mining in the Arctic tell of favorable conditions, such as increasing resource prices on the world market and the minimal political risk of meeting resistance among inhabitants of the Arctic, while climate change is only "a minor and locally variable factor" (Haley et al. 2011, p. 37). In respect to tourism, mining is understood as a potential threat to all nature-based activities that rely on relatively undisturbed environment. Therefore, a lot of effort has gone into masking the effects of mining, while simultaneously using transport connections for goods, resources, and people as they travel to and from tourist destinations as much as sites of extractions.

The mining town of Kiruna in Northern Sweden has become a well-known, hybrid, attraction, especially since the town, its people, and its houses are being moved to make space for more extraction; a recent TV series, *Midnight Sun*, displayed its scenery abundantly from a bird's eye view, promoting film-induced tourism (Beeton 2005). People's perceptions of the place have changed over time and industrial processes of boom and bust have given meaning and modern lifestyle to its populace (Granås 2012; Nyseth and Granås 2007). The active gold mines in Kittilä and, perhaps even more so, Lemmenjoki, in Finland with their programs for visitors, connect with the imaginaries of the Klondike tourism in Yukon territory, where towns like Dawson City invite visitors to commemorate the great gold rush of the 19th century. Abandoned mines and mine towns are part of an industrial heritage (Conesa et al. 2008). The best-known example is Svalbard, where former sites have remained in place and have transformed extraction into outdoor exhibitions (Roura 2011). As places like these are readily accessible on established transport routes, many of the former mining communities have decided to exploit their heritage for leisure visits, such as Kennicott/McCarthy and Whittier in Alaska, while other former mine towns are to be discovered by independent travelers without the comfort and infrastructure of a professional tourism industry.

Creative ideas are needed to turn the lunar-like landscapes left behind by open-pit mining into less frightening spaces for leisure activity. Reclaimed areas could be turned into skating parks or similar attractions where projects are feasible and funds are available. If mining in the Arctic is to increase, comprehensive solutions are needed that entail respectful arrangements beyond extraction. When careful attention is being devoted to balance tourism and resource extraction, this certainly represents an achievement; however, it continues to neglect local inhabitants for whom a once meaningful landscape may simply be gone when economic benefit is prioritized.

5.7 Conclusion

Tourism is a powerful global force that is turning the north into a *Global North*—tied into the Global South by facing the same global problems. It is reasonable to think that it is in the best interests of the northern regions to be able to preserve Arctic land and landscapes rather than transform them with bulldozers and mines causing pollution in air, water, and land. Thus, tourism plays on the side of efforts towards ecological advancements and less stress through human presence on nature. Tourism can also support the opportunities for many of the Global North's indigenous peoples to take control over their future and revive some of their past in forms of living with and of nature without leaving a permanent, damaging trace on it.

Table 5.1 Contesting paradigms of Arctic development through tourism

Focus on environment	Focus on extraction of economic resources	Focus on arctic tourism	Focus on cultures and communities of the arctic	Focus on geopolitical governance
Energy solutions	**Industrial rationality**	**Seasons in nature**	**Indigenous peoples and other local inhabitants**	**Land-use planning**
Logistics and infrastructure; e.g., air travel	Quantifying logics	Responsible once-in-a-lifetime experiences	Indigenous use of land	The role of reindeer herding in tourist products
Renewable energy solutions	Commodification of the tourist experience	High-season-driven services, facilities, and jobs allowing nature and local people periods annually for resting	Battles over cultural copyrights and local values	The affordances of forestry for visitation in nature
Clean air as eco-luxus	Growth sought through major investments in building projects	Gender politics of livelihoods	Indigenous arts and business overcoming old and new forms of colonialism	The risks in neighborhoods of extractive industries

By way of listing a number of global challenges that tourism is actively part of, we wish to identify alternative routes for developing the Arctic, with an emphasis on peoples' creativity and responsibility in making a living while respecting local culture and the nonhuman world. We consider ways of living with nature in the north as a renewable, responsible tourist product that is truly worth the label "once-in-a-lifetime", while perhaps also leading to a permanent transformation in lifestyle and consumption.

To sum up the dimensions of Arctic development in the context of tourism, we have formulated the suggestive table (Table 5.1).

Locals—both indigenous and non-indigenous peoples—are crucial agents in future Arctic tourism. If they engage actively with tourist companies and visitors themselves and steer the tourist practices into responsible ones, their choices and abilities will enable them to make the best out of it and set priorities and limits. For this to happen, it is crucial to compile and review past experiences and ongoing developments around the world research and development institutions in order to have a lively debate and effective planning around tourism.

Tourism presents an opportunity to re-create Arctic livelihoods, with seasons and their affordances, and to invite people from around the world to not only see, but also to learn from and respect Arctic realities. Visitors, strangers, and new locals are needed in order for Arctic societies to actively engage in and shape Arctic future through co-evolving and co-habitation, following an ethics of mobile neighboring (Veijola and Falin 2016) with other peoples and species. The younger generations must be given a chance to take care of the Arctic as well.

We should not forget that mobility goes both ways; in other words, the same globalized processes that bring people into the Arctic through work and leisure also enable the people of the Arctic to leave their homes. Agencies in the Arctic are both rooted and mobile, and so are its images and belongings.

Acknowledgements We wish to thank Pasi Satokangas and Nuccio Mazzullo for valuable comments on the manuscript.

References

Adkins L (2005) The new economy, property and personhood. Theor Cult Soc 22(1):111–130
Abram S, Lund KA (2016) Green ice? Tourism Ecologies in the High North. In: Abram S, Lund KA (eds) Green ice: tourism ecologies in the High North. Palgrave Macmillan, London, pp 1–24
Aikio E-M (2015) Tourism development in Saami communities in Finland. Sustain Dev Circumpolar North. From Tana, Norway to Oktemtsy, Yakutia, Russia 1:167–175
Beeton S (2005) Film-induced tourism. Channel View Publications, Claveton
Brotons L, Mönkkönen M, Huhta E, Nikula A, Rajasärkkä A (2003) Effects of landscape structure and forest reserve location on old-growth forest bird species in Northern Finland. Landscape Ecol 18(4):377–393
Butler RW (1990) Alternative tourism: pious hope or trojan horse? Jour Travel Research 28(3):40–45
Christensen T (1992) Greenland wants tourism. Polar Record 28(164):62–63
Conesa HM, Schulin R, Nowack B (2008) Mining landscape: a cultural tourist opportunity or an environmental problem?: The study case of the Cartagena—La Uni6n Mining District (SE Spain). Ecol Econ 64(4):690–700
Cunningham PA, Huijbens EH, Wearing SL (2012) From whaling to whale watching: examining sustainability and cultural rhetoric. J Sustain Tourism 20:143–161
Denzin NK (2013) Indians on display: global commodification of Native America in performance, art, and museums. Left Coast Press, Walnut Creek, CA
Descola P, Pálsson G (eds) (1996) Nature and Society. Anthropological perspectives. Routledge, London
Dredge D (2010) Place change and tourism development conflict: Evaluating public interest. Tour Manag 31(1):104–112
Dressler WH, Berkes F, Mathias J (2001) Beluga hunters in a mixed economy: managing the impacts of nature-based tourism in the Canadian western Arctic. Polar Record 37(200):35–48
Einarsson N (2009) From good to eat to good to watch: whale watching, adaptation and change in Icelandic fishing communities. Polar Res 28:129–138
Gauriloff K (2007) Huuto tuuleen—A shout into the wind. Oktober
Gesler WM (1993) Therapeutic landscapes: theory and a case study of Epidaurus, Greece. Environ Plan D: Soc Space 11(2):171–189
Granås B (2012) Ambiguous place meanings: living with the industrially marked town in Kiruna, Sweden. Geogr Ann: Ser B Human Geogr 94(2):125–139
Grenier AA (2011) Introduction. In: Grenier AA, and Müller DK (eds) Polar tourism. A tool for regional development. Presses de l'Université du Québec, Québec, pp 1–22

Haley S, Klick M, Szymoniak N, Crow A (2011) Observing trends and assessing data for Arctic mining. Polar Geogr 34:37–61
Hall CM (2016) Loving nature to death: tourism consumption, biodiversity loss and the Anthropocene. In: Gren M, Huijbens EH (eds) Tourism and the Anthropocene. Contemporary geographies of leisure, tourism and mobility. Routledge, Oxon, pp 52–73
Heimtun B, Viken A (2015) Nordlysturismen – En institusjonelt betinget utvikling? In: Viken A (ed) Turisme destinasjonsutvikling. Gyldendal Akademisk, Oslo, pp 175–195
Icelandic Tourist Board (2016) Tourism in Iceland in figures (p 27). Reykjavik
Ilkevich S (2015) Arctic tourism in Russia: major challenges for economic, community and environmental sustainability. In: 2nd international multidisciplinary scientific conference on social sciences and arts, SGEM, vol 3. SGEM, Wien, pp 63–70
Ilkevich S, Strömberg P (2017) Arctic tourism in Russia: attractions, experiences, challenges and potentials. In: Lee Y-S, Weaver DB, Prebensen NK (eds) Arctic tourism experiences. Production, consumption and sustainability. CABI, Wallingford, pp 169–180
Ingold T (2000) The perception of the environment. Essays on livelihood, dwelling and skill. Routledge, London
Johnston ME (2011) Arctic tourism introduction. In: Maher PT, Stewart EJ, Lück M (eds) Polar tourism. Human, environmental and governance dimensions. Cognizant Communication Corporation, New York, pp 17–32
Kaae BC, Råhede MK (2011) Tourism Strategies and Environmental Labeling in Greenland. In: Maher PT, Stewart EJ, Lück M (eds) Polar tourism. Human, environmental and governance dimensions. Cognizant Communication Corporation, New York, pp 81–101
Koikkalainen S, Valkonen J, Huilaja H (2016) Drunken sled dogs, alcohol use and teamwork in nature tourism guiding. J Comp Res Anthropol Sociol 7(1):s. 41–58
Kramvig B, Flemmen AB (2016) What alters when the traditional sámi costume travels? A study of affective investments in the sápmi. In: Frykman J, Povrzanović Frykman M (eds) Sensitive objects. affect and material culture. Nordic Academic Press, Lund, pp 179–197
Kramvig B, Kristoffersen B, Førde A (2016) Responsible cohabitation in Arctic waters: the promise of a spectacle tourist whale. In: Abram S, Lund KA (eds) Green ice: tourism ecologies in the European High North. Palgrave Macmillan, London, pp 25–47
Kristoffersen B, Norum R, Kramvig B (2016) Arctic whale watching and Anthropocene ethics. In: Gren M, Huijbens EH (eds) Tourism and the Anthropocene. contemporary geographies of leisure, tourism and mobility. Routledge, Oxon, pp 94–110
Kumpula T, Pajunen A, Kaarlejärvi E, Forbes BC, Stammler F (2011) Land use and land cover change in Arctic Russia: ecological and social implications of industrial development. Spec Issue Polit Policy Carbon Capture Storage 21(2):550–562
Lasserre F, Têtu P-L (2015) The cruise tourism industry in the Canadian Arctic: analysis of activities and perceptions of cruise ship operators. Polar Rec 51(1):24–38
Lemelin H, Dawson J, Stewart EJ, Maher P, Lück M (2010) Last-chance tourism: the boom, doom, and gloom of visiting vanishing destinations. Curr Iss Tourism 13(5):477–493
Lemelin RH, Koster R, Youroukos N (2015) Tangible and intangible indicators of successful aboriginal tourism initiatives: a case study of two successful aboriginal tourism lodges in Northern Canada. Tour Manag 47:318–328
Lemelin RH, Wiersma EC (2007) Gazing upon Nanuk, the polar bear: the social and visual dimensions of the wildlife gaze in Churchill, Manitoba. Polar Geogr 30(1–2):37–53
Länsman A-S (2004) Väärtisuhteet Lapin matkailussa. Kulttuurianalyysi suomalaisten ja saamelaisten kohtaamisesta. Inari, Kustannus-Puntsi
Maher PT, Gelter H, Hillmer-Pegram K, Hovgaard G, Hull J, Jóhannesson GÞ, Pashkevich A (2014) Arctic Tourism: Realities and Possibilities. In: Heininen L, Exner-Pirot H, Plouffe J (eds) Arctic yearbook: human capital in the north. Northern Research Forum, Akureyri, pp 290–306
Mackenzie A, Vurdubakis T (2011) Codes and codings in crisis. Signification, performativity and excess. Theory Cult Soc 28(6):3–23

Mazzullo N, Ingold T (2008) Being along: place, time and movement among Sámi people. In: Bærenholdt JO, Granås B (eds) Mobility and place: enacting Northern European peripheries. Ashgate, Aldershot, pp 27–38
No author (2016, 28 December) Norway's tourism boom shows no signs of slowing down. The Local
Nyseth T, Granås B (eds) (2007) Place reinvention in the north. Dynamics and governance perspectives (Nordic Research Programme 2005–2008 No. 1). Stockholm
Nuttall M (2002) Protecting the Arctic. Indigenous peoples and cultural survival. Routledge, London
Nuttall M (2008) Climate change and the warming politics of autonomy in Greenland. Indigenous Aff 1–2:44–51
Olsen K (2003) The touristic construction of the "emblematic" Sámi 1. Acta Borealia 20:3–20
Örnberg JC, Room R (2014) Impacts of tourism on drinking and alcohol policy in low- and middle-income countries: a selective thematic review. Contemp Drug Probl 41(2):145–169
Pashkevich A, Stjernström O, Lundmark L (2016) Nature-based tourism, conservation and institutional governance: a case study from the Russian Arctic. Polar J 6(1):112–130
Qu J, Villumsen A, Villumsen O (2015) Challenges to sustainable Arctic tourist lodging: a proposed solution for Greenland. Polar Record 51(5):485–491
Roura RM (2011) The footprint of polar tourism: tourist behaviour at cultural heritage sites in Antarctica and Svalbard, vol 7. Arctic Centre, University of Groningen, Groningen
Saarinen J (1999) Representations of indigeneity: Sami culture in the discourses of tourism. In: Sant PM, Brown JN (eds) Indigeneity: constructions and re/presentations. Nova Science Publishers, New York, pp 231–249
Salmela T, Valtonen A, Miettinen S (2017) An uncanny night in a nature bubble: designing embodied sleeping experiences. In: Fesenmaier DR, Xiang Z (eds) Design science in tourism: foundations of destination management. Springer International Publishing, Cham, pp 69–93
Sejersen F (2007) Entrepreneurs in Greenland. In: Dana L-P, Anderson RB (eds) International handbook of research on indigenous entrepreneurship. Edward Elgar Publishing, Cheltenham
Smith LT (2012) Decolonizing methodologies. Research and Indigenous Peoples. Zed Books Ltd, London, New York
Snyder JM (2007) Pioneers of polar tourism and their legacy. In: Snyder JM, Stonehouse B (eds) Prospects for polar tourism. CABI, Oxon, pp 15–31
Staalesen A (2017) Tourism outdoes coal in Svalbard settlement. Barents Observer, April 18
Stammler F, Ivanova A (2016) Resources, rights and communities: extractive mega-projects and local people in the Russian Arctic. Europe-Asia Stud 68(7):1220–1244
Statistics Faroe Islands (2016) Faroe Islands in figures 2016 (p 40). Argir
Statistics Iceland (2017) Overnight stays in 2016 just over 8.8 million, April 12
Statistics Norway (2016) This is Svalbard 2016. What the figures say (p 28). Oslo
Stewart EJ, Draper D, Dawson J (2011) Coping with change and vulnerability: a case study of resident attitudes toward tourism in Cambridge Bay and Pond Inlet, Nunavut, Canada. In: Maher PT, Stewart EJ, Lück M (eds) Polar Tourism. Human, Environmental and Governance Dimensions. Cognizant Communication Corporation, New York, pp 33–53
Tuulentie S, Mettiäinen I (2007) Local participation in the evolution of ski resorts: the case of Ylläs and Levi in Finnish Lapland. Forest Snow Landscape Res 81(1/2):207–222
Tyrväinen L, Uusitalo M, Silvennoinen H, Hasu E (2014) Towards sustainable growth in nature-based tourism destinations: clients' views of land use options in Finnish Lapland. Landscape Urban Plan 122:1–15
Tzekina M (2014) Estimation of tourism potential of Russian Far North. PhD thesis, Moscow State University, Moscow
Urry J (2000) Sociology beyond societies. Mobilities for the twenty-first century. Routledge, London
Valkonen J, Veijola S (eds) (2008) Töissä tunturissa. Ajatuksia ja kirjoituksia matkailutyöstä. Suomen Akatemia/Lapin yliopistokustannus
Valtonen A, Veijola S (2011) Sleep in tourism. Ann Tourism Res 38(1):175–192
Veijola S, Falin P (2016) Mobile neighboring. Mobilities 11(3):382–399

Veijola S, Jokinen E (2008) Towards a hostessing society? Mobile arrangements of gender and labour. NORA—Nord J Feminist Gend Res 16(3):166–181
Visit Greenland (2016) Tourism report Q3–Q4 2015 (p 15). Statistics Greenland, Nuuk
Wilson K (2003) Therapeutic landscapes and First Nations peoples: an exploration of culture, health and place. Health and Place 9(2):83–93

Chapter 6
Arctic Shipping: A Contrasted Expansion of a Largely Destinational Market

Frédéric Lasserre

6.1 Introduction

In the frame of climate change, sea ice conditions are changing; the length of the navigable season, depending on the vessel ice class, has already expanded and is expected to increase further (Stephenson et al. 2013, 2014). This reduction in sea ice extent and volume has triggered scenarios of fast expansion of maritime trade along Arctic sea routes. The impact of climate change on melting Arctic sea ice has been widely discussed in the scientific literature, as well as in the media. The media largely reported two events that fuelled these narratives on the advent of Arctic shipping. In 2009, the now-defunct German heavy-lift shipping company Beluga sent two multipurpose ships from South Korea to deliver construction material to Yamburg in Siberia, and the ships then loaded steel pipes in Arkhangelsk before proceeding to the Netherlands, thus performing what was actually destinational shipping.[1] This voyage was hailed, incorrectly, as the first commercial transit along the Northeast Passage.[2] Then, in 2013, the bulk cargo ship *Nordic Orion* transited the Northwest Passage from Vancouver to Pori (Finland).

The fast melting of sea ice and the important media coverage of these shipping experiments have triggered debates among scholars, government officials, and journalists about the potential development of commercial shipping in the Arctic,

[1] As they unloaded in Yamburg, and loaded in Arkhangelsk (DNV USA 2010). Transit shipping implies no stopover along the way.

[2] In 1997, the Finnish tanker *Uikku* had crossed the Northeast Passage. Several Soviet/Russian cargo ships had also completed the transit earlier. The media probably meant that this was the first *foreign* cargo transit when they reported the 2009 Beluga semi-transit, but they had forgotten about the *Uikku*. Source: Lawson Brigham, quoted in the *New York Times*, "Commercial Arctic Passage Nearing Goal", Sept. 4, 2009; Lasserre (2010c).

F. Lasserre (✉)
Université Laval, Québec, Canada
e-mail: frederic.lasserre@ggr.ulaval.ca

M. Finger and L. Heininen (eds.), *The GlobalArctic Handbook*,
https://doi.org/10.1007/978-3-319-91995-9_6

particularly along the fabled Northwest and Northeast Passages, which offer much shorter distances between Europe and Asia than the classical itineraries through Panama or Suez and Malacca. The abrupt changes experienced in the Arctic fuelled sustained interest from the media (Christensen 2013) that developed a narrative in which climate change would bring about major environmental transformations, but also geopolitical tensions over resources, continental shelves and strategic seaways where traffic was bound to expand quickly (Lasserre 2010a, b).

These debates often turned to dramatic reports or assertions about the oncoming surge in commercial traffic in the Arctic (Beveridge et al. 2016). For instance, Yang Huigen of the Polar Research Institute of China predicted that 5–15% of China's international trade would use the Northern Sea Route (NSR) by 2020 (*The Economist* 2014). Jong-Deog Kim, head of the Polar Policy Research Center at the Korean Maritime Institute in Seoul, predicted that traffic between Europe and Asia along the Northern Sea Route would grow by 6.5% a year and could potentially account for a quarter of all cargo traffic by 2030 (Koranyi 2013). Didier Schmitt (2013) estimated that, by 2030, the proportion of global traffic that would pass through the Northern Sea Route would be 15%. Similarly, reportedly, the 2013 "transit of the MV *Nordic Orion* marks an important place in maritime history. Arctic shipping is now a reality. The advent of Arctic shipping could usher in a new era of trade" (O'Leary 2014). In turn, the reported strong expansion of shipping traffic underlines uncertainties about safety at sea—what happens if there is an accident?—and about sovereignty claims—to what extent are States going to respect Russia's and Canada's claims of sovereignty over the Northeast and the Northwest passages, respectively? (Fig. 6.1).

Several dozen papers have been published about the future of Arctic shipping since the turn of the century, mostly focusing on the idea of climate change and shorter distances as the driver for the expansion of shipping: as sea ice retreated and opened shorter routes, Arctic shipping was bound to expand quickly (Borgerson 2008; Howard 2009; Lasserre 2010a; Maurette 2010; Emmerson 2011; Young 2011; Rahman et al. 2014; Keupp 2015). The idea of a developing Arctic shipping continues to attract attention from the scientific community. A few have looked at destinational shipping (Thorez 2008; Brigham 2010; Pelletier and Guy 2012, 2015), but the majority have focused on transit shipping, which probably reflects media, governmental, and academic interest in the opening of these potential Arctic routes and joins the dominant narrative about the impacts of climate change in the region.

Since the impact of climate change was outlined, climate change has caused ice to melt at an unabated rate. Shipping, or vessel traffic, has definitely increased in the Arctic; however, transit shipping still shows very limited traffic. Arctic shipping has developed, but presents contrasted rates of growth depending on the region and on the market. This chapter analyzes the evolution of commercial Arctic shipping (cargo and cruise) and underlines the major differences between Arctic regions, hinting that the receding ice may be a facilitating factor, but not a driver of the development of Arctic shipping.

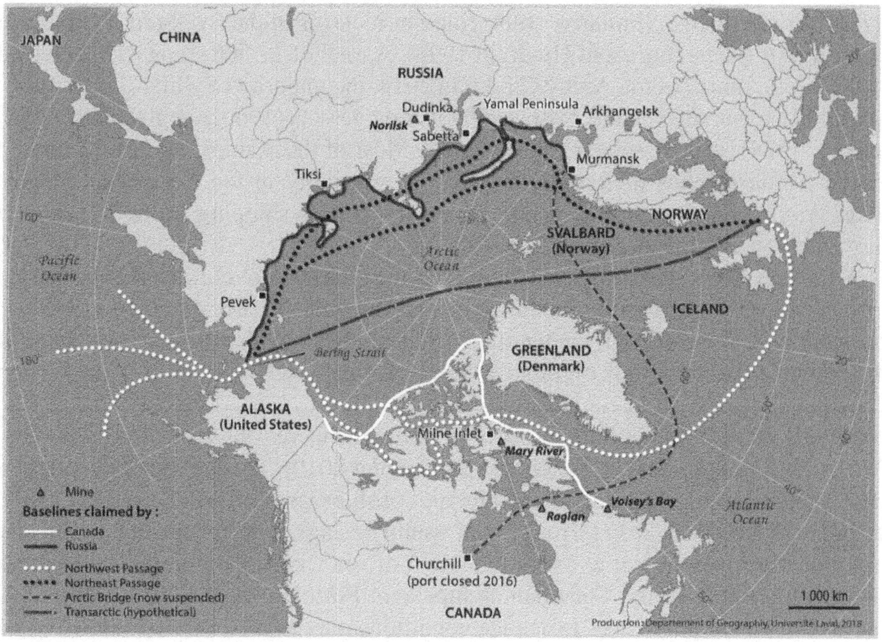

Fig. 6.1 Present and potential Arctic sea routes

6.2 Arctic Shipping Is not New

The strong media, governmental, and scientific attention to Arctic shipping over the past 15 years largely rests on the idea that climate change, resulting in the melting of sea ice, will open up much shorter routes, which will free up new and strategic seaways for world transit traffic (Howard 2009; Lasserre 2010c; Emmerson 2011). The narrative of climate change opening up new shipping routes contributed to the idea that Arctic shipping was a new feature of the region. This is not true, of course, as access to the Arctic has long rested solely on sea transport. Despite the heavy ice concentration, shipping developed whenever government or business ventures were set up. In Svalbard, the mining frenzy that began around the turn of the 20th century and underlined the issue of sovereignty over the archipelago, largely relied on shipping for supply and coal shipment (Arlov 1994). The Soviet government established the Northern Sea Route Administration in 1932 in order to assert its sovereignty and to provide support for the planned economic development of Siberian natural resources (Mulherin et al. 1994; Thorez 2008; Lasserre 2010c). Thanks to the warming effect of the Gulf Stream, which leaves it ice-free nearly year-round, the port of Murmansk has been busy since 1916. Cargo shipping has been present in the Canadian Arctic since the 1920s, albeit with limited numbers, for community supply, the construction of the DEW line, the development of natural resources project, and for grain export after the construction of the port of Churchill in Hudson's Bay in

1931 (Wright 2016). Similarly, cruise tourism took off in the Svalbard archipelago in the 1970s (Kaltenborn and Hindrum 1996). A total of 11,066 ships were detected in the Arctic (north of the Arctic Circle) in 2014, the majority of which were supply, research, and survey vessels, followed by fishing vessels (1960), cargo ships (1892), tankers, (524), and passenger (308) vessels. Most of these ships were concentrated in the Arctic reaches of the North Atlantic, especially in the North and Barents Seas. Arctic shipping represented 9.3% of the world's shipping traffic in terms of movement in 2014 (Eguíluz et al. 2016).

So, while Arctic shipping is not new, after the aftermath of climate change started to become apparent for the media as sea-ice melted, debate about its scope and possible development took center-stage. Indeed, in parallel with the movement of the retreat of sea-ice, it took off rapidly in several regions where it had previously been constrained by ice. In Greenland, cruise tourism has expanded quickly since 1994, especially since 2003 with the number of port calls increasing from 164 in 2003 to 375 in 2008 (Snyder 2007; Stonehouse and Snyder 2010). The number of voyages in the Canadian archipelago, including all types of ships, grew from 121 in 2005 to 416 in 2017 (NORDREG 2017). However, it would be an over-simplification to directly link the expansion of traffic to the reduction of sea-ice. In their statistical analysis of shipping traffic in the Canadian archipelago, Pizzolato et al. (2014) underlined the fact that external factors to sea ice conditions may also be driving the observed increases in shipping traffic in the Canadian Arctic. Contrasting images can be drawn from the observation of traffic in different regions of the Arctic.[3]

6.3 Cruise Tourism Is Expanding in Niche Markets

The main cruise tourism markets are Greenland and Svalbard, but not Canada—-contrary to a popular image—although this market could develop in the future. The sizes of these three markets are currently very different (Table 6.1).

The year 2016 witnessed a jump in passenger traffic in the Canadian Arctic, thanks to the transit of the large ship *Crystal Serenity* that had 1619 people on board. However, the cruise shipping market in the Canadian archipelago remains dominated by smaller vessels of less than 380 people on board: in 2017, the *Crystal Serenity* transited again with 1391 people on board, but no other large ship sailed Canadian Arctic waters.

However, figures show that the cruise market is much more developed in Greenland and in Svalbard, although these markets remain modest when compared with other destinations (Lasserre and Têtu 2015). Figures also suggest that the market is not expanding constantly. In Svalbard, the number of passengers increased from 5000 in 1975 (Kaltenborn and Hindrum 1996) to 24,000 in 1994, before retreating to 19,736 in 2003, peaking at 40,265 in 2007 and stabilizing around this figure. In

[3]Iceland, for the purposes of this research, is not included in the extent of maritime Arctic studied here.

Table 6.1 Number of cruise passengers, Canadian Arctic, Greenland and Svalbard, 2003–2016

Destinations	2003	2004	2005	2006	2007	2008	2009
Canadian Arctic			1239	1936	2131	2633	1406
Greenland	9655	15,654	16,446	22,051	23,506	28,891	26,976
Svalbard	19,736	21,206	29,224	34,908	40,256	38,737	38,269
Destinations	2010	2011	2012	2013	2014	2015	2016
Canadian Arctic	2307	1629	1496	*3760*	*2469*	*4587*	*6036*
Greenland	30,271	29,826	23,399	21,496	20,214	25,049	24,244
Svalbard	35,448	33,896	42,363	36,257			41,000

Sources Lasserre and Têtu (2015); *Greenland in Figures* 2016; NORDREG (2017); Statistics Norway, *This is Svalbard* 2016

Note For the Canadian Arctic, figures displayed after 2013 inclusively report people on board; i.e., passengers and crew

Table 6.2 Number of cruise voyages, Canadian Arctic, Greenland and Svalbard, 2003–2016

Destinations	2003	2004	2005	2006	2007	2008	2009	2010
Canadian Arctic	10	17	12	15	17	20	11	18
Greenland				121	81	99	77	89
Svalbard	43	45	50	41	42	45	54	33
Destinations	2010	2011	2012	2013	2014	2015	2016	2017
Canadian Arctic	18	11	10	17	11	18	20	19
Greenland	89	64	72	88	89	90	104	
Svalbard	33	34	36	33				

Sources Lasserre and Têtu (2015); *Greenland in Figures* 2016; NORDREG (2017); Statistics Norway, *This is Svalbard* 2016

Greenland, after rapid expansion in the first decade of the century, the market receded sharply to 21,496 in 2013 before stabilizing at around 24,000 passengers per year.

The number of voyages also underlines the fact that Canada lags behind Greenland and Svalbard in terms of cruise traffic (Table 6.2). In Canada, the number of cruises has not yet been greater than 20 voyages per year. In Greenland, the number reached 104 in 2016 after the traffic experienced ups and downs, just like for Svalbard. Comparing Norwegian and Greenlandic figures reveals that much larger ships are used for cruises in Svalbard: the average number of passengers per cruise is 812 over the period 2003–2013, compared to 283 in Greenland (2006–2016). This underlines that the Greenlandic cruise market is structured around small ships as in Canada, whereas much larger ships ply Svalbard waters.

It also appears that, if the market did expand at the beginning of the 21st century, possibly helped by ice melt, the expansion stalled afterwards, hinting that the receding ice was probably not the sole driver. Research shows that local regulations, market conditions in consumer countries, and above all transport infrastructure are also decisive factors that condition the variable growth pace for the industry (Lasserre and Têtu 2015; Keil 2017).

6.4 Major Arctic Routes: A Contrasted Evolution

As underlined above, commercial shipping is not new to the Arctic. Moreover, depending on the area, the traffic is not negligible, especially in ice-free zones such as those north of Iceland and in the Barents Sea. Most narratives about Arctic shipping had high expectations about the opening up of new sea routes, the Northwest and the Northeast Passages, especially with a transit function (Borgerson 2008; Howard 2009; Lasserre 2010c; Emmerson 2011; Christensen 2013). These new potential shipping routes were believed to offer high potential for transit shipping between the North Atlantic and Asia or the North American Pacific seaboard. The Arctic bridge, a sea route between the Russian port of Murmansk and the Canadian port of Churchill (see Fig. 6.1) was also believed to have a bright future (Lasserre 2010a, b, c). To what extent were these expectations fulfilled?

6.4.1 Greenland

Given that cruise shipping is expanding in Greenland, and that Arctic waters are receding fast along Greenlandic coasts, one could surmise that cargo traffic has experienced similar growth. Royal Arctic Lines, a Danish shipping line, has long developed a commercial network servicing Greenlandic communities. As port infrastructure is quite developed in Greenland when compared to other places like Canada, with real docks facilitating mooring and the loading/unloading of cargo, even container cargo services are offered to Greenland. However, traffic figures indicate that receding ice is not a determining factor (Tables 6.3 and 6.4).

Table 6.3 contains the number of voyages to Greenland (cabotage or domestic traffic is not included), passenger ships, and consolidated with cargo and container ships. This traffic expanded until 2008 (206 voyages), then contracted gradually until 2013 (143). Tanker traffic also peaked in 2010 (58) and receded sharply in 2013 (24). However, fishing vessels and cruise ships are witnessing an expansion of their activity, from 49 voyages in 2004 to 124 in 2013 for fishing vessels, from 83 in 2005 to a peak in 2010 (193), and then 130 in 2013. Total traffic expanded, in contrasting ways to those just exposed, from 390 voyages in 2004 to a 2010 peak of 855, then contracted to 507 in 2013.

Cargo traffic (Table 6.4) first experienced rapid growth between 2000 and 2008, from 649,000 to 906,000 m^3 (an increase of 39.6%). This traffic was more fueled by exports (growth of traffic from Greenland 67.4%) than by imports (28.2%). Traffic

Table 6.3 Number of recorded sailings (voyages) for the whole of Greenland, by ship category, 2004–2013

Ship Category	2004	2005	2006	2007	2008	2009	2010	2011	2012	2013
Transport ships (passengers, cargo, container)	142	192	159	240	206	171	162	184	155	143
Tankers	47	51	39	42	42	57	58	60	54	24
Fishing vessels	49	65	58	54	44	54	169	145	101	124
Research vessels	44	44	48	37	77	62	71	44	63	20
Cruise ships	84	83	86	87	124	96	193	113	106	130
Government vessels	8	27	13	21	24	12	16	17	25	12
Other ships	16	36	23	35	74	59	786	134	86	54
Total	390	498	426	516	591	511	855	697	590	507

Based on Arctic Command (2014), in Christensen et al. (2018). Each sailing (voyage) corresponds to one entry into the Greenland Exclusive Economic Zone

Table 6.4 Cargo traffic to and from Greenland, 2000–2016

	2000	2005	2006	2007	2008	2009	2010	2011	2012	2013	2014	2015	2016
To Greenland	347	418	404	426	445	401	386	423	389	353	340	373	402
From Greenland	184	307	329	314	308	291	310	300	295	269	266	252	288
Internal	118	115	144	139	137	129	118	111	127	103	107	108	113
special project	6	7	6	7	16	7	22	20	17	20	20	30.5	26.5
Total	649	840	877	879	906	828	836	854	828	745	733	763.5	829.5

Unit Thousands of cubic meters
Source Royal Arctic Lines Annual Reports; *Greenland in Figures* 2014, 2016, Nuuk: Statistics Greenland

then receded to 733,000 m^3 in 2015; in this decline, imports (−16.2%) are less to blame than a weakening of traffic from Greenland (−18.2%) and within Greenland (−21.2%) and was not compensated by the expansion of traffic generated by special projects. Imports are largely affected by the construction industry and domestic consumption, while exports rely largely on fish products and thus depend on world markets. These traffic figures show that even though the receding ice may have facilitated operations in Greenlandic ports, especially in spring and fall, that did not translate into a continuous and steady expansion of commercial traffic; here again, the melting of the ice is an enabler, not a driver, and the paramount factor remains markets opportunities.

6.4.2 The Northwest Passage

Although there is a long continuous history of a few transit passages through the Northwest Passages, this traffic was mainly fueled by government icebreakers or pleasure crafts. Transit traffic remains very modest and mainly fueled by tourism and pleasure-crafts (see Table 6.5). As for destinational traffic—that is, ships going to the Canadian Arctic to load, unload, or perform an economic activity there—it is apparent that this segment is experiencing significant growth, fueled by fishing but also commercial cargo traffic. This market is driven by the servicing of local communities, by natural resources exploitation and by the Arctic Bridge traffic up 2016, when the port of Churchill was closed down by its owner, Omnitrax.

The above figures underline several facts:

- Traffic is indeed increasing in the Canadian Arctic, with 416 voyages to the Arctic region in 2017, compared to 121 voyages in 2005 (a 3.4-fold increase).
- Commercial cargo ships represented 188 voyages (45.2%) of these voyages in 2017, compared to 121 (53.7%) in 2005. The expansion of this traffic does not appear to have been strongly affected by the closing down of the port of Churchill and the end of the Arctic Bridge.
- Despite a general trend towards the expansion of traffic, some submarkets may stagnate or contract. For instance, the Arctic bridge never really took off as traffic declined after 2010 without OmniTrax being able to develop alternate markets. The 2012 closure of the Canadian Wheat Board was not good for the economic fortunes of Churchill or the Hudson Bay Railroad, since it had provided the vast majority of the port's traffic through grain exports (Financial Post 2017). Traffic dropped from 650,000 tons of grain in 2010 to 190,000 in 2015 and then to zero in 2016 when the port was shut down. Shipping companies Desgagnés and Royal Arctic Lines had tried to develop a regular service between Churchill and Nuuk, but the profitability never materialized as the project coincided with the economic downturn of 2008 (Brooks and Frost 2012).
- Fishing vessel experienced expansion, from 20 voyages in 2005 (16.5% of voyages) to 138 (33.2%) in 2017.
- Pleasure-crafts and adventurers also increased their presence in Canadian Arctic waters, from 10 ships (8.3%) in 2005 to 32 (7.8%) in 2017.
- These figures attest to a growing destinational traffic. Transits remain scarce, having peaked at 31 in 2012.
- Pleasure crafts or adventurers conducted most of the transits (Table 6.6) along the Northwest Passage. In 2012, they were responsible for 23 (74.2%) of the transits; 13 (59%) in 2013; 10 (58.8%) in 2014; 19 (70.4%) in 2015, 15 (65.2%) in 2016 and 21 (65.6%) in 2017.
- Canadian government ships come second in terms of traffic, with between one and four ships per year (the same number as passenger vessels).
- Commercial cargo vessels, including general cargo, bulk and tankers, account for a very small share of transit traffic: for most of the period, their traffic is nil, then 1 ship between 2011 and 2016 but none in 2015, then 2 vessels in 2017. The media

Table 6.5 Number of voyages in the Canadian Arctic

	2005	2006	2007	2008	2009	2010	2011	2012	2013	2014	2015	2016	2017
Ships in the Canadian Arctic (no. of voyages)	121	135	181	209	185	257	317	314	349	301	315	347	416
Of which													
Fishing vessels (number of voyages)	20	26	39	52	44	78	136	114	137	119	129	131	138
Cargo ships or barges (number of voyages)	65	67	101	105	100	124	126	124	127	108	120	147	188
Of which													
General cargo	16	17	28	30	23	34	38	32	35	32	34	36	50
Tanker	17	16	24	29	23	28	30	31	28	25	27	23	24
Bulk	21	17	27	25	27	27	23	26	27	33	36	53	72
Tugs and Barges	11	17	22	21	27	35	33	35	36	18	23	35	42
Pleasure-crafts and Adventurers	10	6	9	7	13	13	15	27	32	30	23	22	32
Cruise/Passenger vessels	12	15	17	20	11	18	11	10	17	11	18	20	19
Government vessels (Navy, Coast Guard)	9	9	9	10	10	13	20	16	17	23	16	20	22
Icebreakers											2	2	2
Research vessels	6	12	9	12	7	11	11	23	20	10	9	6	13
Others											1	1	4

Source NORDREG, figures compiled by author from data compiled by NORDREG Iqaluit
Traffic north of 60°N

Table 6.6 Transits of the Northwest Passage

Year	Canadian government ships	General cargo	Tankers	Bulk carriers	Passenger	Tugs	Pleasure-crafts and Adventurers	Research vessels	Foreign government	Others	Total
2017	2	1	1	0	3	0	21	1	0	2	32
2016	3	1	0	0	3	0	15	0	0	1	23
2015	4	0	0	0	2	0	19	0	0	2	27
2014	4	0	0	1	2	0	10	0	0	0	17
2013	2	0	0	1	4	0	13	2	0	0	22
2012	2	0	1	0	2	2	23	1	0	0	31
2011	4	0	1	0	1	0	15	0	0	0	21
2010	4	0	0	0	3	2	11	0	0	0	20
2009	3	0	0	0	2	2	10	0	0	0	17
2008	3	0	0	0	1	0	7	1	0	1	13
2007	3	0	0	0	2	0	4	0	0	0	9
2006	4	0	0	0	2	2	3	0	0	2	13
2005	4	0	0	0	2	0	2	1	0	2	11
2004	3	0	0	0	1	0	2	0	0	0	6
2003	3	0	0	0	2	6	2	0	1	0	14
2002	4	0	0	0	2	2	2	2	0	0	12
2001	2	0	0	0	2	0	2	0	0	0	6
2000	1	0	0	0	2	0	2	0	1	0	6

Source NORDREG Iqaluit

The Canada Shipping Act, 2001 defines a pleasure-craft as a "vessel that is used for pleasure and does not carry passengers". Ministry of Justice. Canada Shipping Act, 2001 (S.c. 2001, c.26). http://laws-lois.justice.gc.ca/eng/acts/C-10.15/. Adventurers are pleasure-crafts not reporting to NORDREG

placed much emphasis in 2013 on the bulk carrier *Nordic Orion* transiting the NWP between Vancouver and Finland, but subsequent years proved the route to remain poorly attractive for commercial cargo transit shipping (Guy and Lasserre 2016).

Therefore, it appears that if Canadian Arctic waters are definitely busier, this increasing traffic is largely fueled by ships that go to the Arctic to perform their economic objective (destinational traffic), whereas transit traffic remains tiny. Moreover, most of this transit traffic is fueled by pleasure boats; the commercial cargo component for now remains very limited. Destinational traffic is experiencing a real, if not steady growth. This traffic is fueled by the expansion of traffic from the exploitation of natural resources and by the servicing of local communities. Bulk traffic has benefited from the exploitation of Arctic or subarctic mines like Voisey's Bay (Labrador), Raglan (Quebec), and Mary River (Baffin Island, Nunavut); this traffic has largely made up for the drying up of traffic to and from Churchill since its port closed down in 2016. For instance, Baffinland Iron Mines shipped 4.1 million tons of ore from its mine in Mary River through its port of Milne Inlet in 2017 (Maritime Magazine 2018). The company eventually intends to reach annual volume of 12 million tons. The first shipment took place in 2015 (Maritime Magazine 2016) and traffic that year reached 920,000 tons. As for community servicing, cargo companies like Desgagnés, NSSI, and NEAS could take advantage of the receding ice to tap into a real demand for cheaper consumer goods. The shipping season now extends over five months instead of four; NEAS could set up 16 voyages instead of 12 in 2017 (Ryan 2018).

6.4.3 The Northern Sea Route

The Northern Sea Route is the section of the Northeast Passage between the Kara Gate and the Bering Strait, and administered by the Northern Sea Route Administration (NSRA). This specification is important inasmuch as it accounts for the fact that ships sailing from Murmansk, for instance, to Asia are considered as transiting ships by the NSRA, whereas, compared to the Northwest Passage, they are performing a destinational voyage, as they departed from a Russian Arctic port.

The origins of Russian thinking on the Northeast Passage route go back to the Russo-Japanese War of 1905: the Russian General Staff considered sending vessels to the Pacific via this route, as it was shorter than the journey around the Cape of Good Hope that the Russian fleet was forced to complete before its final defeat at the battle of Tsushima. An economic dimension was later added to this reflection with the advent of the USSR: the Soviet economic development policy was initially based largely on the desire to maximize the development of own resources, due to a political desire not to rely on foreign supply of raw materials. In 1920, the Committee of the Northern Maritime Route was formed to "equip, improve, and study" (Mulherin et al. 1994: 10) the passage over its entire length. In 1932, Moscow decided to link development of exploitation of the natural resources of the North and active support to the navigation of the Maritime Route, creating the Central Administration of the Northern Sea Route (Lasserre 2010c). Traffic gradually increased as ports were set up and Arctic mines, as in Norilsk, were opened, and peaked in 1987 at 6.6 million tons, then collapsed after the demise of the Soviet Union in 1991, reaching a post-Soviet low of 1.45 million tons in 1998 (Thorez 2008).

In 1987, according to the government's policy of openness, USSR President Mikhail Gorbachev suggested that the Northern Sea Route should be opened for foreign vessels. The official opening took place on July 1, 1991 (Ragner 2000; Blunden 2012; Pastusiak 2016). However, largely due to complex formalities, until 2009 Soviet (then Russian) ships carried out most transit voyages along the NSR.[4] Transit traffic held out until it collapsed in 1993 (see Table 6.7).

[4]In 1997, the Finnish tanker Uikku traveled the commercial route from northern Murmansk to Pevek before exiting through the Bering Strait. It took up fuel and made the journey in the opposite direction to Murmansk. The Russian authorities state that a Latvian tanker also made a full transit in 1997. See Brigham, Lawson, quoted in Revkin (2009) and Lasserre (2010c).

Table 6.7 Evolution of Northern Sea Route transits, 1991–1997

	1991	1992	1993	1994	1995	1996	1997
Number of vessels	15	12	22	7	8	3	2
Tonnage (1000 tons)	210	186	226	10	120	38	30

Source CHNL Information Office, 2013. *NSR transits before 2011*, www.arctis-search.com/NSR%2bTransits%2bbefore%2b2011, a. Jan. 24, 2018

Table 6.8 Number of official transits, Northern Sea Route, 2010–2017

Ship type	2007	2008	2009	2010	2011	2012	2013	2014	2015	2016	2017
Icebreaker					2	3	2	2	1	2	
Government ship					1	0	1	1	3	1	
Cruise or passenger ship				1	1	0	1	3	1	1	
Tug, supply vessel	1	1		4	4	5	1	1	4	4	
Commercial ship	1	2	5	6	31	38	64	24	15	11	
Research ship				2	2	0	2	0	0		
Total official transit	2	3	5	13	41	46	71	31	18	19	25?

Source CHNL Information Office, Transit Statistics, www.arctic-lio.com/nsr_transits, and NSR transits before 2011, www.arctis-search.com/NSR+Transits+before+2011&structure=Arctic+Sea+Routes, a. Jan. 26, 2018. Tentative figures for 2017

Within the frame of climate change narratives, the attention paid by a few shipping companies, and the Russian desire to promote traffic through reformed administrative frameworks and tariffs (Keupp and Schöb 2015; Gritsenko and Kiiski 2016), traffic picked up after 2009 (see Table 6.8).

Figures show that transit traffic began picking up in 2010, expanding rapidly to 71 in 2013 only to drop sharply afterwards to 18 in 2015 and 19 in 2016. This decline, and later stagnation at low levels, in transit traffic along the Northern Sea Route, is clearly out of step with the media forecasts announcing the advent of heavy traffic along Arctic routes. This is due to several factors (Balmasov 2016; Humpert 2016; Doyon et al. 2017a, b):

- The decline in oil prices and fuel prices, which makes the search for possible reductions in transit costs less attractive for shipping companies.

Table 6.9 Number of transit permit application for navigation in the NSR, 2013–2017

	2013	2014	2015	2016	2017
Permit applications	718	661	730	721	664
Refusals	83	30	15	3	2
Permit application, foreign-flagged vessels	141	124	130	142	108

Source Northern Sea Route Administration, *List of applications*, www.nsra.ru/en/rassmotrenie_zayavleniy/perechen_zayavlenii.html, a. Jan. 22, 2018

- The decline in commodity prices, which makes Arctic resources less attractive, both for exploitation and for initial investment for transport with specialized vessels.
- The continuing decline in both bulk and container freight rates, which discourages shipping companies facing overcapacity from investing in new ice-bound vessels.
- The reorientation of certain export routes for raw materials, including natural gas with the opening of the Russian terminal at Ust-Luga on the Baltic Sea, carrying volumes previously shipped via Vitino in the White Sea (Pettersen 2014).
- The priority deployment of Russian icebreakers to infrastructure projects, notably the Sabetta port linked to the gas project on the Yamal Peninsula. The lower availability of buildings has dissuaded some carriers from hiring their vessels for lack of guarantee escort.
- A tariff schedule for the services of the Northern Sea Route, sometimes considered opaque by the maritime carriers.

Here again, it appears that economic factors, some of which, like world commodity prices or freight rates, have nothing to do with the Arctic, have a much greater impact on the development of Arctic shipping than the mere melting of sea ice that continued unabated during the period of transit decline.

As for Canada, a detailed review of transit traffic compared to destinational traffic underlines the fact that the two categories of activity are not evolving on par. The NSR appears attractive given the high number of applications for transit (see Table 6.9), with figures oscillating between 661 and 730. It seems either the NSR Administration is less demanding or that shipping companies have adapted to regulations, as the number of refusals dropped from 83 in 2013 to two in 2017. Interestingly, most applications were made for Russian-flagged vessels, as the number of applications for foreign-flagged vessels range between 142 and 108. The trend towards an even greater dominance of Russian shipping companies is likely to be enhanced following the Russian Parliament (Duma) recent decision to ban foreign-flagged vessels from the transportation of oil, gas and coal along the NSR, except if operations are already under way (Staalesen 2017). This move was reportedly designed to boost Russia's shipbuilding industry in order to tap into an expanding market (RT News 2017), but that could represent a risky bet.

Tonnage traffic figures confirm that destinational shipping is developing fast, while transit stagnates at low levels. Transit tonnage peaked at 1.261 million tons in

Table 6.10 Traffic along the NSR, total and transit, in metric tons, 2010–2017

	2010	2011	2012	2013	2014	2015	2016	2017
NSR, transit tonnage, metric tons	111,000	820,789	1,261,545	1,176 454	274,103	39,586	214,513	194,364
NSR, total tonnage, metric tons	2,085,000	3,225,000	3,750,000	3,914,000	3,982,000	5,432,000	6,060,000	9,737,000

Source CHNL Information Office, *Transit Statistics*, www.arctic-lio.com/nsr_transits, a. Jan. 26, 2018; Staalesen (2018)

2012, then collapsed in 2014 to 274,103 and 39,586 tons in 2015, only to recover to 194,364 tons in 2017 (Table 6.10).

However, total tonnage in the NSR increased from 2.09 million tons (Mt) in 2010 to 6.06 Mt in 2016 and then 9.74 Mt in 2017, according to the Russian Federal Agency for Maritime and River Transport, the biggest annual volume ever, surpassing the previous record of 6.6 Mt set in 1987. In particular, traffic seems to have exploded in 2017. Figures from the Association of Russian Sea Ports show that the Russian Arctic seaports in 2017 handled a total of 74.2 Mt, an increase of 49.1% over 2016. The growth is rooted in a significant hike in Russian Arctic shipments (Staalesen 2018). According to data from the Seaport Association, Murmansk accounted for almost two-thirds of the total port turnover. The harbormasters in the Arctic city saw a total of 51.7 Mt of goods being shipped through the port in 2017, an increase of 54.5% over 2016. The growth comes as several major Arctic industrial projects related to the exploitation of natural resources are in the making. Among them are the Yamal LNG and the projected Arctic LNG 2, both of which will have major effects on regional shipping. Also, oil shipments from new projects like the Novy Port, as well as the terminal at Varandey, are leading to higher volumes.

The Varandey terminal handled 8.2 Mt of oil in 2017, an increase of 3.4% year-on-year. Of the volumes handled in Murmansk, as much as 29 Mt were oil products, while the volumes of minerals and ores traditionally handled by the port remained stable. Forecasts expect volumes to continue to grow over in subsequent years, possibly to more than 70 Mt after 2020. Sabetta, the new port on the Yamal Peninsula, recorded huge growth in 2017. According to the Federal Agency for Maritime and River Transport, as much as 7.99 Mt of goods were handled at Sabetta, up from 2.85 million tons in 2016 (Staalesen 2018).

6.5 Conclusion

In Canada, as well as in Russian Siberia, destinational traffic dominates a very real expansion of Arctic shipping: ships come to the Arctic to perform an economic activity, rather than just to transit. Along the NSR, the number of transits and trade volume both increased from 2011 to 2013 and declined in 2014 and 2015, before recovering slightly in 2017, indicating an unstable and vulnerable shipping environment up to now (Zhang et al. 2016). The NSR seems to be more appealing to liquid, bulk, and general cargo transportation, while container shipping companies have not carried out any voyage—this may change as Maersk has announced it would carry out an experimental trip.[5] Most activities are still domestic and destinational in nature. This expansion of destinational traffic is partly sustained by the expansion of community resupply in Canada; however, in both countries, especially in Russia, it is the construction of infrastructure and the development of natural resources exploitation that fuels the present strong growth.

Similarly, traffic is dominated in Siberia by Russian shipping companies, and in the Canadian Arctic by Canadian companies, for natural resource transportation as well as for the community resupply market. Canadian shipping companies in particular have adapted to the poor infrastructure of the Canadian Arctic villages and to the numerous barriers of entry to this niche market (Giguère et al. 2017). If natural resource exploitation picks up in Canada, it could prove more attractive for foreign corporations.

This is in line with past Russian declarations to the effect they did not expect transit traffic along the NSR to develop to large volumes (Pettersen 2013; Barents Observer 2015). It is also in line with previous analyses of shipping companies' strategies, which showed that cost per transit may prove to be lower, depending on the origin/destination couple, but that this factor is not paramount in the decision to develop shipping in the Arctic: this business decision rests on the perceived strategic opportunity perceived by shipping companies.

The melting of sea ice may act as an enabler, but it is not in itself sufficient to trigger the development of massive traffic along Arctic seaways, nor of single-voyage cost-effectiveness: it does not drive the expansion of Arctic shipping since its evolution is contrasted between regions and between market segments, except for a few market niches like community resupply in Canada where demand is consequent. Shipping companies display a very limited interest for transit traffic, being more interested in the natural resources market. They stress that entering the Arctic market is a strategic diversification move that implies much broader considerations. It continues to be seen as a risky choice, both operationally and commercially, and implies business strategy choices that involve the global picture of the positioning of the company in its regional or global market (Lasserre and Pelletier 2011; Lee and Kim 2015; Beveridge et al. 2016; Lasserre et al. 2016).

References

Arctic Command (2014) Datasæt (Greenpos). [Datasets (Greenpos)]. Made available by Naalakkersuisut of Arktisk Kommando, 2014

Arlov TB (1994) A short history of Svalbard. Norwegian Polar Institute, Oslo

[5]Maersk announced that it is considering conducting test trips for transits along the NSR with a 3000 TEU container carrier in September 2018. "NSRA met with the representatives of Maersk company", NSR Administration, Nov. 9, 2017, www.nsra.ru/en/glavnaya/novosti/n14.html, and Jan. 11, 2018.

Balmasov S (2016) Head of the NSR Information Office, Murmansk, personal correspondence with Frédéric Lasserre, 25 Nov 2016
Barents Observer (2015) Northern Sea Route—focus on domestic projects. The Barents Observer, 17 Oct, http://barentsobserver.com/en/arctic/2015/10/northern-sea-route-focus-domestic-projects-17-10. Accessed 22 Jan 2018
Beveridge L, Fournier M, Lasserre F, Huang L, Têtu P-L (2016) Interest of Asian shipping companies in navigating the Arctic. Polar Sci 10(3):404–414
Blunden M (2012) Geopolitics and the Northern Sea Route. Int Aff 88(1):115–129
Borgerson S (2008) Arctic meltdown: the economic and security implications of global warming. Foreign Aff 87(2):63–77
Brigham LW (2010) Think again: the Arctic. Foreign Policy 181:71
Brooks MR, Frost JD (2012) Providing freight services to remote arctic communities: are there lessons for practitioners from services to Greenland and Canada's northeast? Res Transp Bus Manag 4:69–78
Christensen M (2013) Arctic climate change and the media. The news story that was. In Christensen M, Nilsson A, Wormbs N (eds) Media and the politics of climate change. Palgrave McMillan, New York, pp 26–51
Christensen T, Lasserre F, Dawson J, Guy E, Pelletier J-F (2018) Shipping. In AMAP (ed) Adaptation actions for a changing Arctic–Baffin Bay and Davis Strait. AMAP, forthcoming
DNV USA (2010) Beluga Shipping masters first commercial transit of the Northeast Passage. www.dnvusa.com/industry/maritime/publicationsanddownloads/publications/updates/arctic/2010/01_2010/BelugaShippingmastersfirstcommercialtransitoftheNortheastPassage.asp. Accessed 7 June 2016
Doyon J-F, Lasserre F, Pic P (2017a) Perceptions et stratégies de l'industrie maritime de vrac relativement à l'ouverture des passages arctiques. Cahiers de l'Institut EDS 33 (January), 43 p, online at www.ihqeds.ulaval.ca/fileadmin/fichiers/fichiersIHQEDS/Publications/CahiersInstitutEDS/CA17-01_TransportMaritime.pdf
Doyon J-F, Lasserre F, Pic P, Têtu P-L, Fournier M, Huang L, Beveridge L (2017b) Perceptions et stratégies de l'industrie maritime de vrac relativement à l'ouverture des passages arctiques. Géotransports 8:5–22
Eguíluz VM, Fernández-Gracia J, Irigoien X, Duarte CM (2016) A quantitative assessment of Arctic shipping in 2010–2014. Nat Sci Rep 6:30682
Emmerson C (2011) The future history of the Arctic. Random House, London
Giguère MA, Comtois C, Slack B (2017) Constraints on Canadian Arctic maritime connections. Case Stud Transp Policy 5(2):355–366
Gritsenko D, Kiiski T (2016) A review of Russian ice-breaking tariff policy on the northern sea route 1991–2014. Polar Rec 52(2):144–158
Guy E, Lasserre F (2016) Commercial shipping in the Arctic: new perspectives, challenges and regulations. Polar Rec 52(3):294–304
Howard R (2009) The Arctic gold rush: the new race for tomorrow's natural resources. Continuum, London
Humpert M (2016) Arctic shipping on the Northern Sea Route in deep freeze? High North News, 1 June 2016, www.highnorthnews.com/arctic-shipping-on-the-northern-sea-route-in-deep-freeze/. Accessed 24 Nov 2016
Kaltenborn B, Hindrum R (1996) Opportunities and problems associated with the development of arctic tourism—a case study from Svalbard. Directorate for Nature Management, Norway, DN-1996
Keil K (2017) More and more Arctic tourists—but where exactly? High North News, 2 Feb 2017, http://www.highnorthnews.com/more-and-more-arctic-tourists-but-where-exactly/. Accessed 17 Jan 2018
Keupp M (2015) The Northern Sea Route: introduction and overview. In Keupp M (ed) The Northern Sea Route. A comprehensive analysis. Springer, Wiesbaden, pp 7–20

Keupp M, Schöb R (2015) The potential of container vessel operation on the Northern Sea Route: nautical, regulatory, and operative issues. In: Keupp MM (ed) The Northern Sea Route. Springer, Wiesbaden, pp 53–68

Koranyi B (2013) Ice levels, rule changes to boost Arctic northern sea route. Reuters, 29 May, online: www.reuters.com/article/2013/05/29/shipping-arctic-idUSL5N0EA0RF20130529. Accessed 3 June 2016

Lasserre F (2010a) Arctic shipping routes: from the Panama myth to reality. Int J 66:793–808

Lasserre F (2010b) Géopolitiques arctiques: pétrole et routes maritimes au cœur des rivalités régionales? Critique Int 49:131–156

Lasserre F (2010c) Vers une autoroute maritime? Passages arctiques et trafic maritime international. In Lasserre F (ed) Passages et mers arctiques. Géopolitique d'une région en mutation. Presses de l'Université du Québec, Québec, pp 449–478

Lasserre F, Pelletier S (2011) Polar super seaways? Maritime transport in the Arctic: an analysis of shipowners' intentions. J Transp Geogr 19(6):1465–1473

Lasserre F, Têtu P-L (2015) The cruise tourism industry in the Canadian Arctic: analysis of activities and perceptions of cruise ship operators. Polar Rec 51(1):24–38

Lasserre F, Beveridge L, Fournier M, Têtu P-L, Huang L (2016) Polar seaways? Maritime transport in the Arctic: an analysis of shipowners' intentions II. J Transp Geogr 57(2016):105–114

Lee T, Kim HJ (2015) Barriers of voyaging on the Northern Sea Route: a perspective from shipping Companies. Mar Policy 62:264–270

Maritime Magazine (2016) First Baffinland iron ore shipping season totals 920 000 tonnes. Marit Mag 79:94

Maritime Magazine (2018) Baffinland Iron Mines ships record tonnage in 2017. Marit Mag 87:98–99

Maurette F (2010) Rate, impact and scope of climate change in the Canadian Arctic: Synthesis Report. DRDC CORA CR 2010–190. Defense R&D Canada, Ottawa

Mulherin N, Sodhi D, Smallidge E (1994) Northern Sea Route and icebreaking technology. Cold Regions Research & Engineering Laboratory, Fairbanks

NORDREG (2017) Ships movements, Canadian Arctic. Data compiled since 2005. NORDREG, Iqaluit

O'Leary C (2014) New ice age: the dawn of Arctic shipping and Canada's fight for sovereignty over the Northwest Passage. Univ Miami Inter-Am Law Rev 46(1):117–135

Pastusiak T (2016) The Northern Sea Route as a shipping lane. Springer, Basel

Pelletier JF, Guy E (2012) Évaluation des Activités de Transport Maritime en Arctique Canadien. Cahiers Scientifiques du Transport 61:3–33

Pelletier JF, Guy E (2015) Supply and demand for the Eastern Canadian Arctic Sealift. Marit Policy Manag 42(7):669–681

Pettersen T (2013) Northern Sea Route no alternative to Suez–Deputy Minister. Barents Observer, 19 Dec, http://barentsobserver.com/en/business/2013/12/northern-sea-route-no-alternative-suez-deputy-minister-19-12. Accessed 15 Jan 2018

Pettersen T (2014) Northern Sea Route traffic plummeted. Barents Observer, 16 Dec, http://barentsobserver.com/en/arctic/2014/12/northern-sea-route-traffic-plummeted-16-12. Accessed 15 Nov 2016

Pizzolato L, Howell SE, Derksen C, Dawson J, Copland L (2014) Changing sea ice conditions and marine transportation activity in Canadian Arctic waters between 1990 and 2012. Clim Change 123(2):161–173

Ragner CL (2000) Northern Sea Route cargo flows and infrastructure—present state and future potential. The Fridtjof Nansen Institute, Report 13

Rahman NA, Saharuddin AH, Rasdi R (2014) Effect of the Northern Sea Route opening to the shipping activities at Malacca straits. Int J e-Navig Marit Econ 1:85–98

Revkin A (2009) Commercial arctic passage nearing goal, New York Times, September 4, 2009. https://dotearth.blogs.nytimes.com/2009/09/04/commercial-arctic-passage-nearing-goal/, c. August 15, 2016.

RT News (2017) Putin wants to keep foreign shipping out of Russia's Northern Sea Route. RT News, 17 Nov, https://www.rt.com/business/410138-putin-northern-sea-route/. Accessed 18 Jan 2018

Ryan L (2018) Vital contribution of NEAS to Arctic sealift. Marit Mag 87:87–89

Schmitt D (2013) Arctic: the challenges of opening up the northern route. Horizon, the EU Research & Innovation Magazine. 31 Jan. http://horizon-magazine.eu/article/arctic-challenges-opening-northern-route_en.html. Accessed 3 June 2016

Snyder J (2007) The economic role of Arctic tourism. In: Snyder J, Stonehouse B (eds) Prospects for polar tourism. Cabi, Cambridge, pp 102–122

Staalesen A (2017) Russian legislators ban foreign shipments of oil, natural gas and coal along Northern Sea Route. The Independent Barents Observer, 26 Dec, https://thebarentsobserver.com/en/arctic/2017/12/russian-legislators-ban-foreign-shipments-oil-natural-gas-and-coal-along-northern-sea

Staalesen A (2018) Arctic seaports bustle as shipping on Northern Sea Route reaches new high. The Independent Barents Observer, 16 Jan, https://thebarentsobserver.com/en/arctic/2018/01/arctic-seaports-bustle-shipping-northern-sea-route-reaches-new-high

Stephenson SR, Smith LC, Brigham LW, Agnew JA (2013) Projected 21st-century changes to Arctic marine access. Clim Change 118(3–4):885–899

Stephenson SR, Brigham LW, Smith LC (2014) Marine accessibility along Russia's Northern Sea route. Polar Geogr 37(2):111–133

Stonehouse B, Snyder J (2010) Polar tourism: an environmental perspective. Channel View Publications

Thorez P (2008) La Route maritime du Nord. Les promesses d'une seconde vie. Le Courrier des Pays de l'Est 2008/2, no. 1066:48–59

Wright C (2016) Arctic cargo. A history of marine transportation in Canada's North. Christopher Wright, Digby

Young OR (2011) The future of the Arctic: cauldron of conflict or zone of peace? Int Aff 87(1):185–193

Zhang Y, Meng Q, Zhang L (2016) Is the Northern Sea Route attractive to shipping companies? Some insights from recent ship traffic data. Mar Policy 73:53–60

Chapter 7
Boreal Forests of the Circumpolar World

Hayley Hesseln

7.1 Introduction

The boreal forest accounts for about one-third of all global forest resources and is found in seven of the eight Arctic countries. The boreal biome, also known as the taiga, lies south of the Arctic Circle and runs through Iceland, Norway, Sweden, Finland, Russia, the United States, and Canada (out of the eight Arctic countries), as well as Japan, Mongolia, and Scotland, making it one of the largest biomes in the world.

While the word "Arctic" does not typically invoke images of lush green forests, forest resources play a critical role in sustaining economic and social development in northern regions. The boreal forest is also vital to the carbon cycle, helping to maintain the world's carbon balance by both storing and releasing CO_2.

The resource is also critical to sustainability in its broadest sense. The literature pertains to three types of sustainability: economic, ecological, and social. The forest provides goods and services that support local and global economies and provides employment and income to society. The boreal forest also plays a critical role to society's wellbeing, affording people recreation opportunities and natural experiences. Finally, maintaining or increasing biodiversity is essential to highly functioning ecosystems that, in turn, support economic and social well-being.

As demand for forest resources increases globally, so too will pressure to use, conserve, and protect the boreal forest, which will force people to make trade-offs. Furthermore, while the resource is largely homogeneous across the world, ownership, management regimes, and policies are not. How do such similarities and differences affect forest sustainability? Is it possible to manage forest resources in a way that addresses all aspects of sustainability? This chapter reviews the state of the boreal forest across the Circumpolar North and examines differences between the regions,

---

H. Hesseln (✉)
University of Saskatchewan, Saskatoon, Canada
e-mail: h.hesseln@usask.ca

M. Finger and L. Heininen (eds.), *The GlobalArctic Handbook*,
https://doi.org/10.1007/978-3-319-91995-9_7

including forest ownership, management, policy, and innovation in the context of social, economic, and environmental sustainability.

This chapter begins by putting the boreal forests into perspective, describing the physical resource across the Arctic countries. I then discuss variations in ownership regimes and how this affects management and sustainability. Next, I discuss forest management and governance by country. Finally, I examine global forces that have increased the pressure for resource use, and how society has responded to ensure future sustainability.

7.2 The Boreal Forest in Perspective

The boreal forest constitutes a vast eco-zone that lies below the Arctic Circle and spans the Circumpolar North. Also known as the taiga, the boreal covers approximately 16.6 million square kilometers (Runesson 2016) and accounts for approximately 33% of the world's total forested area. Given its subarctic location, and the fact that it is the coldest terrestrial ecosystem on earth, the forest is largely a function of extremes. In general, the eco-zone is characterized by short summers that can be hot; long and dark winters with extremely low temperatures; and low moisture resulting in relatively short growing seasons.

Despite these extreme climatic conditions, the boreal is highly productive and is the source of a wide range of timber and non-timber forest products and services that support complex economies. This biome is often described as a mosaic of forest types that include species such as black and white spruce, larch, poplar, birch, and aspen, many of which are used for commercial purposes. Additionally, shrubs such as highbush cranberry, alder, and Labrador tea (Fig. 7.1) characterize the forest and provide commercial opportunities to produce medicines and energy, for example. Groundcover plants include mosses, grasses, sedges, lowbush cranberry, and lichen, the latter of which is an essential source of nutrition for caribou and reindeer. Figure 7.2 illustrates the extent of the boreal forest in green, which spans North America, northern Europe, and Asia.

It is difficult to distinguish boreal forest images by location given that the forest is similar in structure and species across its range. This is an important consideration given that forests are managed differently: the implication being that sustainable forest use and management is a direct function of ownership and management.

In terms of land area, Russia is home to the largest area of the boreal forest, followed by Canada and Alaska. Additionally, Russia's forests account for about 20% of the world's forests, which means that sustainable management is important not only to Russia, but also to the rest of the world (FAO 2012). As a proportion of landmass, Finland (73%), Sweden (68%), and Russia (50%) have the highest percentages of forest, which means that the forest is clearly a highly important resource in terms of environmental, social, and economic sustainability in the Barents region. Although Iceland has the lowest proportion of forests, at only 1% of land area, there is a small industry in that country that makes products for niche markets. Table 7.1

Fig. 7.1 Labrador tea as depicted by William Miller (London 1817–1833)

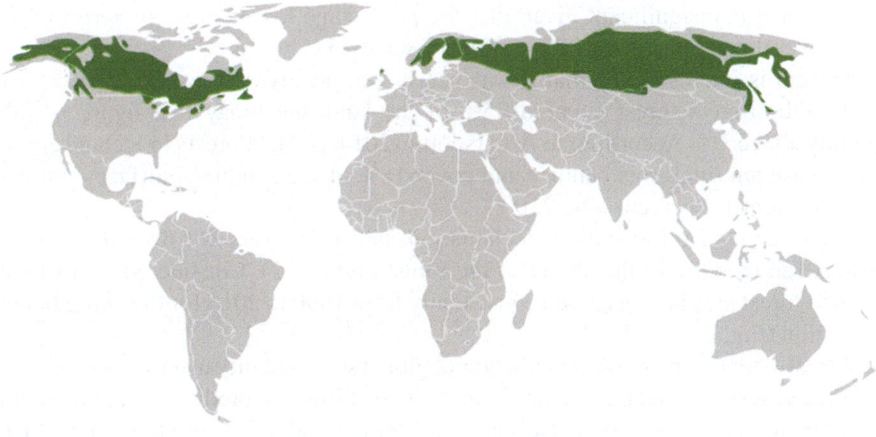

Fig. 7.2 The global extent of the boreal forest/taiga. Mark Baldwin-Smith CC BY-SA 3.0

describes the amount of forested area in each country by hectares and percentage of total land area categorized as forest, other wooded land, and other land. In Canada, for example, the boreal forest accounts for 38% of the total land area.

While the United States shows 34% forest cover, that figure refers to all states, including Hawaii. Alaska's forest area is estimated to be 52 m ha, which constitutes approximately 17% of total forested area. Finally, although deforestation is a problem

Table 7.1 Forested and other wooded land—2015 (*source* FAO 2015)

Country	Land area						Inland water (1000 ha)	Country area (1000 ha)
	Forest		Other wooded land		Other land			
	1000 ha	% of land area	1000 ha	% of land area	(1000 ha)	Of which with tree cover		
Canada	347,069	38	40,866	4.5	521,416	8499	89,116	909,351
Finland	22,218	73	801	3	7371	231	3452	30,390
Iceland	49	1	144	1	9832	10	275	10,025
Norway	12,112	40	2012	7	16,303	–	1951	30,427
Russia	814,931	50	74,925	5	747,832	20,328	72,137	1,637,687
Sweden	28,073	68	2432	6	10,529	864	3996	41,034
US*	310,095	34	21,279	2	584,818	27,668	47,011	916,192

*United States figures include all forested areas in the lower 48 states, Alaska and Hawaii

in many parts of the world, the boreal forest among the circumpolar countries is not in decline. Countries such as Canada and Norway show a slight decrease in forest cover, but it is insignificant given that the percentage loss is close to zero in both cases over the period from 1990 to 2015 (FAO 2015).

Iceland is interesting in that it was nearly completely deforested. When it was settled, the boreal forest covered 25% of the land, but today that figure is only slightly above 1%. According to Arnarsdóttir (2014, p. 1), "Forests in Iceland are set to increase ten times this century and cover 12% of the country's surface—instead of the current 1.2 percent—by 2100."

As a result of reforestation, reclamation, and forest renewal overall, statistics indicate an increase in the extent of the forest (Table 7.2). Considered as a whole, planted forest area has increased consistently from 1990 to 2015 for the entire boreal ecoregion (Fig. 7.3).

Forest extent is an important indicator of global sustainability in terms of providing ecological services, such as aquatic function, air filtration, biodiversity, and habitat for flora and fauna, as well as carbon sequestration. Table 7.3 provides an estimate of the carbon stock and the annual percentage change in stock between 1990 and 2015. The boreal forest plays a critical role in mitigating climate change given its role in capturing carbon in living biomass, but also in the soils and forest floor litter.

The boreal forest is biologically homogeneous across the circumpolar north. The differences lie in ownership, management, policy, and use, all of which are dictated by culture and beliefs. To what extent is use of the boreal forest sustainable, and what can we learn by different management approaches? Ownership has clear implications for how resources are used and managed. Additionally, global economic pressures

Table 7.2 Extent of the forest 1990–2015 (*source* FAO 2015)

Country	Forest area (1000 ha)					Annual change rate							
	1990	2000	2005	2010	2015	1990–2000		2000–2010		2010–2015		1990–2015	
						1000 ha/year	%	1000 ha/year	%	1000 ha/year	%	1000 ha/year	%
Canada	348,273	347,802	347,576	347,302	347,069	−47.1	0	−50	0	−46.6	0	−48.2	0
Finland	21,875	22,445	22,143	22,218	22,218	57	0	−22.7	−0.1	0	0	13.7	0.1
Iceland	16	29	37	43	49	1.3	6	1.4	4	1.3	2.9	1.3	4.6
Norway	12,123	12,113	12,092	12,102	12,112	−1.9	0	−1.1	0	2	0	−0.8	0
Russia	808,950	809,269	808,790	815,136	814,931	31.9	0	586.7	0.1	−41	0	239.2	0
Sweden	28,063	28,163	28,218	28,073	28,073	10	0	−9	0	0	0	0.4	0
USA	302,450	303,536	304,757	308,720	310,095	108.6	0	518.4	0.2	275	0.1	305.8	0.1

Table 7.3 Carbon stock in living forest biomass 1990–2015 (*source* FAO 2015)

Country	Living forest biomass (million tonnes)						Annual change rate							
	1990	2000	2005	2010	2015	By area 2015	1990–2000		2000–2010		2010–2015		1990–2015	
						Tonnes/ha	1000 tones/year	%	1000 tones/year	%	1000 tones/year	%	1000 tones/year	%
Canada	14,427	14,408	14,136	13,992			−1900	0	−41,600	−3				
Finland	633	716	745	780	780	35	8260	1	6460	1	0	0	5888	1
Iceland						13	6	3	17	6	50	10	19	6
Norway	332	337	409	443	476	39	4500	1	6600	2	6600	1	5760	2
Russia	32,504	32,157	32,210	32,500	32,800	40	−34,700	0	34,300	0	6000	0	11,840	0
Sweden	950	1016	1091	1103	1114	40	6570	1	8680	1	2300	0	6560	1
USA	1448	15,711	16,393	17,067	17,330	56	123,600	8	135,600	1	52,600	0	115,280	1

Fig. 7.3 Image of the boreal forest in autumn (Photo by Phil, CC BY-NC-ND 2.0.)

have influenced environmental sustainability, societies and cultures, and economic prosperity. Sustainable forest management is becoming increasingly important as populations grow and, consequently, so too do demands on forests. I now look at forest ownership and management regimes and compare the state of the forest industry in different countries.

7.3 Forest Ownership

Boreal forest ownership in the circumpolar north falls into two categories: public or private. Russia, Canada, and the United States have the highest percentage of public ownership, where federal, state, or provincial governments manage forests, whereas private ownership is prevalent in the Scandinavian countries, Iceland, and Finland (see Table 7.4).

In Canada, the majority of forests are owned and managed by 10 provincial and three territorial governments. The federal government, which owns only 4% of the forestland, is responsible for managing national parks, national defense reserves, and land held in reserve for indigenous peoples. Private holdings account for 6% of total forest ownership and are made up of timber companies and small family woodlots

Table 7.4 Forest ownership (*source* FAO 2015)

Country	Ownership pattern			Private ownership		
	Public	Private	Other	Individuals	Business entities and institutions	Local, indigenous and tribal communities
Canada*	94	6	n.s.	84	16	0
Finland	32	68	0	84	16	0
Iceland	30	70	0	73	27	0
Norway	14	86	0	89	8	3
Russia	100	0	0	–	–	–
Sweden	24	76	0	63	29	8
Alaska/USA**	43	57	0	69	31	0

*NRCan (2016)
**Statistics for Alaska were compiled from Alaska Forest Association (2015)

located predominately in the eastern part of Canada and British Columbia in the west (NRCan 2016).

In the US, 76% of land is owned and managed by the government, with 24% owned by indigenous corporations and less than 1% privately owned by individual landowners. Public ownership is further divided into the federal government (51%) and a combination of state lands, university holdings, and local governments (25%) (Alaska Forest Association 2015). Russia is currently in a period of forest policy transformation and might consider reforming private ownership rights. Russia owns 100% of its forests (FAO 2012)

The predominant form of ownership in the Scandinavian[1] countries is private: in Norway, the public–private split is 86–14%, followed by Sweden (76/24), Iceland (70/30), and Finland (68/32). Private individuals make up the greatest percentage of owners in each jurisdiction, followed by corporations/businesses. This point is particularly interesting in that the number of owners is vastly greater than the number of individual forest owners in North America and Russia, and individual forest holdings are relatively small in size, making forest management and long-term access to timber supply more complex.

In terms of productive forest, the FAO reports growing stock for both commercial forestland and other wooded land. Russia's growing stock of all commercial species is the greatest, at 81 billion m^3, followed by Canada at 47.3 billion m^3 (Table 7.5).

With the exception of Iceland, forestry and the production and sale of forestry products are relatively important to each nation in terms of its contribution to GDP.

[1]For the purposes of this article, Iceland and Finland are included.

Table 7.5 Growing stock in forest and other wooded land 2015 (*source* FAO 2015)

Country	Forest				Other wooded land	
	Total (million m^3)	m^3/ha	Coniferous (million m^3)	Broadleaf (million m^3)	Total (million m^3)	m^3/ha
Canada	47,320	n/a	n/a	n/a	n/a	n/a
Finland	2320	104	1850	569	8	10
Iceland	1	10	n/a	n/a	n/a	n/a
Norway	1157	96	841	316	8	4
Russia	81,488	100	57,536	23,952	1534	20
Sweden	2989	106	2481	507	7	3
USA	40,699	131	23,282	17,416	414	19

7.4 Forestry Management and Governance

Although ownership differs by nation state, as do authorizing agencies within states, governance and policy characteristics are similar across the biome, with a strong focus on all aspects of sustainability. This section provides an overview of forest governance and how it is related to industrial development and the forest products industry. I will focus first on North America and Russia, where most forestland is owned and managed by the government. I will then focus on Scandinavia, where the predominant form of ownership is private.

7.4.1 Russia

Russian forests represent significant potential for economic development, but are highly under-utilized. The forest industry accounts for only 1.3% of GDP and 1% of the country's employment although forest resources cover 50% of Russia's landmass, and make up approximately 20% of the world's forests (Ulybina 2014a, b). The forest resource is regulated and managed by the Federal Forest Agency, which has recently undergone a great degree of administrative change, reporting to the Ministry of Natural Resources up to 2008, then to the Ministry of Agriculture from 2008 to 2010, and directly to the Government of the Russian Federation from 2010 to 2012. Currently, the Federal Forestry Agency falls under the purview of the Ministry of Natural Resources and Ecology (see Petrov and Lobovikov 2012).

Forestry policy has also undergone changes that provide new opportunities for forest stakeholders, including both industry and timber-reliant communities. Prior to 2013, forest policy was fraught with bureaucratic difficulties that hampered the development of the Russian forest economy and increased the incidence of illegal logging and forest mismanagement. Forest Stewardship Council International, in conjunction with the World Wildlife Fund (WWF), Greenpeace, and a host of FSC-certified companies, began working together in 2011 with the Ministry of Natural Resources to develop new policies. The outcome has been a new focus on modernizing Russia's forest policy that seeks greater stakeholder involvement in decision making, on improving forest management practices to reflect natural processes and preserve ecosystem integrity, and on harmonizing Russian forestry practices to bring them in line with international standards (FSC 2013). Sustainable forest management, as defined by the FSC, requires that forest owners and managers follow 10 principles that address all aspects of sustainability, including the environment (monitoring and assessment of the environment based on high conservation values), society (observing legislation regarding human rights and community relations), and economics (providing long-term values and benefits) (FSC 2015).

According to the Russian Government website,[2] the Federal Forestry Agency is currently responsible for overseeing all forestry issues with the exception of those related to specially protected nature reserves. The agency also provides forest management services (Russian Government 2016). More specifically, it oversees contracting and administrative services, forest inventory, forest pathology, forest fire protection, research, and education (Petrov and Lobovikov 2012).

As in North America, the private sector in Russia secures timber resources through contracts that specify rights to forest resources that can be secured from one to 49 years depending on use. Forest companies must follow the Forest Code of 2006, which constitutes the primary legislation governing all management and use of the forest. The Forest Code outlines key principles of forest legislation and provides guidance on a wide range of factors affecting the disposition of forest resources, including property rights, harvesting practices, and environmental management (see Russian Government 2006).

While reforms are being implemented, reports suggest that there has not been much practical change. According to Ulybina (2014a, b), who closely examined the recent forest policy reforms, the forest continues to be viewed by the government as an economic resource and is managed without sufficient consideration for society. Furthermore, despite stakeholder involvement being formally articulated, the reality is that decision-making continues to be top-down without regard for the professional foresters, timber-dependent communities, or NGOs that are affected by such decisions. Ulybina suggested that the reason for this is the continuance of informal institutions that continue to govern society at large. The WWF in Russia echoes these concerns and cautions that, without significant change in the legislation, sustainable forest management is jeopardy (WWF 2016).

[2]Russian Government, retrieved from: http://government.ru/en/department/245/.

7.4.2 *Alaska*

Forests in Alaska are governed by four entities including the federal government (USDA Forest Service), the State of Alaska and other local governments including the university system, Native corporations, and to a small degree, private landowners. Forest management and use are guided by national legislation such as the National Forest Management Act (NFMA), the Wilderness Act, the Endangered Species Act of 1973, and the National Environmental Policy Act (NEPA). Additionally, the Alaska Division of Forestry, in the Alaska Department of Natural Resources, oversees the Forestry Strategic Plan, which protects environmental values, manages wildfire, provides access to timber resources, and engages in forest education and stewardship (Alaska Department of Natural Resources 2016).

There are three state laws that govern forest management on state lands. These laws dictate that the timber industry will (i) follow the principles of sustained yield and multiple-use; (ii) prepare forest management plans that assess commercial harvests, forest harvest practices, and industrial effects on the environment; and (iii) fully address the multiple-use aspects of the forest, including community needs and values. It is also worth noting that securing the highest economic return does not outweigh the production of non-market values. Finally, the State of Alaska has the power to regulate forest harvest regulation on private land (Alaska Department of Natural Resources 2016).

Historically, the forest industry in Alaska, as in the rest of North America, produced a wide variety of forest products and played a significant role in Alaska's economy. As a result of rising production costs, increasing competition from lower-cost producers, and stricter environmental regulations, the pulp and paper producers closed their mills. Today, the industry is much smaller and largely produces lumber and wood products for bioenergy.

The National Forests have come under increasing pressure over the last half-century to increase forest access for multiple-use, to designate more area as wilderness and habitat for endangered species. The result has been a significant decline in the annual allowable cut (AAC), which is the annual amount of stumpage available for sale to industry. The lack of timber supply is manifested in business closures, lost jobs, and a weak timber economy.

7.4.3 *Canada*

Forests played a central role in the westward expansion, settlement, and economic development of Canada and the United States, providing the resources needed to build those nations. According to Natural Resources Canada, the country is second only to the US as the largest exporter of primary forest products worldwide. Domestically, the forest sector is within the top five in terms of contributions to net trade. Additionally, the Canadian forest industry is an important employer particularly in northern and remote regions.

Canada's forest legislation is developed and enforced by provincial and territorial governments. While legislation and regulations are specific to each jurisdiction, the overarching goal is to practice sustainable forest management (SFM), which considers ecosystem and environmental functioning, community, and social well-being, and economic development (NRCan 2016).

Provincial crown forests are managed through tenure agreements that were originally designed to provide economic and social benefits. The tenure system grants access to forest companies to secure wood for commercial purposes. Tenure agreements vary in purpose and length and are defined by unique sets of rights and responsibilities. For example, Saskatchewan uses three types of tenure instruments: forest management agreements (FMAs), term supply licenses (TSLs), and forest product permits (FPPs). These vary in length, with the longest being 20 years for FMAs, and the shortest being up to one year for FPPs. Agreements specify either the volume of timber permissible to harvest, or the geographic area. The type of tenure granted has a direct effect on timber supply security.

The tenure system writ large has been criticized for its outdated disposition of rights to timber supply and to meet economic and social goals (Haley and Nelson 2007). British Columbia (BC) began a process of tenure reform in 2014 to change the nature and duration of rights to timber resources. A key concern was the duration of agreements: the timber industry wanted to secure longer tenures to provide incentives for research and development and to have more secure access to timber supply. Today, BC has 13 types of tenure, the most secure of which is issued for no less than 25 years and no more than 99 years, but replaceable every 10 years (Government of British Columbia 2012).

Tenure reform has also taken place in Ontario in response to the needs of Aboriginal people, the timber industry, and the economy in general. Tenure modernization includes new legislation to establish local forest management corporations (LFMCs) and to enhance timber supply licenses to reflect the need to respond to fluctuations in forest products markets (Ontario Ministry of Natural Resources and Forestry 2016).

The forest industry in Canada is economically significant, contributing 1.25% of the country's GDP in 2013, which is valued at CAD19.8 billion. In terms of global production, Canada has the largest forest product trade balance (NRCan 2016). For this reason, it is imperative that the timber industry and local communities have secure access to a long-term wood supply.

7.4.4 Iceland

Forests historically covered about 25% of Iceland but were felled to clear land for agriculture. The climatic region supports growth of the boreal forest and there has been a strong push to increase afforestation, not only to restore forestlands, but also to develop a forest industry. The majority of forests are owned by private individuals and managed by forestry societies run by volunteers for the purpose of land reclamation and afforestation (see Eysteinsson 2016). Funding cuts caused by the global financial crisis have slowed afforestation significantly (Lange 2015). Today, only about 1% of Iceland is forested, but the country has the highest afforestation rate, some of which is being planted in native birch (Icelandic Forestry Association 2016).

Forest protection and legislation largely follows EU directives and focuses on ecological restoration as directed by the National Forest Programme established in 2006 (Eysteinsson 2016). Directives focus on sustainability in terms of developing a commercially valuable resource; managing the forest for ecological outcomes that maintain soil and water conservation, as well as carbon sequestration; and providing social access for recreation and well-being (Eysteinsson 2016).

The forest industry is small but is growing and serves local niche markets. The sector employs 30 people and produces fuelwood, wood products such as fence posts and lumber, as well as landscaping products from waste wood (Eysteinsson 2016).

7.4.5 Norway

Private individuals own approximately 80% of productive forests in Norway, with 43,000 of these individuals belonging to the Norwegian Forest Owners Association. Although the average property size is relatively small, given the large number of non-industrial private owners, the timber industry secures most of its wood through private individuals.

Forest policy in Norway, as in the other Scandinavian countries, promotes all aspects of sustainability, including social, economic, and ecological goals. Furthermore, Norway has committed to pursuing such goals through the ratification of international agreements such as the Rio Convention. Those who choose to use the forest for commercial purposes must comply with regulations set out in Norway's Forest Act of 2005. While owners are free to use their forests as they see fit, if they engage in timber production they must consider environmental values and long-term sustainability as part of their harvesting practices (Nordic Timber 2016). Specific regulations apply to harvesting methods, as well as the effects of harvesting on wildlife, recreation opportunities, social values and water resources, for example.

The Forestry Act uses the "Living Forest" standard, which follows forest certification regulations set out by the Programme for the Endorsement of Forest Certification (PEFC). Any of the private non-industrial forest owners wishing to sell timber must be certified. Because costs are shared through forest associations, nearly all of Norwegian commercial timber is certified (Håbesland et al. 2015).

To ensure private owners are compensated for providing public goods and to manage forests sustainably, a percentage of harvest income is held in reserve (between four and 40%) in the Forest Trust Fund. Forest owners are permitted to use such funds to support long-term forest planning, reforestation, improved environmental values, and to provide and maintain infrastructure such as roads (Nordic Timber 2016, Håbesland et al. 2015).

The forest products industry in Norway was also negatively affected by the global crisis, but continues to produce sawn wood, roundwood, paper products, and bioenergy. Observations from forest landowners and the timber industry suggest that the bioenergy sector has cooled considerably as a result of low oil prices. Also, concerns are rising among society about the state of the environment, thus adding pressure to increase forest use for carbon sequestration and the production of environmental goods and services (Sjølie et al. 2016).

7.4.6 Sweden

Family enterprises own approximately half of the productive forestland in Sweden. As of 2015, there were about 200,000 family enterprises, with average holdings of 50 ha. To help manage and operate forest businesses, many families work within co-operatives, thus gaining economic advantages to sales and marketing. Additionally, many family enterprises operate mills and are engaged in several aspects of forest production all along the supply chain. The second largest category is industrial private, of which there is a small number of companies that control 25% of forested land in Sweden. The state and other government owners account for the remainder (Royal Swedish Academy of Agriculture and forestry 2015).

Long-term sustainability is of primary importance to all forest owners, which is reflected in Sweden's forest policy and stems from a history of heavy cutting in the 18th and 19th centuries, followed by a period of strict regulations to improve forest stewardship. Forest policy was restructured in 1994 to liberalize business using a "freedom under responsibility" philosophy while ensuring biodiversity is protected (reflecting a balance between economic development and biodiversity; see the Swedish Forestry Model 2009). Because social values are highly important to Swedish society, the government established the National Forest Programme, which enables greater public participation in decision making (Royal Swedish Academy of Agriculture and Forestry 2015).

The Swedish Forest Agency (SFA) oversees forest operations to ensure regulations are followed and provides support services for training, advice, and information. Additionally, the Swedish Environmental Protection Agency works in conjunction with the SFA to establish nature reserves and to oversee environmental legislation. Key components of the forest act include oversight regarding harvesting methods by species and age, adherence to reforestation, and stewardship of the forest to manage socio-cultural and environmental values (Royal Swedish Academy of Agriculture and Forestry 2009).

The forest products industry in Sweden is highly innovative and is an important contributor to its economy, accounting for 9–12% of employment, exports, and added value. The industry contributes significantly to world demand for sawn timber (11%), paper exports (close to 9%), and pulp (about 6%) (Royal Swedish Academy of Agriculture and Forestry 2015). In fact, Sweden is the world's second-largest exporter when all forest products are considered.

The vision in Sweden is to have a bio-based economy by 2035, in which the forest industry will play a substantial role, providing a renewable resource to produce building materials and energy, added value products (such as chemicals, textiles, and medical products) in a sustainable manner. To that end, the pursuit of sustainability addresses all three aspects: certified forests that are diverse and ecologically robust, the protection and development of community values and social well-being, and economic development that supports jobs and economic growth (Swedish Forest Industries Federation 2012).

7.4.7 Finland

Forests in Finland are highly important contributors to the economy, to social well-being, and are managed to protect the environment and biodiversity. Private families and businesses own most forestland (68%): 737,000 owners with forest land holdings greater than two hectares and average holdings of 30 ha. Of the total private owners, about 74% are families (Lier and Parviainen 2013).

The Finnish Ministry of Agriculture and Forestry is the primary governing body responsible for forest policy legislation and regulations. However, the Ministry of Trade and the Ministry of the Environment also play a role in terms of imports and exports of forest products, and environmental sustainability (FAO 2010a, b). Private landowners, while responsible for their own forests and free to meet their own needs, cooperate with Forest Centers that report to the Ministry of Agriculture and Forestry, to ensure forests are managed according to forest regulations that promote socio-cultural, economic and environmental sustainability. As in the other Scandinavian countries, the Forest Centers also provide professional advice and monitoring.

In 2015 the Ministry of Agriculture and Forestry announced the new National Forestry Strategy 2025 that is influenced by both strategies regarding the bioeconomy, and biodiversity. The main principle and objectives are:

> The vision is "Sustainable forest management is a source of growing welfare."
>
> The three strategic objectives to make the vision come true are: (1) Finland is a competitive operating environment for forest-based business, (2) Forest-based business and activities and their structures are renewed and diversified and (3) Forests are in active, economically, ecologically and socially sustainable, and diverse use.
>
> (Finnish Ministry of Agriculture and Forestry 2015)

Forest management and policy seeks a balance between commercial production and environmental protection. Thus, the legislation requires reforestation after harvest and that practices focus on maintaining biological diversity and natural forest processes. Additionally, about three million ha (13%) of the forest are designated for restricted use, of which 9% is strictly protected making it the highest percentage in Europe (Parviainen and Lier 2015).

Because forests are of great important to Finns, the government uses an open and transparent process to ensure public participation in all large forestry projects.

This is of particular importance given increasing demand for forest resources from non-timber interests. Collaboration is further enhanced through Forest Management Associations, the Central Union of Agricultural Producers and Forest Owners, the Federation of Forest Owners, The Finnish Forest and Park Service, and the Finnish Reindeer Owners' Association (FAO 2010a, b).

The forest products industry is crucial to Finland's economy, contributing approximately 4% of the country's GDP and employing about 65,000 people. Because of its importance, Finland is known for its forestry expertise and innovation. Modern forestry practices produce not only traditional forest products such as sawn timber and roundwood, but increasingly also produce bioproducts from wood such as cosmetics, pharmaceutical and fiber packaging. Additionally, sectors such as pulp and paper use waste products to produce increasingly more bioenergy bringing the total energy consumption of wood-based fuels up to 25% (Parviainen and Lier 2015).

7.4.8 Implications

How does management of the boreal forest compare across the Circumpolar North? There are some differences among the Arctic countries when it comes to forest policy and management. Additionally, there are differences in rights associated with ownership (public or private), management regimes, and policies that range from having a great degree of autonomy for private owners in the Nordic countries to heavy regulations that guide timber management in Russia and North America. Also, the boreal plays a relatively different role given the extent to which each country relies on the forest as an economic base. For example, the extent of the forest in Finland is 73% of land area, whereas in Iceland it is only 1% of land area.

Despite significant differences in management and ownership regimes, the similarities are greater. The resource itself is relatively homogeneous across the Circumpolar North, meaning that harvesting regimes, reforestation, wildlife habitat and ecosystem functioning, for example, are similar and that countries can collaborate to learn from each other. Additionally, timber products are somewhat homogeneous: sawn wood, round wood, pulp and paper, etc., and now the development of new wood fiber products that are used to make everything from cosmetics to clothing.

One of the most prominent similarities among nations is the focus on sustainability of the environment, communities, and economies. This view on sustainability is not only enshrined in national policies and agreements, but is also reflected by forestry practitioners in their actions to certify significant tracts of forestland. Even Russia has adopted legislation based on the principals set out by the FCS and, over time and with increasing pressure from the global communities, will be more likely to implement practices to achieve such goals.

7.5 The Global Forces of Change

The fact that the boreal forest is a vast resource might lead one to think that increasing demands can easily be met. The ability to meet demands might exist in the short run, but an increasing global population and the need for food security will result in an increase in conversion of the world's forests, thereby displacing wood production and creating pressure for forest products derived from the boreal forest and other regions (FAO 2016). In order to achieve sustainability in the truest sense of the word, it is imperative that forest managers and owners, whether private or public, continue to reevaluate policy and promote innovation considering global changes. How have the Arctic countries dealt with such change to date?

Increasing demand for forest products will be driven primarily by demographic change, including a projected increase in the world's population to nine billion people by 2050 and continued economic growth estimated to be US$100 trillion by 2030 (FAO 2009). This means an increase in demand for wood products, but also for other non-timber forest products and of course energy. Countries such as Sweden and Finland have been leading the global forest industry in this area with innovations in building materials using wood rather than steel (for example), the production of chemicals, pharmaceuticals, other wood-based fiber products, and bioenergy.

The demand for non-timber uses of the forest is also rising, as evidenced by increasing conflict among different societal groups with competing interests. For example, Saami reindeer herders in the Barents Region have long had the right for their herds to move freely over the land from the high-elevation calving grounds down through the forests to lower elevations in the winter. However, more intense forest harvesting, and expanding infrastructure such as roads, dams, and wind farms are creating increasingly more conflict, which demands innovative solutions (see Korosuo et al. 2014; Sandström et al. 2012).

Similar conflicts are arising in Canada between indigenous peoples and industry in the boreal forest, where competing interests want different benefits from the same land base, including those derived from mines, dams, and forest harvesting, which often conflicts with traditional uses, the routes of migratory species such as caribou/reindeer, and other wild food sources such as salmon. A lot of research has been conducted in this area to develop methods for resolution and mitigation, including benefit-sharing agreements, the assignation of property rights, and exploration of how to manage for multiple-use needs (see, for example, Wyatt et al. 2013, 2016; Horstkotte et al. 2014; Langston 2013).

Global impacts are unavoidable in today's economy given the great degree of connectedness among world commodities and financial markets. As a result, many countries' forest industries were adversely affected as a result of the global financial crisis, which negatively affected forest-dependent communities, particularly with respect to the production of traditional wood products. Additionally, increasing competition from lower-cost pulp and paper producers in nations such as Brazil and Indonesia, combined with decreasing global demand for paper and newsprint, contributed to mill closures and substantial layoffs. While many countries have recovered, the size of industry has contracted greatly in some places. The timber industry contributes approximately 4% to global GDP today, up from a low of about 3% in 2009, but significantly down from highs of over 10% in the 1980s (Parviainen and Lier 2015). Similarly, the timber industry in Canada has suffered. Between 2004 and 2014, employment in the logging industry fell by more than one-third, from over 300,000 workers to 190,000. Forest products that historically contributed three percent per year to the Canadian GDP fell to only one percent by 2013. The number of paper mills in Canada in 2000 was 50 and now is fewer than 30 (NRCan 2016).

While the economic crisis has had long-lasting effects in many areas, the downturn forced the restructuring of the industry, which resulted in the emergence of a stronger bio-economy that produces new and diverse forest products. More efficient use of forest resources led not only to innovation, but also to a level of economic diversity not seen prior to the crisis. As a result, many timber-based economies have developed new trading partners, which has helped secure long-term stability and reduce the negative effects of future economic shocks. Both the Swedish and Finnish forest industries have been leaders in this regard.

One of the greatest sources of change stems from ecological disturbance that not only transforms ecosystem dynamics, but also affects human systems. While ecological processes are natural, human intervention often creates substantial ecological change whereby forest systems move from a stable dynamic equilibrium to one that is unstable. For example, a long history of wildland fire suppression in North America, combined with climate change, is resulting in changes in fire regimes from ones that were high frequency and low intensity to those that are low frequency and high intensity and, most recently, to high frequency and high intensity. The result is increasing suppression costs and losses, a higher incidence of tragic fires where human lives and structures are lost, and where ecosystem diversity is threatened. Such consequences have long been predicted for the boreal forest (see for example Flannigan and Wagner 1991; Stocks et al. 1998; Robinne et al. 2016). The Nordic countries have also been successful in total fire suppression, yet continued climate change poses increasing risk of loss with the threat of both increased wildfire activity and drought (Lehtonen et al. 2014; Lidskog and Sjödin 2016). Additionally, deteriorating forest health has been linked to climate change; an example is the vast expanses of beetle-killed forests in the western part of Canada (Cullingham et al. 2011).

7.6 Conclusion

We live in a world where change is the only constant, meaning that we must continually adapt our use and management of forest resources to meet demand and satisfy increasingly diverse wants, in the face of economic crises and ecological disturbance. As global society becomes more environmentally conscious and resources become scarcer, there will be more pressure to use forestland for the production of non-timber forest products that range, for example, from mushrooms and medicinal products to recreation and wilderness experiences, to wind farm installations where reindeer have traditionally migrated. At the same time, forest owners and timber producers are adversely affected by external forces such as the global economic crisis that occurred between 2007 and 2009, the effects of which are still being felt in 2017. Finally, there is a vast amount of literature on the effects on the boreal forest of climate change and subsequent effects that range from vast areas of pine forests killed by the pine beetle in Canada, increasingly erratic wildlife behavior in North America and Russia, and possible drought in Scandinavia.

The response to all types of change and global effects has been extensive. Scientific research has focused on all aspects of boreal forest resiliency, from genetics and biological processes, to social vulnerability and policy effectiveness largely to increase environmental, social, and economic sustainability through adaptive change. Through scientific discovery and technological advance, forest owners and users have been successful in adapting to change to strengthen forest sustainability. If this trend persists, regardless of change, the boreal forests will continue to provide the ecological foundation that supports other human desires.

Box 7.1: The Role of the Forest Sector

The forestry sector in 1990 played a much greater role as a contributor to GDP in all of the circumpolar countries. By 2011, however, contributions to GDP had fallen by roughly half (see table below). Finland's forest industry contributes the greatest proportion, at 4.3%, followed by Sweden at 2.9% and Canada at 1.2%. The forest sector contributes only 0.6% to the US economy, meaning that the percentage contribution to total GPD from Alaska's forest industry would be close to zero.

Contribution of the forestry sector to total Gross Domestic Product, 1990–2011 (%)

	1990	2000	2001	2002	2003	2004	2005	2006	2007	2008	2009	2010	2011
Canada	2.6	3.2	3	2.8	2.6	2.6	2.4	2	1.8	1.4	1.2	1.3	1.2
Finland	7.2	7.9	7.5	6.7	6.1	5.8	5.1	5.3	5.5	4.6	3.7	4.4	4.3
Iceland	0.6	0.4	0.3	0.4	0.4	0.3	0.3	0.3	0.3	0.2	0.2	0.2	0.2
Norway	1.9	1.3	1.4	1.2	1.1	1.1	1	0.9	0.9	0.7	0.6	0.6	0.6
Russia	1.4	1.2	1.1	1.1	1.1	1	1	1	1	0.8	0.8	0.8	0.8
Sweden	4.3	4.2	4.2	3.8	3.7	3.4	2.8	3.1	3.2	2.9	3	3.1	2.9
USA	1.2	1	0.9	0.9	0.8	0.8	0.8	0.8	0.7	0.6	0.7	0.6	0.6

Source Data retrieved from the FAO, Lebedys and Li (2014)

The greatest contributing sectors to GDP in each country are services and manufacturing. Forestry is grouped within agriculture.

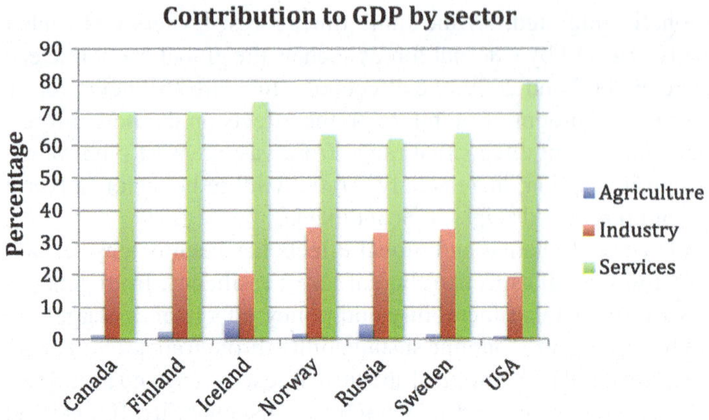

Source: Data from the World Fact Book produced by the CIA (2014).

Box 7.2: A Comparison of Global Forest Value

The global financial crisis has taken a toll on forest production in the North. A look at how gross value added has changed from 1990 to 2011 reveals a general decrease in value added in the northern countries with the exception of Sweden. Value added (VA) is the value of output produced net of purchases from other sectors. It is also used to calculate GDP. The table below shows gross VA in 2011 prices, adjusted for exchange rates for the Arctic states as well as the top four forest-producing countries in the world—China, Japan, India, and Brazil.

Gross value added in the forestry sector (ISIC Rev.4 Divisions 02, 16 and 17), 1990–2011 (in million USD at 2011 prices and exchange rates)

	1990	2000	2001	2002	2003	2004	2005
Canada	26,392	43,339	41,168	39,019	36,505	38,812	35,858
Sweden	13,150	15,408	15,594	14,669	14,541	13,815	11,896
Russia	19,564	12,086	11,216	12,345	12,422	12,688	13,589
Finland	10,959	14,868	14,536	13,186	12,170	11,950	10,923
Norway	4926	4714	5184	4571	4188	4273	3976
Iceland	44	35	34	35	36	32	33
USA (50 states)	110,346	135,498	119,250	118,179	114,607	119,348	117,134
China	17,434	30,834	32,595	36,236	41,877	52,719	56,898
Japan	63,397	51,356	47,301	43,508	43,197	45,041	44,197
India	18,803	22,964	23,687	23,224	23,700	22,733	24,736
Brazil	24,732	19,928	20,166	21,315	24,605	26,086	22,510
Total North America	136,740	178,840	160,419	157,201	151,115	158,163	152,994
Total Europe	203,489	190,928	190,832	185,523	179,492	180,655	173,926

	2006	2007	2008	2009	2010	2011
Canada	31,975	28,482	23,075	19,539	20,435	19,789
Sweden	13,469	14,753	12,919	12,657	13,967	13,841
Russia	13,724	15,278	12,486	11,992	12,789	13,075
Finland	11,830	12,997	10,943	7973	9800	9645
Norway	3856	3961	3164	2642	2755	2434
Iceland	34	32	30	30	29	29
USA (50 states)	118,119	109,329	98,367	100,138	98,697	95,664
China	64,752	79,082	88,420	96,308	112,689	124,622
Japan	40,846	39,335	39,205	37,853	39,858	39,999
India	27,640	26,180	27,330	28,744	30,129	30,958
Brazil	23,595	23,654	23,318	20,979	21,942	22,513
Total North America	150,096	137,813	121,444	119,678	119,134	115,454
Total Europe	178,659	188,646	171,540	154,587	161,799	164,147

The most valuable forest sector in the world in 2011 was China at US$124 trillion; that is approximately eight percent more than what is produced in North America, and approximately 75% of that produced in all of Europe. China also shows the greatest growth in VA, seemingly impervious to the global slowdown that affected most other forest producing nations. India's VA from the forest sector has also increased in the last two decades from $18 trillion to almost $31 trillion

It is interesting to note that Sweden, Finland, Iceland, and Norway account for almost 16% of total VA in Europe (excluding Russia). Similarly, Canada accounts for 17% of the total North American VA, which comes primarily from the boreal forest

Source Data retrieved from the FAO, Lebedys and Li (2014)

References

Alaska Department of Natural Resources (2016) Alaska division of forestry overview. In: Anchorage AK (ed). Retrieved from: http://forestry.alaska.gov/overview. Accessed 20 Oct 2017

Alaska Forest Association (2015) Alaska forest facts. Retrieved from: http://www.akforest.org/facts.htm. Accessed 20 Oct 2017

Arnarsdóttir, Eygló Svala (2014) Forests in Iceland to increase tenfold by 2100. Iceland Review, July 30th, 2014. Retrieved from: http://icelandreview.com/author/258. Accessed 20 Oct 2017

Berg EC, Gale CB, Morgan TA, Brackley AM, Keegan CE, Alexander SJ, Christensen GA, McIver CP, Scudder MG (2014) Alaska's timber harvest and forest products industry, 2011. Gen. Tech. Rep. PNW-GTR-903. US Department of Agriculture, Forest Service, Pacific Northwest Research Station, Portland, 39 p

Central Intelligence Agency (2009) The CIA world factbook 2010. Skyhorse Publishing Inc. Updated in 2014. Retrieved from: https://www.cia.gov/library/publications/the-world-factbook/ . Accessed 20 Oct 2017

Cullingham CI, Cooke JE, Dang S, Davis CS, Cooke BJ, Coltman DW (2011) Mountain pine beetle host-range expansion threatens the boreal forest. Mol Ecol 20(10):2157–2171

Eysteinsson Þ (2016) Forestry in a treeless land, 4th edn. Division Chief National Forests, Iceland

FAO (2009) State of the world's forests 2009. Society, forests and forestry: adapting for the future. Rome

FAO (2010a) Forest management—managers, Finland. Retrieved from: http://www.fao.org/forestry/country/61584/en/fin/. Accessed 20 Oct 2017

FAO (2010b) Forest management. Finland. Retrieved from: http://www.fao.org/forestry/country/61584/en/fin/. Accessed 20 Oct 2017

FAO (2012) The Russian Federation forest outlook study to 2030

FAO (2015) Global forest resources assessment 2015: desk reference. Rome, 253 pp

FAO (2016) State of the world's forests 2016. Forests and agriculture: land-use challenges and opportunities. Rome

Finnish Forest Research Institute (2012a) Finnish forests and forest management in a nutshell. Retrieved from: http://www.metla.fi/metinfo/sustainability/SF-1-forest-industry.htm. Accessed 20 Oct 2017

Finnish Forest Research Institute (2012b) Retrieved from: http://www.smy.fi/en/forest-fi/forest-facts/finnish-forests-owned-by-finns/. Accessed 20 Oct 2017

Finnish Ministry of Agriculture and Forestry (2015) National forest strategy 2025. Ministry of Agriculture and Forestry, Helsinki. Retrieved from: http://mmm.fi/en/nfs. Accessed 20 Oct 2017

Flannigan MD, Wagner CV (1991) Climate change and wildfire in Canada. Can J For Res 21(1):66–72

FSC (2013) FSC helps to shape New Russian Forest policy. Wednesday, 30th October 2013. Retrieved from: https://ic.fsc.org/en/news/id/541. Accessed 20 Oct 2017

FSC (2015) FSC(r) International Standard: FSC principles and criteria for forest stewardship. FSC-STD-01-001 V5-2 EN. Forest Stewardship Council International Secretariat, Bonn, Germany, 32 pp

Government of British Columbia (2012) Timber tenures in British Columbia: managing public forests in the public interest. Lands and Natural Resource Operations, British Columbia Ministry of Forests. 17 pp. Retrieved from: https://www.for.gov.bc.ca/ftp/hth/external/!publish/web/timber-tenures/timber-tenures-2006.pdf. Accessed 20 Oct 2017

Håbesland D, Kilgore M, Snyder S, Becker D, Solberg B, Sjølie HK, Lindstad BH (2015) An assessment of Norwegian family forest owner interest in carbon offset programs and comparisons to the lake states. Staff paper series. No. 237, Department of Forest Resources, University of Minnesota (237)

Haley D, Nelson H (2007) Has the time come to rethink Canada's Crown forest tenure systems? For Chronicle 83(5):630–641

Horstkotte T, Sandström C, Moen J (2014) Exploring the multiple use of boreal landscapes in northern Sweden: The importance of social-ecological diversity for mobility and flexibility. Hum Ecol 42(5):671–682

Icelandic Forestry Association (2016) Icelandic forest info. Retrieved from: http://www.skog.is/forest/index.php?option=com_content&view=article&id=18&Itemid=38. Accessed 20 Oct 2017

Korosuo A, Sandström P, Öhman K, Eriksson LO (2014) Impacts of different forest management scenarios on forestry and reindeer husbandry. Scand J For Res 29(sup1):234–251

Lange K (2015) Forestry in Iceland? Knots and Bolts, Winter 2015. Retrieved from: http://northernwoodlands.org/knots_and_bolts/forestry-iceland. Accessed 20 Oct 2017

Langston N (2013) Mining the boreal north. Am Sci 101(2):98–102

Lebedys A, Li Y (2014) Contribution of the forestry sector to national economies, 1990–2011. FAO, Forest Finance Working Paper FSFM/ACC/09

Lehtonen I, Ruosteenoja K, Venäläinen A, Gregow H (2014) The projected 21st century forest-fire risk in Finland under different greenhouse gas scenarios. Boreal Environ Res 19:127–139

Lidskog R, Sjödin D (2016) Extreme events and climate change: the post-disaster dynamics of forest fires and forest storms in Sweden. Scand J For Res 31(2):148–155

Lier M, Parviainen J (2013) Integration of nature protection in forest policy in Finland. INTEGRATE Country Report. FICENT-OEF, Freiburg, 65 pp

Natural Resources Institute Finland (2012) State of Finland's forests 2012: Finnish forests and forest management in a nutshell. Retried from: http://www.metla.fi/metinfo/sustainability/SF-1-forest-industry.htm. Accessed 20 Oct 2017

Nordic Timber (2016) Norwegian forest policy. Retrieved from: http://www.nordictimber.org/norwegian-forest-policy. Accessed 20 Oct 2017

Norway's Forest Act (2005) Retrieved from: http://faolex.fao.org/cgi-bin/faolex.exe?rec_id=053322&database=faolex&search_type=link&table=result&lang=eng&format_name=@ERALL. Accessed 20 Oct 2017

NRCan (2016) Overview of Canada's forest products industry. Retrieved from: http://www.nrcan.gc.ca/forests/industry/overview/13311. Accessed 20 Oct 2017

Oester B, Parviainen J (2008) Finnish forestry—a role model for Switzerland? English version retrieved at: http://www.tandfonline.com/doi/full/10.1080/02827581.2014.902152. Accessed 20 Oct 2017

Ontario Ministry of Natural Resources and Forestry (2016) Ontario Forest Modernization Act, 2011. Government of Ontario. Retrieved from: https://www.ontario.ca/page/forest-tenure-modernization. Accessed 20 Oct 2017

Parviainen J, Lier M (2015) Finlands forest facts—2015. LUKE Natural Resources Institute, Finland. Retrieved from: http://www.metla.fi/metinfo/kestavyys/pdf/finlands-forests-facts-2015-small.pdf. Accessed 20 Oct 2017

Petrov A, Lobovikov M (2012) The Russian federation forest sector: outlook study to 2030. FAO, Rome

Robinne FN, Parisien MA, Flannigan M (2016) Anthropogenic influence on wildfire activity in Alberta, Canada. Int J Wildland Fire. https://doi.org/10.1071/wf16058

Royal Swedish Academy of Agriculture and Forestry (2009) The Swedish forestry model. Stockholm, Sweden, p 16

Royal Swedish Academy of Agriculture and Forestry (2015) Forests and forestry in Sweden Stockholm, Sweden, 24 pp

Runesson UT (2016) Boreal forests. Lakehead University, Thunder Bay, ON Canada. Retrieved at: http://www.borealforest.org. Accessed 20 Oct 2017

Russian Government (2006) Forest code of the Russian Federation, 2006. (Translated and retrieved from: http://www.fire.uni-freiburg.de/GlobalNetworks/BalticRegion/ForestCode-3rdReading-061108-eng.pdf). Accessed 20 Oct 2017

Russian Government (2016) Federal agency for forestry. Retrieved from: http://government.ru/en/department/245/. Accessed 20 Oct 2017

Sandström P, Sandström C, Svensson J, Jougda L, Baer K (2012) Participatory GIS to mitigate conflicts between reindeer husbandry and forestry in Vilhelmina model forest Sweden. For Chronicle 88(3):254–260

Sjølie HK, Becker D, Håbesland D, Solberg B, Lindstad BH, Snyder S, Kilgore M (2016) Willingness of nonindustrial private forest owners in Norway to supply logging residues for wood energy. Small-scale For 15(1):29–43

Stocks BJ, Fosberg MA, Lynham TJ, Mearns L, Wotton BM, Yang Q, Jin JZ, Lawrence K, Hartley GR, Mason JA, McKenney DW (1998) Climate change and forest fire potential in Russian and Canadian boreal forests. Clim Change 38(1):1–13

Swedish Forest Industries Federation (2012) The forest industry—the driver for a sustainable bioeconomy. Swedish Forest Industries Federation Sustainability Report 2012. English Abridged Version, 28 pp

Ulybina O (2014a) Interaction, cooperation and governance in the Russian forest sector. J Rural Stud 34:246–253

Ulybina O (2014b) Russian forests: the path of reform. For Policy Econ 38:143–150

World Wildlife Fund (2016) Russian forest program. Retrieved from: https://www.wwf.ru/about/what_we_do/forests/eng. Accessed 20 Oct 2017

Wyatt S, Fortier JF, Natcher DC, Smith MAP, Hébert M (2013) Collaboration between aboriginal peoples and the Canadian forest sector: a typology of arrangements for establishing control and determining benefits of forestlands. J Environ Manage 115:21–31

Chapter 8
Climate Change and China's Rise to Great Power Status: Implications for the Global Arctic

Sanna Kopra

8.1 Introduction

In the last few decades, the world has faced two significant changes that have fueled the globalization of the Arctic. In the first, with climate change now a key threat, the Arctic has become a showcase of climate governance. It is now the fastest-warming region on earth, and the melting of ice in the Arctic is accelerating climate change and altering ecosystems globally (ACIA 2004). In the second change, the People's Republic of China has undergone a profound identity change from an isolated communist state to an emerging great power, a transition that has dramatically transformed its sphere of interests. China has recently become interested in Arctic affairs and wants to be recognized as a legitimate stakeholder in governance of the region. This ambition has sparked speculation about whether China is challenging the rights and interests of the Arctic states and whether its engagement entails a heightened risk of military conflict in the region (see, for example, Cassotta et al. 2015; Rainwater 2013; Wright 2011).

This chapter investigates the interplay of these two changes by examining the emergence of China's great power status and its implications for the Global Arctic. I focus particularly on the country's climate policies in particular, because climate change is a central issue in its increasing role in the Arctic. China argues that it has special interests in the Arctic due to adverse effects of climate change and that it must have the opportunity to be involved in governance of the region. Another salient consideration is that China is the largest emitter of greenhouse gases (GHG) in the world, which means that the fate of the Arctic—and therefore of the whole world—hinges a substantially on China's climate change policies.

S. Kopra (✉)
University of Helsinki, Helsinki, Finland
e-mail: sanna.kopra@helsinki.fi

M. Finger and L. Heininen (eds.), *The GlobalArctic Handbook*,
https://doi.org/10.1007/978-3-319-91995-9_8

8.2 China's Growing Great Power Status

China's era of reform began in the late 1970s, following the death of Premier Mao Zedong. Economic and political reforms brought rapid growth, which also gradually transformed the country's national identity during the 1980s and 1990s. Although the change in identity was largely the result of economic developments, it also encouraged China to take a more active part in the international community than it had before (Qin 2004). Moreover, when the Cold War ended and the Soviet Union collapsed in the early 1990s, the international community begun to regard China as a potential great power. However, at the same time as other states sought to integrate China into the community, "the China threat" theories started to develop, especially after the Tiananmen incident in 1989; countries like the United States, Japan, and Taiwan started to speculate about whether a rising China posed a threat to international society (see, for example, Deng 2006). Since then, the Chinese government has paid special attention to the state's international image (see, for example, Deng 2008; Gries 2004), while Chinese intellectuals have debated the international role and responsibilities of a rising China (Shambaugh 2013; Xia 2001). In 2005, then-US Deputy Secretary of State Robert Zoellick gave a famous speech in which he urged China to become a responsible stakeholder, indicating that the country was not yet fulfilling the global responsibilities commensurate with its emerging great power status (Zoellick 2005). Because China's national identity is in flux,[1] however, it has been very difficult for the Chinese leadership to agree on the scope of China's global responsibility. In general, China accepts that its permanent seat on the UN Security Council entails special responsibilities (Wang 2013).

China has recently started to identify itself as a great power and to define and implement its "grand strategy" in a more determined way than ever before. In particular, since President Xi Jinping took office in March 2013, the country has departed from the "keep a low profile" foreign policy that Deng Xiaoping set in the 1990s and has expressed its policy objectives in an unprecedentedly explicit manner. Notably, in his inauguration speech, Xi introduced his vision of the "China dream" (中国梦): "Realizing the great renewal of the Chinese nation is the greatest dream for the Chinese nation in modern history" (as cited in Xinhua 2012). The dream was quickly adopted as one of the hallmarks of Xi's regime and can now be interpreted as China's "grand strategy". This strategy draws on two "centenary goals", which the country aims to achieve by the 100th anniversary of the founding of the PRC in 2049. The first is "doubling the 2010 GDP and per capita income of urban and rural residents and finishing the building of a society of initial prosperity in all respects", and the second is "turning China into a modern socialist country that is prosperous, strong,

[1] Jin (2011, 251) has distinguished four identities for China, all of which all are currently in flux. First, China's identity as the largest developing country is being challenged by its economic miracle. Second, China's identity as a regional power is no longer very accurate, as contemporary China has wide global interests and influence. Third, China's identity as a socialist country is being eroded by its capitalist market system and consumerism. Fourth, China's identity as a Confucian civilization is being lost in modernization and market economy.

democratic, culturally advanced and harmonious". Although these aims were originally put forward in 1997 by then-President Jiang Zemin (Xinhua 2014a) at the 15th National Congress of the Communist Party of China (CPC) (see Jiang 1997), Xi Jinping's dream has elevated them to the status of China's strategic priorities. This means that although the government's ultimate goal officially continues to be the realization of communism, in practice its key priorities are economic development and building a harmonious society. Despite its growing international status, China's core interests lie in domestic and regional issues and include state sovereignty, national security, safeguarding China's political system established by the Constitution, territorial integrity and national reunification, overall social stability, as well as ensuring sustainable economic and social development. The country's foreign policy must support and promote these interests.

Guided by the "China dream", China increasingly identifies itself as a great power in the international arena. In February 2012, then-Vice President Xi Jinping referred to China as a great power by introducing the concept of the "new type of great power relationship" (新型大国关系) and highlighted the need to expand shared interests and mutually beneficial cooperation between the US and China (Xi 2012). In June 2013, China's Foreign Minister Wang Yi gave an unusually comprehensive statement of China's foreign policy entitled "Exploring the Path of Major-Country Diplomacy with Chinese Characteristics" (Wang 2013) and in November 2014 President Xi Jinping presented the concept of "major-country diplomacy with Chinese characteristics" at a high-level foreign policy conference in Beijing (Xinhua 2014a). Xi's speech established guiding principles for Chinese foreign policy for the coming years. On the one hand, he emphasized China's increased interdependence with the rest of the world and the need for global cooperation; on the other, he stressed—in extraordinarily explicit terms—the growing importance of foreign policy in ensuring the realization of the "China dream" and achievement of the "two centenary goals" (see also Swaine 2015).

As a result, China is now taking a more assertive approach to foreign policy than ever before. Advised by Xi, China will "never give up [its] legitimate rights and will never sacrifice [its] national core interests" (as cited in Xinhua 2013a). The Chinese government has also attempted to set international agendas proactively: it has begun to develop and promote its own concepts and ideas, such as "harmonious world" (和谐世界), "the Asia-Pacific dream" (亚太梦) and a "new type of international relations" (新型国际关系) in order to organize international society. Time will tell if these new concepts succeed in transforming international discourses and practices such that they become "less Westernized" and better accommodate Chinese values and interests. The purpose of these concepts is to reform international society in a "responsible manner", not to replace the existing practices from which China has benefitted. In some policy sectors, however, China has suggested alternative sources of global governance, proposing new foreign policy initiatives such as One Belt, One Road, or the New Silk Road and establishing new multilateral financial institutions such as the Asian Infrastructure Investment Bank and the BRICS New Development Bank.

Along with its other efforts to advance its great power status, China has significantly prioritized maritime issues on its agenda and is now endeavoring to become a "maritime power". This objective was declared by then-President Hu Jintao in his report to the 18th National Congress of the CPC in November 2012. According to Hu, China should enhance its "capacity for exploiting marine resources, develop the marine economy, protect the marine ecological environment, resolutely safeguard China's maritime rights and interests, and build China into a maritime power" (Hu 2012). According to Devitt (2016, 2), China's conception of a maritime power not only emphasizes strong naval power, but also includes civilian capabilities such as a good coast guard and port infrastructure, advanced shipbuilding capacity, high-level merchant shipping and fishing fleets, as well as the technology and know-how to utilize maritime resources. The objective of becoming a maritime power was incorporated into China's 2012 white paper on armed forces, which was published after Xi Jinping took office in spring of 2013. That white paper states:

> China is a major maritime as well as land country. The seas and oceans provide immense space and abundant resources for Chinas sustainable development, and thus are of vital importance to the people's wellbeing and China's future. It is an essential national development strategy to exploit, utilize and protect the seas and oceans, and build China into a maritime power. (Information Office of the State Council 2013)

Although China has not officially made reference to the Arctic when outlining its efforts to become a maritime power, some Chinese scholars have made strong linkages between the two (Jakobson 2015, 161). Since becoming a maritime power now constitutes an essential part of Xi Jinping's "China dream", which in turn is a key guideline of China's foreign policy, it can be assumed that China's increasing activities in the Arctic Ocean are informed by this objective.

International climate politics is an especially interesting window to China's emerging great power status, because China has increasingly started to call itself a major power in international negotiations on climate change while continuing to emphasize that it is a developing country (Kopra 2016). Although China resists formal talks on climate change in the United Nations Security Council, it seems to acknowledge that great powers should take the lead in international efforts to combat the problem. For example, China's special envoy at the September 2014 UN Climate Summit, Zhang Gaoli, declared: "responding to climate change is what China needs to do to achieve sustainable development at home as well as to fulfill its due international obligation as a responsible major country" (Zhang 2014). Moreover, China has published its key climate commitments in joint statements with the US, indicating that this was done with a view to the great power context (see White House 2014, 2015, 2016a, b). Yet, while *China's National Climate Change Plan (2014–2020)* confirms the country's responsibility as a great power in climate change mitigation, the document also defends the state's "legitimate development rights and interests" as a developing country (National Development and Reform Commission 2014, 4–5).

8.3 China's Role in International Climate Politics

There is clear evidence that human activities such as burning fossil fuels have increased the concentration of GHGs in the atmosphere and are thus contributing to climate change (Intergovernmental Panel on Climate Change 2014). Historically, climate change has largely been caused by the industrialization of developed countries, but the rapidly growing proportion of GHG emissions in emerging economies is also a contributing factor. In particular, China's total GHG emissions have grown at an extraordinary rate due to its rapid economic growth: between 1990 and 2013, the country's carbon emissions increased by 80% (PBL Netherlands Environmental Assessment Agency 2015, 10). In 2006, China's total emissions surpassed those of the US, and it is now the world's largest carbon dioxide emitter. In 2015, China accounted for 30% of total global carbon dioxide emissions (PBL Netherlands Environmental Assessment Agency 2015, 12). Although the Arctic produces essentially no GHG emissions, the region has warmed more rapidly than any other region on earth during the last 30 years. The impacts of climate change in the Arctic, such as the reduction in sea ice and loss of snow cover, not only harm the region's vulnerable natural environment and livelihood of the local populations, but also have significant global impacts (ACIA 2004).

Since climate change is a complex global problem, global efforts are necessary to address it; it cannot be solved by Arctic regional governance. China's engagement in international climate politics is undoubtedly imperative: without its participation, any global effort to combat climate change is going to fail. Since the early 1990s, international negotiations on climate change have been conducted under the UN Framework Convention on Climate Change (UNFCCC). For a long time, China's attitude towards the UNFCCC was marked by a reluctance to take any measures to mitigate climate change. Instead, it underlined the historic responsibility of developed countries and demanded that they shoulder all the responsibility for tackling the problem. At present, however, the Chinese leadership recognizes that climate change is a "challenge faced by the entire world" and can therefore only be solved by "extensive international cooperation" (Information Office of the State Council of the People's Republic of China 2008). China has gradually changed its attitude in international climate politics since the 2009 UN Climate Conference in Copenhagen, where it came under harsh international criticism. Although China continues to refuse to commit itself to legally binding emissions targets at the international level, it can no longer be seen as "irresponsible".

China played an influential role at the UN Climate Conference in Paris in 2015, where the most recent global climate agreement was concluded. For the first time ever, the president (instead of the premier) represented the country, portraying it as a responsible stakeholder and committed facilitator of the agreement. After the conference, Foreign Ministry Spokesperson Hong (2015) praised "China's sense of responsibility as a major country in tackling climate change". For China, the "bottom-up" approach of the Paris Agreement was appealing: it does not impose a top-down obligation on any of the parties, but instead allows them to commit to voluntary, domestically formulated mitigation plans. China obviously prefers to make moderate voluntary commitments without legal international obligations, as they involve no fear of failure but rather offer it an opportunity to "gain face"[2] by exceeding global expectations.

In its Intended Nationally Determined Contribution (INDC) to the UNFCCC, published in 2015, China committed to peak its carbon emissions around 2030 and pledged to reduce its carbon intensity—that is, the amount of carbon dioxide per unit of GDP—by 60–65% from its 2005 level by 2030. China also undertook to increase the share of non-fossil fuels in primary energy consumption to around 20% and to increase its forest stock volume by around 4.5 billion m^3 over the 2005 level (National Development and Reform Commission 2015, 5). Notably, China was among the first countries to ratify the Paris Agreement, in September 2016, an action which undoubtedly heightened the willingness of other states to ratify it.

The Paris Agreement entered into force in early November 2016, just a few days before climate skeptic Donald Trump was elected as US president. His election raised serious concerns worldwide regarding the US commitment to the agreement, as he had repeatedly threatened to withdraw the US from it if elected. On June 1, 2017 President realized his plans and announced that the United States would cease all commitments to the Paris Agreement (White House 2017). As Trump seems to vitiate all the climate policies put in place by the previous administration under the leadership of Barack Obama, his policies have highlighted China's new emerging leadership in international climate politics. Indeed, it seems that China might well be ready to assume that position: since Trump's election, the Chinese leadership has declared several times that it will not dilute its climate commitments despite the US withdrawal from the Paris Agreement (see, for example, China Daily 2016). Notably, at the 2017 World Economic Forum in Davos, Xi Jinping urged that "[a]ll signatories should stick to [the Paris Agreement] instead of walking away from it as this is a responsibility we must assume for future generations" (World Economic Forum 2017). Furthermore, in a speech at the United Nations Office at Geneva, Xi (2017) sent a clear message to the world—and especially to President Trump—about China's unwavering climate commitments.

Talk of China's new responsibility is not only empty rhetoric; rather, it seems that the Chinese political leadership and the general public widely agree that they—and the planet—cannot afford to follow the Western model of industrialization based on a "pollute first, clean up later" mentality. China's new development policies acknowledge that going green is the only way to realize the "China dream". Since the early 2000s, China has examined how it could modernize in a more sustainable way and hence alleviate social and environmental problems caused by its development model (Dent 2014, 57). To ensure an energy supply, the government launched a series of

[2]In the Chinese context, the concept of "face" is often used to describe human concerns over honor and respect. It explains why national image-building plays such an important role in Chinese foreign policy. The Chinese government seeks to maintain face because it means "maintaining authority" and the state's national honor; losing face would mean "losing status and the ability to pursue instrumental goals" (Gries 2004, 29). In China, others can also "give face" to a person or social group.

policies and measures to decrease dependency on (imported) fossil fuels and to promote the production of non-fossil energy, especially hydropower and nuclear energy.

In addition to economic interests, China has strong human security incentives to respond to climate change, given that it is geographically very vulnerable to climate-related disasters. The Chinese government is increasingly aware of the adverse effects of climate change on the country. Advised by the first *National Assessment Report on Climate Change*, published in 2006, China's first *National Climate Change Program*, released in 2007, recognized that the climate was already changing in China: the average surface temperature had already increased by between 0.5 and 0.8 °C during the 20th century; mountain glaciers were melting at an accelerated rate; the frequency and intensity of heat waves had increased in the northern provinces; and heavy precipitation had increased in the southern provinces (National Development and Reform Commission 2007).

Forecasts estimate that China's average temperatures will rise between 1.3 and 5.0 °C by the end of the century (China Climate Change Info-Net 2015). Therefore, it is assumed that China will suffer the most from a "business as usual" scenario—in which no effective emission cuts are implemented and the global average temperature will increase to 4 °C—and has the most to gain from limiting warming to 2 °C (Strauss et al. 2015, 10). Although China is not an Arctic state, climate change in the Arctic poses tremendous risks to the country. For example, many of China's mega-cities, such as Shanghai, Tianjin, and Hong Kong, are located in coastal areas and are at high risk of flooding due to rising sea levels caused by the melting of sea ice in the Arctic. Furthermore, climate change in the Arctic will alter many global natural processes, causing changes that are likely to hamper China's agricultural production. Food security is not only important for feeding China's large population, but is also a critical aspect of the legitimacy of the Communist Party. Given these problems, it is no surprise that scientific research on climate change is currently one of the key interests of China's Arctic policies.

Unfortunately, it is now becoming clear that states have failed to stop climate change. The Earth's climate is changing regardless of any actions states may be taking, and the Arctic is already experiencing rapid changes as result. In this light, adaptation is important, alongside mitigation. China's first nationwide climate change adaptation strategy, published in 2013, warned that Chinese society in its entirety is ill-prepared to deal with the serious threats posed by climate change. Taking "significantly enhanced adaptation capacity" as the ultimate goal, the plan outlines a wide range of measures to be implemented by 2020 in order to protect water, forest, and soil resources, safeguard agricultural output, strengthen infrastructure, improve risk management systems, increase public awareness, and establish institutional mechanisms (National Development and Reform Commission 2013). Engagement in the Arctic offers China a chance to gain scientific knowledge that can be utilized to formulate plans allowing the country to adapt to climate change. Accordingly, the primary interest of Chinese polar research is to gain better understanding of climatic changes in the Arctic and their impacts on China. In particular, the Chinese government is interested in learning more about linkages between climate change in the Arctic and extreme weather, floods, and impacts on agriculture in the country. In contrast to sensitive issues such as sovereignty and resource exploitation, the issue of climate change is regarded as sufficiently uncontroversial for China to encourage enhanced collaboration with the Arctic states.

China has made major structural changes in its economy and energy sectors in an effort to reduce its carbon emissions. In fact, forecasts suggest that the emissions peak could occur earlier than 2030 because China's total coal consumption seems to have already peaked in 2013 (see, for example, Buckley and Sanzillo 2015) and the growth of its total emissions has slowed since 2012 (PBL Netherlands Environmental Assessment Agency 2015, 5). In 2014, China's *Energy Development Strategy Action Plan (2014–2020)* included, for the first time, a cap on national coal consumption by 2020 and pledged to raise the share of non-fossil fuels in the total primary energy mix to 15% by 2020 from 9.8% in 2013 (Xinhua 2014b). Looking forward, China has issued many other ambitious strategies, such as *China's National Climate Change Plan (2014–2020)*, published in 2014 and its *Integrated Reform Plan for Promoting Ecological Progress*, published in 2015, to reduce emissions in the energy and industrial sectors, the construction industry, transportation and agriculture, as well as to develop high value-added manufacturing and a less energy-intensive service sector. Moreover, 11 Chinese cities and provinces have committed themselves to peaking their carbon emissions before the national target of 2030, with two of the largest cities, Beijing and Guangzhou, even promising to do so by the end of 2020 (US–China Climate Leaders' Declaration 2015). Most importantly, China's *13th Five-Year Plan (2016–2020)* integrated the state's international climate pledges into the regime's most authoritative domestic development objectives and measures. The plan aims to reduce carbon intensity by 18% from 2015 levels. It seeks to not only implement, but also *strengthen*, the country's INDC and reiterates the above-mentioned national energy consumption cap of 5 billion metric tons of standard coal equivalent by 2020 (National People's Congress & Chinese People's Political Consultative Conference 2016). All in all, there are strong hopes that China will fulfill its international climate commitments in the coming years while simultaneously growing its economy.

8.4 China's Grand Strategy and the Arctic

As China is not an Arctic country, engagement in the Arctic is not one of China's political priorities. Nevertheless, it has become increasingly interested in the Arctic in recent years (see, for example, Chen 2012; Jakobson 2010; Jakobson and Peng 2012). In general, the country's involvement in polar affairs is nothing new: it acceded to the Svalbard Treaty as far back as in 1925 and started its own Antarctic research program, with the support of the US, New Zealand, Japan, and Australia, in the late 1970s and early 1980s (Brady 2012,104). In the early 1990s, Chinese scientists started to conduct research and participate in multinational projects in the Arctic, and since 1994 they have conducted expeditions in the both Polar Regions on board the research ice breaker *Xuelong* (Snow Dragon) (see Jakobson 2010,

3–5). However, the main focus of China's polar research, in terms of funding and expeditions, has been—and still is—the Antarctic (Jakobson 2015, 156).

In addition to the traditional scientific interest in the Arctic, an increasing number of Chinese analysts are now working on geopolitical, economic, legal, and other aspects of the Arctic. On the one hand, this growing interest is seemingly motivated by the globalization of the Arctic; as demonstrated by the present volume, there are a number of political, economic and environmental changes going on in the Arctic that have global repercussions, including impacts on China. On the other hand, as China's global status has dramatically changed, it is no surprise that the country has started to pursue economic and political interests at the global level. The melting of the Arctic icecap certainly provides economic opportunities that will support China's long-term development strategy. The northern sea routes are of particular interest to China, because they not only provide it with faster and shorter access to Western markets, but also decrease its heavy dependence on the narrow and politically insecure Strait of Malacca (the so-called "Malacca dilemma"). Furthermore, with its growing hunger for oil and other natural resources, the large unexplored areas in the Arctic containing petroleum and untapped minerals resources are also of interest. Since China does not have the advanced technology required to extract oil or gas from the Arctic continental shelf, it has increased cooperation with transnational oil corporations and the Arctic states—and will probably continue to do so. In political terms, the Arctic offers another platform for China to take part in international governance and enhance the country's international status (For more detailed discussion of China's interests in the Arctic, see, for example, Chen 2012; Jakobson 2010; Jakobson and Peng 2012; Kopra 2013; Li and Bertelsen 2013).

China's growing international status has raised global concerns that it will challenge the existing international norms and institutions. Classic realists believe that because the world is "condemned to perpetual great-power competition" (Mearsheimer 2001, 2), the rise of China will inevitably lead to a war for hegemony. China's increasing interest in the Arctic has generated much public unease, although many of the concerns are based on myths and misconceptions of China's Arctic diplomacy. These include Chinese Rear Admiral Yin Zhuo's misinterpreted statement that the Arctic does not belong to any country, China's involvement in the mining of Greenland's rare earth elements, a Chinese businessman's plans to buy a large area of land in Iceland, and the Chinese government's alleged plans to establish a "super-embassy" with a massive number of personnel in Iceland (see Ping and Lanteigne 2015, 2–13). For their part, critics of realism the "China threat" view as simplistic, because circumstances and ideas both influence how great powers behave. On the one hand, a rising China is faced with a very different international system and very different political problems than previous rising powers have encountered. On the other hand, national identities, norms, and values influence how great powers see the world and how they behave. Regarding the Arctic, the globalization of the region and the causes and effects of climate change largely link the interests of states, including the great powers.

China has attempted to alter regional power relations in the East Asian region, especially in the South China Sea, but this does not prove that it is a revisionist power on the global stage. There is no evidence that China would pose a threat in the Arctic region (Cassotta et al. 2015). It has never disputed the Arctic states' sovereign rights over their exclusive economic zones (EEZs), nor has it questioned the role of the Arctic Council, the key regional forum established in 1996 to enhance cooperation and coordinate interaction amongst the eight Arctic states and Arctic indigenous communities on sustainable development and environment protection in the region, or suggested an alternative international body to govern Arctic affairs. In order to be granted "permanent observer" status on the Arctic Council, China had to accept the Nuuk criteria defined by the Council in 2011. By so doing, it agreed to "recognize Arctic States' sovereignty, sovereign rights and jurisdiction in the Arctic" as well as the Law of the Sea and regional indigenous peoples' culture and rights (Hong 2013; Arctic Council 2011).

As part of its efforts to gain access to Arctic governance, China has increased its cooperation with the Arctic countries and started to identify itself as a "near-Arctic country" and an "Arctic stakeholder". This discursive identification is motivated in particular by the threats that climate change is expected to pose to China's environment, agriculture and coastal urban areas. However, it should be noted that China is not the only non-Arctic country that has made efforts to create an Arctic country "brand". Other observer states on the Arctic Council, such as Japan, United Kingdom, France, South Korea, and Singapore, have published, or are in the process of preparing, policy papers outlining their identities, visions, and interests with respect to the Arctic. Nevertheless, China's "branding" efforts have been seen in a far more negative light than those of other non-Arctic states—even though the United Kingdom, for instance, has used a very similar discourse to identify itself as a "nearest neighbour of the Arctic" (United Kingdom Government 2013, ii; see also Ping and Lanteigne 2015, 16).

While the Chinese government has not officially commented on the Nuuk criteria, various Chinese scholars have criticized the Arctic Council's position as the sole decision-maker for the region (Jakobson and Peng 2012, 13–14). For instance, Professor Guo (2011) has argued the following:

> Arctic states announce to the world: The Arctic is an 'Arctic-states' Arctic. They oppose the idea that the Arctic is common property of the whole of humankind and desire to advance their own interests and to impair the participation of non-Arctic states through the Monroe Doctrine.

China is not totally excluded from governance of the Arctic. In 1996, it ratified the UN Convention on the Law of the Sea (UNCLOS), the key international agreement for regulating maritime activities in the Arctic. As a signatory to the 1920 Svalbard Treaty, China has a right of access to Svalbard and as a member of the International Maritime Organization, as well as other international bodies, it can take part in the formulation of international regulations that directly or indirectly influence Arctic governance. As a non-Arctic state, however, China is not eligible to become a full member of the Arctic Council. In 2007, it did acquire ad hoc observer status at Arctic Council meetings. In 2013, after two failed attempts, China's status was elevated to "permanent observer", which does not accord it decision-making power (voting rights), but does guarantee access to all Arctic Council meetings and activities. This was an important decision for China, because it ensured that its voice is now heard in the Council and that it will have an opportunity to "get to know the Arctic better, and then [be] able to join effectively international cooperation" (Xinhua 2013b).

It is likely that China will seek to play a more influential role in the Arctic Council. Notably, Chinese Vice Foreign Minister Zhang Ming declared in 2015 that China is a "major stakeholder" in the Arctic (Zhang 2015). Although China has not questioned the Arctic states' sovereign rights in the region, it often emphasizes that Arctic governance concerns itself with regional as well as global issues and therefore the voices of non-Arctic states must be properly heard (see, for example, Xinhua 2013c). Zhang also seemed to express his concern that non-littoral states would not gain full access to Arctic natural resources. He pointed out that "the overall interests of the international community in the Arctic should be respected" (Zhang 2015).

8.5 Conclusions

In step with its growing great power status, China has started to demand access to all international political forums, and Arctic governance is no exception. In the long run, China has such important economic and security interests related to shipping, access to natural resources and climate change in the Arctic that it will likely seek a more influential role in governance of the region. This ambition is also motivated by its "policy of face" and pursuit of honor globally. Given China's own strong doctrine of sovereignty, however, it is not likely that it will challenge the sovereign rights of the Arctic littoral states.

China's position in international climate negotiations has changed dramatically since the early 2010s. It no longer focuses exclusively on the historic responsibility of developed countries and has started to make ambitious efforts of its own to mitigate climate change. Indeed, it seems that China is now increasingly taking on the role of a great power in international climate politics. If the country manages to continue to expand its economy while substantially reducing carbon emissions, it has great potential to act as a role model in climate change mitigation globally. In any event, China has a key role to play in this effort in the Arctic: on one hand, it is in a position to make or break an ambitious global agreement on climate change; on the other, its domestic efforts to peak and reduce GHG emissions are necessary in order to limit the global stock of carbon dioxide. Indeed, the early peak of China's GHG emissions is the most important variable in Arctic climate change due to the country's position as the largest GHG emitter in the world. Currently, however, its climate policies do not make references to the melting ice of the Arctic. In any event, deeper involvement in Arctic affairs will probably increase China's motivation to reduce GHG emissions. Scientific research in the Arctic will also help the country to develop its domestic plans for adapting to climate change. From this perspective, China's active Arctic engagement should be encouraged. For the future of the Arctic—and the planet overall—China must make a firmer commitment to respond to climate change.

References

ACIA (2004) Impacts of a warming Arctic: Arctic climate impact assessment. http://amap.no/acia/. Accessed 22 Oct 2016

Arctic Council (2011) Senior Arctic Officials (SAO) Report to Ministers. Nuuk, Greenland. http://library.arcticportal.org/1251/. Accessed 22 Oct 2016

Brady A-M (2012) The emerging economies of Asia and Antarctica: challenges and opportunities. In Jabour J, Haward M, Press AJ (eds) Australia's Antarctica: Proceedings of a Symposium to mark 75 Years of the Australian Antarctic Territory. Institute for Marine & Antarctic Studies Occasional Paper 2, pp 103–113. http://www.imas.utas.edu.au/__data/assets/pdf_file/0005/249692/11_Brady_Final.pdf. Accessed 1 Aug 2012

Buckley T, Sanzillo T (2015) Past peak coal in China. The Institute for Energy Economics and Financial Analysis. http://ieefa.org/wp-content/uploads/2015/11/IEEFA_Peak-Coal_November-2015.pdf. Accessed 23 Oct 2016

Cassotta S, Hossain K, Ren JZ, Goodsite ME (2015) Climate change and China as a global emerging regulatory sea power in the Arctic Ocean: is China a threat for Arctic Ocean security? Beijing Law Rev 6:199–207. https://doi.org/10.4236/blr.2015.63020

Chen G (2012) China's emerging Arctic strategy. Polar J 2(2):358–371

China Climate Change Info-Net (2015)《第三次气候变化国家评估报告》发布. [The "Third National Climate Change Assessment Report" released] http://www.ccchina.gov.cn/Detail.aspx?newsId=56949&TId=57. Accessed 22 Oct 2016

China Daily (2016) Climate meet stays on track. 11 Nov. http://www.chinadaily.com.cn/china/2016-11/11/content_27343097.htm. Accessed 20 Feb 2017

Deng Y (2006) Reputation and the security dilemma: China reacts to the China threat theory. In: Johnston AI, Ross RS (eds) New directions in the study of China's foreign policy. Stanford University Press, Stanford, pp 186–214

Deng Y (2008) China's struggle for status. The realignment of international relations. Cambridge University Press, Cambridge

Dent CM (2014) Renewable energy in East Asia. Towards a new developmentalism. Routledge, New York

Devitt M (2016) Becoming a great "maritime power": a Chinese dream. The Center for Naval Analyses, Arlington. https://www.cna.org/cna_files/pdf/IRM-2016-U-013646.pdf. Accessed 22 Oct 2016

Gries PH (2004) China's new nationalism: pride, politics, and diplomacy. University of California Press, Berkeley

Guo P (2011) 应对北极门罗主义的挑战. [Response to the challenges of the Arctic Monroe Doctrine] http://news.sina.com.cn/w/2011-10-16/151023311701.shtml. 16 Oct. Accessed 24 Sept 2016

Hong L (2013) Foreign Ministry Spokesperson Hong Lei's remarks on China being accepted as an observer of the Arctic Council. http://www.fmprc.gov.cn/mfa_eng/xwfw_665399/s2510_665401/2535_665405/t1040943.shtml. Accessed 24 Sept 2016

Hong L (2015) Foreign Ministry Spokesperson Hong Lei's remarks on the outcomes of the Paris Climate Conference. http://www.fmprc.gov.cn/mfa_eng/xwfw_665399/s2510_665401/t1323918.shtml. Accessed 24 Sept 2016

Hu J (2012) Hu Jintao's report at 18th Party Congress. http://news.xinhuanet.com/english/special/18cpcnc/2012-11/17/c_131981259.htm. Accessed 24 Sept 2016

Information Office of the State Council (2013) The diversified employment of China's armed forces. http://news.xinhuanet.com/english/china/2013-04/16/c_132312681.htm. Accessed 20 Oct 2016

Information Office of the State Council of the People's Republic of China (2008) White paper: China's policies and actions for addressing climate change. http://www.china-un.org/eng/hyyfy/t521513.htm. Accessed 10 Feb 2017

Intergovernmental Panel on Climate Change (2014) Climate change 2014 synthesis report. https://www.ipcc.ch/report/ar5/syr/

Jakobson L (2010) China prepares for an ice-free Arctic. SIPRI Insights on Peace and Security, No. 2010/2. http://books.sipri.org/product_info?c_product_id=402. Accessed 20 Oct 2016
Jakobson L (2015) China's security and the Arctic. In: Dittmer L, Yu M (eds) Routledge handbook on Chinese security. Routledge, London, New York, pp 155–166
Jakobson L, Peng J (2012) China's Arctic aspirations. SIPRI Policy Paper No. 34. http://books.sipri.org/product_info?c_product_id=449. Accessed 20 Oct 2016
Jiang Z (1997) Hold high the great banner of Deng Xiaoping theory for an all-round advancement of the cause of building socialism with Chinese characteristics' into the 21st century. Report Delivered at the 15th National Congress of the Communist Party of China on September 12, 1997. http://www.bjreview.com.cn/document/txt/2011-03/25/content_363499.htm. Accessed 28 Oct 2016
Jin C (2011) Big power's responsibility. China's perspective (trans: Tu X). China Renmin University Press, Beijing
Kopra S (2013) China's Arctic interests. In: Heininen L (ed) Arctic yearbook 2013. Northern Research Forum, Akureyri, pp 107–122
Kopra S (2016) Great power management and China's responsibility in international climate politics. J China Int Relat 4(1):20–37
Li X, Bertelsen RG (2013) The drivers of Chinese Arctic interests: political stability and energy and transportation security. In: Heininen L (ed) Arctic yearbook 2013. Northern Research Forum, Akureyri, pp 53–68
Mearsheimer JJ (2001) The tragedy of great power politics. W.W. Norton & Company, New York
National Development and Reform Commission (2007) China's national climate change programme. http://www.china.org.cn/english/environment/213624.htm. Accessed 4 Sept 2015
National Development and Reform Commission (2013) 国家适应气候变化战略. [China's national climate change adaptation strategy] http://www.sdpc.gov.cn/zcfb/zcfbtz/201312/W020131209343322750059.pdf. Accessed 4 Sept 2015
National Development and Reform Commission (2014) 国家发展改革委关于印发国家应对气候变化规划 (2014–2020年) 的通知. [China's national climate change plan (2014–2020)] http://www.ccchina.gov.cn/nDetail.aspx?newsId=49211&TId=60. Accessed 4 Sept 2015
National Development and Reform Commission (2015) Enhanced actions on climate change: China's intended nationally determined contribution. http://www4.unfccc.int/submissions/indc/Submission%20Pages/submissions.aspx. Accessed 24 Sept 2015
National People's Congress & Chinese People's Political Consultative Conference (2016) 中华人民共和国国民经济和社会发展第十三个五年规划纲要. [The thirteenth five-year plan for the national economic and social development of the People's Republic of China] http://news.xinhuanet.com/politics/2016lh/2016-03/17/c_1118366322.htm. Accessed 23 Oct 2016
PBL Netherlands Environmental Assessment Agency (2015) Trends in global CO_2 emissions. 2015 Report. http://edgar.jrc.ec.europa.eu/news_docs/jrc-2015-trends-in-global-co2-emissions-2015-report-98184.pdf. Accessed 23 Oct 2016
Ping S, Lanteigne M (2015) China's developing arctic policies: myths and misconceptions. J China Int Relat 3(1). http://dx.doi.org/10.5278/ojs.jcir.v3i1.1144
Qin Y (2004) National identity, strategic culture and security interests: three hypotheses on the interaction between China and international society. In: Yu X (ed) Cultural factors in international relations. Shanghai Institute for International Studies, Shanghai, pp 209–221
Rainwater S (2013) Race to the North. China's arctic strategy and its implications. Naval War Coll Rev 66(2)
Shambaugh D (2013) China goes global: the partial power. Oxford University Press, New York
Strauss BH, Kulp S, Levermann A (2015) Mapping choices: carbon, climate, and rising seas, our global legacy. Climate Central Research Report, pp 1–38
Swaine MD (2015) Xi Jinping's address to the central conference on workrelating to foreign affairs: assessing and advancing major-power diplomacy with Chinese characteristics. China Leadersh Monit 46

United Kingdom Government (2013) Adapting to change. UK policy towards the Arctic. https://www.gov.uk/.../Adapting_To_Change_UK_policy_towards_the_Arctic.pdf. Accessed 24 Sept 2016
U.S.–China Climate Leaders' Declaration (2015) On the occasion of the first session of the U.S.–China climate-smart/low-carbon cities summit Los Angeles, CA, 15–16 September 2015. http://go.wh.gov/cSqu8N. Accessed 23 Oct 2016
Wang Y (2013) Exploring the path of major-country diplomacy with Chinese characteristics. http://www.fmprc.gov.cn/mfa_eng/wjb_663304/wjbz_663308/2461_663310/t1053908.shtml. Accessed 22 Oct 2016
White House (2014) U.S.–China joint announcement on climate change. https://www.whitehouse.gov/the-press-office/2014/11/11/us-china-joint-announcement-climate-change. Accessed 5 Oct 2015
White House (2015) U.S.–China joint presidential statement on climate change. https://www.whitehouse.gov/the-press-office/2015/09/25/us-china-joint-presidential-statement-climate-change. Accessed 5 Oct 2015
White House (2016a) U.S.–China joint presidential statement on climate change. https://www.whitehouse.gov/the-press-office/2016/03/31/us-china-joint-presidential-statement-climate-change. Accessed 22 Oct 2016
White House (2016b) U.S.–China climate change cooperation outcomes. https://www.whitehouse.gov/the-press-office/2016/09/03/fact-sheet-us-china-cooperation-climate-change. Accessed 22 Oct 2016
White House (2017) Statement by President Trump on the Paris climate accord. https://www.whitehouse.gov/the-press-office/2017/06/01/statement-president-trump-paris-climate-accord. Accessed 25 Sept 2017
World Economic Forum (2017) President Xi's speech to Davos in full. https://www.weforum.org/agenda/2017/01/full-text-of-xi-jinping-keynote-at-the-world-economic-forum. Accessed 6 Feb 2017
Wright CD (2011) The dragon eyes the top of the world. Arctic policy debate and discussion in China. China Maritime Institute, U.S. Naval War College. http://www.usnwc.edu/Research—Gaming/China-Maritime-Studies-Institute/Publications/documents/China-Maritime-Study-8_The-Dragon-Eyes-the-Top-of-.pdf. Accessed 22 Oct 2016
Xi J (2012) Work together for a bright future of China–US cooperative partnership. http://www.fmprc.gov.cn/mfa_eng/wjdt_665385/zyjh_665391/t910351.shtml. Accessed 22 Oct 2016
Xi J (2017) Work together to build a community of shared future for mankind. http://news.xinhuanet.com/english/2017-01/19/c_135994707.htm?from=singlemessage. Accessed 6 Feb 2017
Xia L (2001) China: a responsible great power. J Contemp China 10(26):17–25
Xinhua (2012) Xi pledges "great renewal of Chinese nation". http://news.xinhuanet.com/english/china/2012-11/29/c_132008231.htm, 29 Nov. Accessed 22 Oct 2016
Xinhua (2013a) Xi vows peaceful development while not waiving legitimate rights. http://news.xinhuanet.com/english/china/2013-01/29/c_132136202.htm, 29 Jan. Accessed 22 Oct 2016
Xinhua (2013b) China granted observer status in Arctic Council. http://news.xinhuanet.com/english/china/2013-05/15/c_132384455.htm, 15 May. Accessed 22 Oct 2016
Xinhua (2013c) China calls for Norway to improve ties. http://news.xinhuanet.com/english/china/2013-01/22/c_132120252.htm, 22 Jan. Accessed 22 Oct 2016
Xinhua (2014a) Xi eyes more enabling int'l environment for China's peaceful development. http://en.people.cn/n/2014/1130/c90883-8815967.html, 30 Nov. Accessed 22 Oct 2016
Xinhua (2014b) China unveils energy strategy, targets for 2020. http://news.xinhuanet.com/english/china/2014-11/19/c_133801014.htm, 19 Nov. Accessed 23 Oct 2016
Zhang G (2014) Build consensus and implement actions for a cooperative and Win-Win global climate governance system. http://www.fmprc.gov.cn/mfa_eng/zxxx_662805/t1194637.shtml Accessed 24 Sept 2015

Zhang M (2015) Keynote speech by Vice Foreign Minister Zhang Ming at the China Country Session of the Third Arctic Circle Assembly. http://www.fmprc.gov.cn/mfa_eng/wjbxw/t1306858.shtml. Accessed 22 Oct 2016

Zoellick RB (2005) Whither China: from membership to responsibility? Remarks to National Committee on U.S.–China Relations. http://2001-2009.state.gov/s/d/former/zoellick/rem/53682.htm. Accessed 24 Sept 2015

Chapter 9
The Arctic Paradox (and How to Solve It). Oil, Gas and Climate Ethics in the Arctic

Teemu Palosaari

9.1 Introduction

This chapter looks at Arctic oil and gas development from the viewpoint of global climate ethics. The purpose is to analyze how topical issues related to climate justice and responsibility are covered in the current Arctic discourse. The analysis focuses on Arctic discussions on new oil and gas resources that become accessible as the sea-ice melts. It has been argued that the development of oil and gas resources in the Arctic is incompatible with the efforts to limit average global warming to 2 °C (McGlade and Ekins 2015). Consequently, the way in which problems and solutions regarding Arctic oil and gas are defined and promoted has global significance.

The chapter begins with a brief overview on some of the key themes of global climate ethics in light of the related research literature. I then introduce the so-called Arctic Paradox in order to highlight how Arctic oil and gas development is unavoidably an ethical issue and how it is entangled with global climate governance. The purpose is to highlight an ongoing Arctic process of interaction, interpretation, and contextualization, in which competing problem definitions, moral evaluations, and treatment recommendations regarding future oil and gas development are promoted. In doing so, I provide an outline of a future research agenda on Arctic climate ethics. To demonstrate the agenda and its potential, I present some tentative analysis on how climate ethics are articulated in the Arctic discourse. Method-wise, this analysis builds on constructivist discourse analysis and frame analysis. For the purposes of the analysis, I have reconstructed the Arctic discourse from material related to Arctic Circle and Arctic Frontiers conferences in Iceland and Norway.

T. Palosaari (✉)
Tampere University, Tampere, Finland
e-mail: teemu.palosaari@icloud.com

M. Finger and L. Heininen (eds.), *The GlobalArctic Handbook*,
https://doi.org/10.1007/978-3-319-91995-9_9

9.2 Global Climate Ethics

Although the process has perhaps been slow, climate change has made its way from the scientific agenda to the political agenda, with the impacts of the climate change being more visible and better documented. With rising public awareness, policy-makers have recognized issues on the scientific agenda. Although the issue of political responsibility remains sensitive, issues concerning climate-change mitigation and related international cooperation and institutionalization, the distribution of costs and benefits, free-rider dilemmas, and problems of enforcement have all entered the debate. The academic community has discussed related issues under headings such as climate justice (see, e.g., Fisher 2015; Baxi 2016) and climate responsibility (e.g., Frumhoff et al. 2015; also Arnold 2011; *Ethics & International Affairs* 28(3) 2014 Roundtable; see also *Peace Review* 25(4) 2013 Symposium on Climate Change and Peace).

Climate change has been labeled as a "super-wicked" problem that has four key features (Levin et al. 2012). Firstly, time is running out and the risk of harm to human communities and ecosystems increases with each passing year. As with many current environmental issues, an expanding world population that has greater access to more powerful technologies contributes to climate change. Secondly, those who have caused the problem are also seeking to provide a solution; it is currently difficult to live without creating some greenhouse emissions. Additionally, according to Wapner, ExxonMobil, British Petroleum, Royal Dutch Shell, PetroChina, Chevron, Gazprom, and many other powerful companies with vested commitments to a fossil-fuel economy are opposed to doing anything significant about climate change (Wapner 2014, 327). Thirdly, an effective central authority or coordinated global governance system addressing climate change is lacking. Moreover, it is not clear if there are actors who are sufficiently powerful and capable of responding to climate change efficaciously. The state-based structure of international relations also causes hindrances to collective climate action, as sovereign states strive to "provide security for and advance the welfare of their own citizens before they can concern themselves with global problems" (Wapner 2014, 328). In the current international system, it might be difficult to reconcile national interests with a global one. Fourthly, collective policy making tends to favor present consumption over possible future gain. This results in policies with short-term time horizons that ignore evidence of significant or even catastrophic impacts of inaction. Wapner has noted that the reluctance to invest in expensive new technologies or shift current patterns of consumption stands as a barrier to climate action (Levin et al. 2012; Riedy 2013; Wapner 2014, 327).

However, according to Mittler, "there has recently been a revival in arguing for climate action on an ethical basis" and climate change has begun to be framed in moral terms (Mittler 2014, 351; see also Schönfeld 2013). The viewpoints of intergenerational and environmental justice have gained ground in the global climate conversation, with the discussions around the Paris Climate Treaty helping to introduce them to a wider audience.

As Rentmeester summarizes in his article on cross-disciplinary, cross-cultural climate ethics, the first wave in ethics of climate change highlighted the danger that humans inflict upon nature and wild species through greenhouse gas emissions (McKibben 1989; Jamieson 1992). The second wave pointed out that climate change places an unfair burden on the most vulnerable; that is, the people in the least developed countries and future generations (Grubb 1995; Shue 1999; Gardiner 2001). The intergenerational aspect to climate change has perhaps generated the most discussion. The current generation prefers to overexploit the atmosphere because it benefits immediately from this practice, but such overexploitation puts future generations at serious risk (Rentmeester 2014).

Intergenerational justice highlights *displacement* of the harm of current practices into the future. By choosing to burn fossil fuels, "present generations are making a choice to enjoy associated benefits while transferring the costs. They are, in other words, displacing the harm of current practices across time" (Wapner 2014, 328). According to Gardiner the full effects of our current behavior are spread over a very long period of time—carbon dioxide emissions might remain in the atmosphere for centuries, even for hundreds of thousands of years—most of which is beyond the lifetimes of those making the decisions (Gardiner 2014, 301). Future generations cannot voice their opinion on these questions that will have impact on their living conditions.

Climate change also raises transnational issues regarding social justice and sustainability (Mittler 2014; Arnold 2011). From that point of view, the citizens of developed countries are responsible for tackling climate change since they have contributed far more to the enhanced greenhouse effect, have more resources to solve the problem, and are often living beyond the means necessary for basic human flourishing, while others in the developed countries have less than is needed for a decent human life (Rentmeester 2014). This is a spatial dimension of displacement, meaning that marginalized groups, from whom industries grab resources and who lack material protection, have little influence on public affairs (Wapner 2014, 331). The "global North" has historically been the main source of emissions and the benefactor of industrialization, whereas the negative effects are most strongly felt in many of the world's poorest regions.

All in all, the impacts of the climate change are now more visible and better documented. Although the question of the acceptance of political responsibility remains sensitive, issues concerning climate change mitigation and related international cooperation and institutionalization, the distribution of costs and benefits, free-rider dilemmas, and problems of enforcement have entered the debate. At the same time, climate ethics has become a fully fledged field in applied ethics (Rentmeester 2014, 10).

In the following, these issues will be placed in the Arctic context. How do the key questions of climate ethics, such as intergenerational justice and transnational responsibility, appear in the Arctic—the area where the impacts of climate change are stronger than elsewhere and where a significant amount of yet unexploited fossil fuel resources are located? The global sea-level rise to which the melting of Arctic sea-ice and glaciers contribute further increases the pressing nature of the case of Arctic climate ethics.

9.3 Arctic Paradox

The Arctic has become exposed to increasing globalization. In fact, the region has long been 'global' in the sense that Northern fishing grounds, whaling, fur trading, and mining have connected the Arctic to markets around the world (Heininen and Southcott 2010). Today, however, the forces of globalization are boosted by climate change and the Arctic is becoming increasingly integrated into the global economy. As indicated elsewhere in this book, there is growing interest in the Arctic sea routes and natural resources that become available as the sea-ice melts.

A significant share of the world's as-yet unexploited oil and gas resources are at the bottom of the Arctic Ocean. As the sea ice melts, coastal states and energy companies view these northern resources with great interest. However, using them would create emissions and accelerate climate change. A debate has started about whether the new Arctic oil and gas reserves should be utilized or left untouched. At the heart of the Arctic discourse lies the question of exploiting new Arctic oil and gas resources at a time when humankind needs to reduce emissions.

Thus, along with the various global issues, new *ethical* questions have emerged that relate to Arctic oil and gas. They concern the "Arctic Paradox": the faster we use fossil fuels, the sooner we get access to new oil and gas resources. Fossil fuels contribute to climate warming, which makes the Arctic sea-ice melt, making new oil and gas resources available. Using those resources then further accelerates climate warming. This makes Arctic oil and gas development unavoidably an ethical issue. Is it acceptable to explore and exploit new oil and gas in the Arctic at a time when humankind needs to reduce its carbon emissions (Palosaari 2012)? To drill or not to drill, that is the question.

Indeed, key questions of the global climate change ethics debate—such as moral responsibility and distribution of burdens and benefits—have lately found their way into Arctic politics. There are conflicting views that range from supporting unlimited oil and gas development to proposing a drilling ban. Some have stressed economic growth and the right of indigenous peoples and other local population to benefit from natural resources, whereas others have highlighted the environmental risks of the mining and fossil energy industry. There are also varying views regarding the extent to which Arctic states, companies, and people have responsibility to mitigate climate change.

In the Arctic case, there are differences in how the actors perceive and promote the oil and gas development. To some it is an ethical problem, to some it is an question of technical standards, and to others it is not a problem at all. The causal interpretations also differ: some see a connection between Arctic oil and gas extraction and climate change, while others refuse to do so. This is arguably a question of differences in how the issue is *framed*.[1] In the context of global climate ethics, it is interesting to analyze what ethical arguments, if any, are presented in order to build legitimacy for future treatment recommendations regarding Arctic oil and gas.

The Arctic is currently in the midst of an ongoing "process of interaction, interpretation and contextualization"[2] due to the political, social, and economic impacts of climate change. The debate is rife with dynamics such as cooperation versus conflict, environment versus extraction, globalization versus periphery, and indigenous peoples' economic growth versus their traditional livelihoods.

In the previous decade, a convincing conclusion on the Arctic discourse was that "there seems to be a certain inevitability about increased Arctic oil and gas exploration and production. The story lines we have identified all point more or less in the same direction. Oil and gas in the Arctic will be developed, like it or not" (Langhelle and Hansen 2008, 341).[3] Currently, given the ongoing debate on global climate change ethics, the number of people who "don't like it" appears to be increasing. Moreover, issues such as divestment, carbon bubble, cleantech, renewable power, and *Energiewende* have become increasingly visible in the media and policy-making.

In light of the major annual Arctic meetings of politicians, business, academia and NGOs (Arctic Circle Assembly held in Iceland, Arctic Frontiers Conference in Norway), many questions of the global climate ethics debate have recently found their way to the speeches of Arctic politicians, scientific institutions, environmental organizations, and companies. In the 2014 Reykjavík Arctic Circle Assembly, for instance, the Arctic Paradox was mentioned in many speeches. At the same event in 2016, the UN secretary-general was given the first ever Arctic Circle Prize for his role in the negotiations on international climate cooperation. In his speech on the occasion, Ban Ki-moon noted that "Arctic melting affects Miami, Mumbai, Shanghai, coastal cities—and so much else. When the Arctic suffers, the world feels the pain."

Global environmental change and globalization of the Arctic have also been likely to boost the same trend. As Young pointed out, the Arctic is not isolated from global trends and it is becoming increasingly difficult to "carve coherent agendas at the regional level and to pursue them without undue concern for the linkages between regional activities and planetary processes" (Young 2013, 14). Indeed, Arctic issues are not solely Arctic, but have linkages to "planetary processes"—and vice versa.

---

[1] To frame is to select some aspects of perceived reality and make them more salient in a communicating text, in order to promote a particular problem definition, causal interpretation, moral evaluation, and/or treatment recommendation (Entman 1993, 52). The last two issues in Entman's definition of framing—moral evaluation and treatment recommendation—appear particularly suitable for the analysis of the Arctic case.

[2] According to Goffman's classic work on frame analysis, meanings only arise in processes of interaction, interpretation, and contextualization (Goffman 1974, 24; Vliegenthart and van Zoonen 2011, 103).

[3] *Arctic Oil and Gas. Sustainability at risk?*, edited by Aslaug Mikkelsen and Oluf Langhelle, analyzes the expanding oil and gas activities in the Arctic from the perspective of sustainable development and corporate social responsibility, and presents findings on how "sustainable development" is understood and how further oil and gas extraction in the Arctic is often perceived compatible with it.

For instance, the melting of sea-ice and glaciers in the Arctic has contributed to the sea-level rise in the Southern hemisphere.

The following section of this chapter will discuss the attempts to solve (or silence) the Arctic Paradox. My main argument is that the current discourse[4] comprises two competing main ways to perceive and communicate about Arctic oil and gas. Each of these ways builds on different ethical undertones and differs in how it defines the main problem, moral responsibility, and what "treatment recommendations" it promotes. These two ways to frame Arctic oil and gas development also differ significantly in their interpretation of the connections between global and Arctic levels.

Hansen et al. (2008) used eight categories of "research propositions and questions pertaining to oil and gas activity in the Arctic": drilling in the Arctic, sustainability, social issues, environment, power and politics, indigenous peoples and local communities, Arctic institutions and organizations, and narratives in the Arctic (p. 100). I aim to add a new category to the list: climate ethics. I am building on the same theoretical notion coming from constructivist discourse analysis, according to which politics is "a struggle for discursive hegemony in which actors try to secure support for their definition of reality" (Hajer 1995, 59). My area of interest here is the arguments and definitions of reality in the case of Arctic oil and gas in the context of global climate change and climate ethics. The analysis on how Arctic stakeholders are positioned as victims, problem-solvers, perpetrators, or as scaremongers (Hansen et al. 2008, 97) can be linked to ethical arguments in the Arctic discourse on who is to blame for climate change and who is responsible for acting on it.

9.4 Arctic Environmental Change: Regional Versus Global Framing

Until very recently, the dominant way to define the problem regarding Arctic oil and gas development was to focus on the environmental risks of drilling and transport to Arctic environment. This made the problem a limited, regional environmental one that could be solved by environmental protection and by using scientific standards, planning, monitoring, impact evaluation, and innovative technology in Arctic extraction operations. The problem definition acknowledges the potential Arctic environmental risks, such as oil spills, and offers solutions that are already available.

This framing is a sort of a decoupling strategy, in which the harmful impact of Arctic oil and gas on climate is detached from environmental problems. While highlighting the risks towards Arctic flora and fauna, this strategy silences the global environmental impacts of Arctic oil and gas development. In this way, the Arctic environmental issues become solvable with the help of science, standards, and technology. Typical buzzwords in this frame are: "Oil spill preparedness", "oil spill response technology", "offshore pollution containment systems", "meteorological products", and "cleantech". The catchphrase "vulnerable Arctic" resonates with much of the

[4]For more on how the discourse was constructed, see *Notes on the material* in the end of this chapter.

research literature and Arctic Council's working groups' environmental approach. It is a well-established part of the Arctic canon that the region is "environmentally sensitive" and requires "highest environmental standards."[5]

This is a framing that is typically evoked by government and economic elites[6]; to some extent, this problem definition also reverberates in Arctic environmental and academic cultures.[7] It is well institutionalized, since the vulnerability of the Arctic environment has been discussed since the early 1990s and many environmental monitoring systems have been established.

The following excerpts combine the vulnerability of Arctic nature and scientific solutions. Both also define and delimit the issue as a regional, Arctic environmental problem and remain silent on the global environmental impacts of Arctic oil and gas development.

> "First, high-quality research is a must in the Arctic. Science is a cornerstone. We really have to know what we are doing, what we can do and what we cannot do. So, we have to research, analyze and monitor extra carefully in the North." …"We need exactly those technological solutions that enable us to tap the potential while avoiding risks. In a word, we need cleantech which is designed also for these conditions." (President of Finland Sauli Niinistö 2014)

> We want to operate safely, we want to operate in manner that's in compliance with regulations and standards, in a manner that is protective of the environment and respects people. (…) a science programme helps us to understand this place we are trying to operate and makes it possible to do so. We have (…) impact assessments studies that are going on including assessment of discharges and on-shore environmental assessment programme. (Alaska Science Team Lead, Shell 2014)

There is one interesting case in the material when environmental standards and sustainability are presented as solutions to the Arctic Paradox itself. The alleged logic is that the money gained from Arctic oil and gas can be invested into alternative energy resources:

> I could foresee that not only would we apply the highest environmental standards, but also make use of the potential material wealth and dividends to invest further in human capital, necessary infrastructure and alternative energy resources. This way, we can reap the potential economic benefits of extracting oil and gas and, at the same time, address the importance of sustainability and human investment. Hence, the 'Arctic paradox', of contributing to climate change by utilising non-renewable resources does not, if properly handled, necessarily have to involve a contradiction in terms. (Prime Minister of Iceland Sigmundur Gunnlaugsson 2014)

This problem definition has become challenged by a more global viewpoint on the Arctic. In that view, the main problem regarding Arctic oil and gas development is not an Arctic problem, but a global one; namely, the global greenhouse gas emissions that utilization of the new under-ice oil and gas reserves would result in. Therefore, many of the above-mentioned solutions would not be relevant. Instead, they might

[5]For example, President of Iceland, Arctic Circle 2014.

[6]For example, Shell Alaska Science Team Lead, Premier of Quebec, President of Finland, President of Iceland, Prime Minister of Iceland, Chancellor of Germany, Arctia Shipping CEO, Umoe CEO, Team Arctic Finland at the Arctic Circle 2014.

[7]At the Arctic Circle 2014 for instance researchers from Arctic University of Norway and UCLA.

be perceived as "shadow solutions", which not only fail to address the core features of the problem at hand but also create the dangerous illusion of genuine action" (Gardiner 2014, 301). While such voices have had a marginal position in the Arctic discourse, some examples have appeared in the material, emphasizing, for instance, that the Arctic is not only of regional concern but also a global issue of resource challenges, environmental degradation, and especially climate change.[8]

9.5 Arctic or Global Responsibility?

The heads of the Arctic nations have typically stated that tackling the climate change is not only a matter for the Arctic region and is instead a global challenge that requires a global response. While this is undoubtedly true, it is also a way of saying that mitigating climate change is a global responsibility rather than just an Arctic one. In this framing, climate change is presented as a problem that can only be solved by global action. Therefore, the Arctic can only react by adapting and building resilience. Furthermore, especially the indigenous peoples' representatives argue that since the current global climate warming has not been caused by the Arctic occupants, the responsibility lies elsewhere. Instead of a specific Arctic climate responsibility, a moral responsibility to provide welfare to Arctic population is often promoted.

> Save the Arctic campaigns need to first look at the causes of climate change – what is happening closer to their own backyards before setting their sights on our homeland. Drive your cars less, reduce, reuse and recycle, look at where pollutants are actually coming from now - as simple examples. (...) There are groups that would seek to stop hunters from earning a small living from a limited, legal and highly regulated trade, on the basis of climate change predictions extrapolated over the next 50 to 100 years. That's not science. That's using climate change as a weapon to further disadvantage already disadvantaged communities. (President of Inuit Tapiriit Kanatami, Arctic Circle 2014)

In some cases, it is argued that the responsibility lies with the consumer rather than the producer. Most oil consumption takes place outside the Arctic region. In the same vein, Arctic gas is particularly constructed not as part of the problem, but as part of the solution. The argument goes that Arctic gas is cost-efficient to transport, has low emissions and can help to "crowd out" coal. These views underline the growing global demand for gas and argue that Arctic gas resources are relevant in all future energy production scenarios, including the 2-degree scenario.[9]

What facilitates such framing is that a voluntary self-marginalization evoking the region's peripheral status is still readily available in the discourse on the Arctic, despite recent globalization trends. A remote region cannot be expected to solve

---

[8]Prince of Monaco, Arctic Circle 2014; Executive Director of International Centre for Reindeer Husbandry, Arctic Frontiers 2015; Professor Emeritus of Marine and Environmental Affairs University of Washington, Arctic Frontiers 2015.

[9]Executive director, The Research Council of Norway, Arctic Frontiers 2015; also the report of an independent research group presented at the Arctic Frontiers 2015, Prime Minister's Office Publications 2015.

global problems alone. Also, the amount of emissions coming from the Arctic is often considered small compared to other regions.

> The region has already become symbolic in the context of global efforts to curb climate change. Of course whether global warming can be kept below the two-degree ceiling will depend on other places, where greenhouse gas emissions are high. (Chancellor of Germany, video speech at Arctic Circle 2014)

Yet, global attention towards the Arctic is growing. The sea-level rise in the coastal areas around the world is linked to the melting of glaciers in the Arctic. NGOs and non-Arctic states have started to question the ethics of Arctic oil and gas exploration. Arctic glacier and sea-ice melting has huge symbolic value in global climate politics as well as a direct impact on sea-level rise around the world. Consequently, a causal connection between Arctic resources and global climate change appears more and more often in the Arctic discourse. The conclusion is that the Arctic actors economically benefiting from fossil fuels cannot escape the moral responsibility.[10] By enjoying the benefits now, the harms of climate change are displaced to future generations, or to other geographical locations, such as coastal cities in the South that are threatened by sea-level rise.

9.6 Sustainable Extraction or Moratorium?

The two competing framings outlined above also offer different treatment recommendations regarding future oil and gas development in the Arctic. The one that builds on a regional problem definition and highlights global responsibility concludes that oil and gas development can be done sustainably in the Arctic. The established Arctic reading of sustainable development has tended to focus on how the resources can be used (Mikkelsen et al. 2008). The framing proposes striking a balance between environmental concerns and economic gains and highlights the economic opportunities and local benefit. Previous research has shown how mining and energy companies in particular use this rhetoric in engaging the indigenous peoples and other local residents (ibid.); this also seems to be the case in light of the current Arctic discourse analyzed here. Cooperation with the local communities in planning, monitoring and running the extraction operations is, in their experience, a useful way to achieve local approval for the economic activities. This framing is supported by the argument that Arctic oil and gas are needed in the "Global Energy Mix". This is argued to be the case even if there is a transition from fossil fuels to renewables, since oil and gas are needed during the transition period.[11]

On the other hand, when the problem definition rests on the emissions resulting from Arctic oil and gas, the view on responsibility is different, and the conclusion is that oil and gas resources that become available as the ice melts are bet-

---

[10]For example, CEO, DNV GL Group; Leader WWF Global Climate and Energy Initiative, Arctic Frontiers 2015.

[11]For example, Executive vice president, Statoil (Exploration), Arctic Frontiers 2015.

ter left untouched. Such interpretation of sustainable development emphasizes the environmental dimension more than human socio-economic well-being. The emerging argument in the Arctic discourse is that, because of the resultant emissions, Arctic oil and gas development cannot be a sustainable use of natural resources, regardless of how safe the oil and gas extraction and transportation might be. Based on this, the typical proposed solution regarding Arctic oil and gas development is a drilling ban in areas that have historically been covered by sea-ice. In light of the analyzed material, conservation areas and strict requirements for test-drilling permissions are mentioned increasingly frequently in the Arctic debates, in a way that challenges the framing of sustainable fossil fuel extraction that has dominated the Arctic discourse for a long time.

9.7 Conclusions

The purpose of this chapter was to analyze how ideas of global climate ethics might appear in the Arctic. Given that climate change is hitting the Arctic hard, and in the global climate discussions new Arctic fossil fuel resources have been labeled "unburnable" (McGlade and Ekins 2015), this is a topical issue. Climate change has provided access to new oil and gas resources in the Arctic, the utilization of which will contribute to further changes in climate. Therefore, the ethical undertones in the Arctic oil and gas debate are worth analyzing. The focus of this chapter was on the different problem definitions, moral evaluation, and treatment recommendations promoted in the Arctic discourse.

The first located framing builds on mainly regional, Arctic, focus. It acknowledges the impacts of global climate change on the Arctic, but chiefly ignores the climate impacts of Arctic oil and gas, and thus ends up discussing Arctic adaptation and resilience building more than climate mitigation. The framing rests on many established interpretations and buzzwords in the Arctic discussions, such as vulnerable Arctic and sustainable development. It bypasses or silences the ethical issues of Arctic Paradox and concludes instead that Arctic oil and gas development is justifiable since it can be done "sustainably" and since it supports local economic growth and welfare. In other words, it replaces climate responsibility with the responsibility to secure and provide local development.

The second located framing challenges these views and takes note of the global impacts of Arctic fossil fuels. It promotes a different interpretation of sustainable development and questions the sustainability of Arctic oil and gas development. It also discusses the displacement of harms of climate change to future generations, and places Arctic oil and gas in the context of global climate change ethics.

The chapter has pointed out the main competing ways to frame Arctic oil and gas development in the context of global climate change. In doing so, it offers a way to study how climate ethics are articulated in the Arctic discourse. Further research could undoubtedly reconstruct multiple and intersecting frames. For instance, there could be more nuanced differences in how the responsibilities of coastal states, other

Arctic states, non-Arctic states, companies, or indigenous peoples to mitigate climate change are defined and promoted. Furthermore, it could be worthwhile analyzing the effects, success, and level of institutionalization of different framings in greater detail. The material and symbolic resources that different actors possess should also be taken into consideration. Based on the current material analyzed, the intersection of global climate ethics and Arctic energy resources is a promising research area.

References

Arnold D (ed) (2011) The ethics of global climate change. Cambridge University Press
Baxi U (2016) Towards a climate change justice theory? J Hum Rights Environ 7(1):7–31
Entman R (1993) Framing: towards clarification of a fractured paradigm. J Commun 43(4):51–58
Fisher S (2015) The emerging geographies of climate justice. Geogr J 181(1):73–82
Frumhoff P, Heede R, Oreskes N (2015) The climate responsibilities of industrial carbon producers. Clim Change 132:157–171
Gardiner S (2001) The real tragedy of commons. Philos Public Aff 30(4)
Gardiner SM (2014) A call for a global constitutional convention focused on future generations. Ethics Int Aff 28(3):299–315
Goffman E (1974) Frame analysis: an essay on the organization of experience. Harper & Row, New York
Grubb M (1995) Seeking fair weather: ethics and international debate on climate change. Int Aff 71(3):463–496
Hajer M (1995) The politics of environmental discourse. Ecological modernization and the policy process. Clarendon Press
Hansen OH, Langhelle O, Anderson R (2008) Framework and methodology: regulation and discourse analysis as a research strategy. In Mikkelsen A, Langhelle O (eds) Arctic oil and gas. Sustainability at risk? Routledge, Abingdon, pp 87–108
Heininen L, Southcott C (2010) Globalization and the circumpolar North. An introduction. In Heininen L, Southcott C (eds) Globalization and the circumpolar North. University of Alaska Press, Fairbanks, pp 1–22
Jamieson D (1992) Ethics, public policy, and global warming. Sci Technol Human Values 17(2):139–153
Langhelle O, Hansen KF (2008) Perceptions of Arctic challenges: Alaska, Canada, Norway and Russia compared. In Mikkelsen A, Langhelle O (eds) Arctic oil and gas. Sustainability at risk? Routledge, Abingdon, pp 317–349
Levin K, Cashore B, Bernstein S, Auld G (2012) Overcoming the tragedy of super wicked problems: constraining our future selves to ameliorate global climate change. Policy Sci 45(2):123–152
McGlade C, Ekins P (2015) The geographical distribution of fossil fuels unused when limiting global warming to 2 °C. Nature 517:187–190
McKibben B (1989) The end of nature. Random House, New York
Mikkelsen A, Langhelle O (2008) Arctic oil and gas. Sustainability at risk? Routledge, Abingdon
Mittler D (2014) The changing ethics of climate change. Ethics Int Aff 28(3):351–358
Palosaari T (2012) The Amazing Race. On resources, conflict, and cooperation in the Arctic. NGP Yearbook 2011, 13–30. http://urn.fi/URN:NBN:fi:uta-201302011022
Rentmeester C (2014) Do no harm: a cross-disciplinary, cross-cultural climate ethics. De Ethica: J Philos Theol Appl Ethics 1(2)
Riedy C (2013) Climate change is a super wicked problem. Planetcentric, 29 May

Schönfeld M (2013) Introduction: plan B: global ethics on climate change. In: Schönfeld M (ed) Global ethics on climate change: the planetary crisis and philosophical alternatives. Routledge, New York, pp 1–8

Shue H (1999) Global environment and international inequality. Int Aff 75(3)

Vliegenthart R, van Zoonen L (2011) Power to the frame: bringing sociology back to frame analysis. Eur J Commun 26(2):101–115

Wapner P (2014) Ethical enhancement in an age of climate change. Ethics Int Aff 28(3):325–334

Young O (2013) Arctic stewardship: maintaining regional resilience in an era of global change. Ethics Int Aff (online) 7 Jan 2013. http://www.ethicsandinternationalaffairs.org/2013/arctic-stewardship-maintaining-regional-resilience-in-an-era-of-global-change/

Chapter 10
Contemporary Arctic Meets World Politics: Rethinking Arctic Exceptionalism in the Age of Uncertainty

Juha Käpylä and Harri Mikkola

10.1 Introduction[1]

The Arctic region is often seen as an exceptional space. Traditionally, the idea of "Arctic exceptionalism" has referred to a romantic view that highlights the region's exotic and unique properties. The Arctic is seen as a vast desert-like area where the forces of nature challenge human capabilities; as a pristine wilderness whose beauty has made it a focal point of environmental conservation; and as a space where indigenous peoples of the North are contented hunters and gatherers living a simple existence in harmony with the natural environment and uncorrupted by the forces of modernity (Young 1992).

More recently, a distinctly more political vision of the exceptional Arctic as a "zone of peace" (Gorbachev 1987/2012) and a "territory of dialogue" (Lavrov quoted in ITAR-TASS 2014) has emerged. In this sense, the Arctic has become understood as a "distinctive region in international society" (Young quoted in Keskitalo 2007, 195); a region detached from world politics and characterized as an apolitical space of regional governance, functional co-operation, and peaceful co-existence (see Heininen et al. 2013, 25). This chapter will discuss and critically analyze this notion of political exceptionalism of the Arctic.

We start by discussing the recent history of Arctic exceptionalism, before discussing in more detail why the Arctic is often considered to be an exceptional zone of peace and co-operation. While these arguments have validity in avoiding/defusing

[1] This chapter builds upon our previous work of the thematic, including Käpylä and Mikkola (2015), Käpylä et al. (2016).

J. Käpylä (✉) · H. Mikkola
The Finnish Institute of International Affairs, Helsinki, Finland
e-mail: juha.kapyla@fiia.fi

H. Mikkola
e-mail: harri.mikkola@fiia.fi

M. Finger and L. Heininen (eds.), *The GlobalArctic Handbook*,
https://doi.org/10.1007/978-3-319-91995-9_10

intra-Arctic conflict dynamics, we argue that more attention should be paid to external political and conflict dynamics and their impacts on the region. By focusing on the crisis in Ukraine, this chapter illustrates how forces and dynamics external to the region have had varied impacts on co-operation in the Arctic. This analysis is further supported by a discussion of on-going dynamics related to an increasingly unpredictable Russia that have raised growing concern in the West, with potential implications for Arctic co-operation. The last substantive part of the chapter discusses recent changes in US domestic and foreign policy, which have increased general uncertainty and can further problematize the co-operative nature of the Arctic.

10.2 The Emergence of Political Exceptionalism of the Arctic

Today's exceptional political vision of the Arctic emerged with the end of the Cold War. The end of superpower rivalry meant that the region lost most of its geostrategic and geopolitical relevance, even though strategic military assets, such as nuclear capabilities, remained in the region. In fact, the geopolitical dynamics of the Arctic had already started to transform in the latter years of the Cold War. In the famous 1987 Murmansk speech, the Soviet Secretary General Mikhail Gorbachev laid down the vision of the Arctic as a zone of peace and co-operation and initiated the gradual process of "desecuritization" of the Arctic as an element of the broader Soviet re-orientation (Åtland 2008, 289–311).

In 1989, the process took on a more concrete form when Finland convened officials from eight Arctic states to start a discussion on matters of environmental protection, resulting in what came to be known as the "Rovaniemi Process". Two years later, in 1991, the eight Arctic states came to agree on the Arctic Environmental Protection Strategy (AEPS); this arrangement was surpassed seven years later, in 1996, by the establishment of the Arctic Council (AC). From its inception, the AC was an international co-operation forum that relied on consensus in adopting its non-binding political recommendations. The mandate of the AC was limited to the promotion of cooperation among the Arctic States and Arctic indigenous communities in issues of sustainable development and environmental protection. This not only constructed the Arctic as a de facto "internal affair" of the community of Arctic Council states and other representatives, but also excluded "high political" matters with geopolitical implications, most notably military security, from the AC agenda (Pedersen 2012, 147–148; Koivurova 2010, 146–148). Together, these developments solidified the vision of the Arctic as an exceptional, encapsulated zone of peace and co-operation.

During the last decade or so, however, the Arctic has re-emerged on the agenda of world politics. The key driver behind this development was the rapid and exceptional warming of the region. Continuous reduction and thinning of the Arctic Ocean ice cover meant that the previously isolated geopolitical frontier was opening up. Consequently, securing access to, and control of, the Arctic resources and maritime routes heightened strategic interest in the region (Mikkola and Käpylä 2013). The resulting political dynamics were increasingly characterized by geopolitical friction and competition, and fears of a new arms race or even a new Cold War. A number of specific interconnected factors at the time contributed to this understanding of the Arctic:

- In 2007, Russia planted its titanium flag deep into the North Pole seabed in a manner that was initially interpreted as an illegal land-grab (Chivers 2007). At the time, Russia also adopted a more self-assured anti-Western rhetoric and resumed its long-range aviation as well as maritime patrols in the Arctic and beyond (Zysk 2011, 86–88).
- In 2008, the US Geological Survey published its often-cited estimation that there were significant undiscovered oil and gas reserves in the Arctic. This estimate was published at a time when there was a lot of talk about dwindling global oil reserves and a growing demand for oil in the emerging markets (Claes and Moe 2014, 97).
- Scholarly literature and media articles both speculated about a possible conflict or new Cold War as a result of the interstate competition over Arctic natural resources (e.g. Borgerson 2008).

However, the situation started to de-escalate as early as 2008 with the help of confidence-building measures by the Arctic states themselves. Subsequent policy and academic literature also highlighted that the Arctic was not going through a serious militarization, and that the calls for resource conflicts or a new Cold War in the region were misinterpretations of the empirical state of affairs (e.g. Wezeman 2012; Lasserre et al. 2012). Gradually, the discourse of an emerging conflict was overshadowed by the paradigm of continuing, if not intensifying, cooperation in the Arctic.

10.3 The Arctic as an Exceptional Region of Peace and Co-operation

At the core of the paradigm of Arctic exceptionalism lie several key assumptions why the Arctic is—and is likely to remain—a zone of peace and co-operation, and why the potential for Arctic conflict has been exaggerated (see Käpylä and Mikkola 2013).

First, there is an assumption that there is *not that much to fight over*. Land areas above the Arctic Circle are under the uncontested sovereignty of the Arctic states. In the maritime domain, the existing Exclusive Economic Zones (EEZs) of the Arctic states are largely uncontested and well defined (Young 2009, 77). It is estimated that approximately 90% of the undiscovered hydrocarbon reserves in the Arctic are within these undisputed EEZs, leaving little to compete for. In the partly contested area around the North Pole, economic exploitation remains close to impossible due to harsh operating conditions and extremely high costs.

Second, there is an assumption that the Arctic area has *governance structures that foster co-operation and defuse potential conflict dynamics*. As the agenda of issues in Arctic governance is manifold, it is not subject to one single comprehensive treaty regime (like the Antarctic is), nor does it fall under the mandate of any single governance structure or organization. Instead, Arctic governance amounts to what Humrich and Wolf (2012, ii) have called "a fragmented rather than a properly integrated multi-level system" that has evolved incrementally as a response to practical needs; that has been operationalized through multiple federal, national, regional, international and global mechanisms; and which remains divided into partly overlapping sectoral spheres.

The most important governance mechanism in the Arctic is the 1982 United Nations Convention on the Law of the Sea (UNCLOS). It provides a global multilateral legal framework for defining the status of maritime areas and settling intra-Arctic sovereignty and border issues regarding maritime routes and continental shelf extensions (Byers 2013, 5–7; see also Koivurova et al. 2015, 3, 5).[2] Another key governance structure is the Arctic Council (AC), which is recognized as the most important multilateral institutional framework in the region, particularly through its co-operative scientific contributions, policy recommendations, and the three recent international agreements that have been negotiated under its auspices. Other relevant governance structures include the International Maritime Organization (IMO) and the Barents-Euro Arctic Council (BEAC). While these governance mechanisms do not deal with hard security issues, there is broad agreement that they foster and contribute to the spirit of co-operation, the rule of international law, and peace in the region (see, e.g., Byers 2013, 9).

According to the third assumption, *Arctic states have explicitly expressed their interest in international cooperation*. For example, the infamous 2007 Russian flag-planting episode did not lead to intensifying competition. Instead, the five Arctic coastal states decided to defuse the situation through the adoption of the 2008 Ilulissat Declaration, in which they affirmed their commitment to existing international law and co-operation. More practically, co-operation and measures to build confidence have included regular military and emergency exercises among Arctic states (Conley 2014, 56). In addition, the Chiefs of Defense of the Arctic states started to hold regular discussions, and a broader gathering of military officers of the Arctic states began convening regularly in the context of the *Arctic Security Forces Roundtable* (Hilde 2014, 159–160). Moreover, all Arctic Council member states, as well as its permanent observers, have endorsed Arctic multilateral governance, especially the Arctic Council (see e.g. Arctic Council 2013, 2015, 2017).

Finally, and perhaps most importantly, there is an assumption that *Arctic states have little to gain* from conflict dynamics that would create an unfruitful investment and development environment for Arctic exploration and exploitation.

---

[2] The United States is the only Arctic country that has not ratified the UNCLOS. However, it accepts the treaty as customary international law.

10.4 The Crisis in Ukraine and the Arctic

The above arguments have validity when it comes to avoiding intra-Arctic conflict dynamics. However, there is a certain conceptual flaw in the approach. The arguments are problematic because of their regional focus, which tends to bracket out global political dynamics and their impacts on the region. As we have argued previously, "[t]o understand the Arctic today, one needs to have a global perspective. The Arctic is not a closed system and regional development is increasingly intertwined with global dynamics" (Käpylä and Mikkola 2013, 8).

Recent conflictual events in world politics—above all, those related to the ongoing crises in Ukraine and Syria and Russia's changed foreign policy behavior—provide us the first context in which to reconsider the exceptional character of the Arctic, both in terms of its regional as well as its co-operative aspects. The crisis in Ukraine has brought external political and conflict dynamics to the Arctic, which has had four main direct and indirect effects on Arctic cooperation.

First, the crisis has had an effect—albeit a limited one—on *the established discourse on the Arctic*. The crisis has negatively affected Western perceptions of Russia and its intentions. Today, there is a widespread distrust of Russia in the West, particularly given the perceived discrepancy between what Russia says and what it does. While Russia has been seen as a pragmatic foreign policy player that can utilize both hard and soft power to advance its interests (Zysk 2011), unlike the Soviet Union, it has become to be seen as a very unpredictable power.

Given the recent change in Western perceptions of Russia, similar events taking place before and after the crisis have been more often interpreted with caution and worry (see, e.g., Conley and Rohloff 2015; Baev 2015). This is especially the case with Russia's overall military modernization, including in the Arctic. Before the crisis in Ukraine, the increase in Russian Arctic capabilities was easy to interpret as legitimate state behavior to monitor and secure the opening border region and strategic assets therein, and to support civilian activities in a harsh environment. After the crisis, increased Russian capability and activity in the Arctic can be approached with more concern, as a sign of potentially more assertive behavior in a conflictual geopolitical situation. This reading is relevant in the Arctic as there are multiple overlapping claims to extend continental shelves coupled with a growing international interest in the region as a result of opening sea lanes and recoverable natural resources. While not a reality yet, this externally induced change in public discourse and awareness of the Arctic—driven by the changing perception of Russia itself—could become the "new normal", affect the political imagination towards the Arctic in general, and erode the Arctic cooperative spirit.

Secondly, the crisis has affected the *established practices of security co-operation in the Arctic.* Already at an early stage, the crisis resulted in the cancellation of the *Northern Eagle* naval exercise between the Norwegian, Russian, and US navies (Pettersen 2014). The crisis has also resulted in the cancellation of the annual Chiefs of Defense meeting among Arctic states, while the Arctic Security Forces Roundtable has been organized without Russian participation (Le Miére 2014).

While some military-to-military contacts remained, these developments were significant. Military cooperation between Russia and NATO countries has been a distinctive and exceptional Arctic feature, and particularly important in building mutual trust in the region (Hilde 2014, 160). These cases illustrate how an external crisis can cancel out the important work done in building confidence and trust in the sphere of military security between the Arctic states.

Thirdly, *Arctic governance structures* have also been affected by the crisis in Ukraine, even though there have been successful efforts to limit the spillover. Prior to the crisis, there was wide agreement that the Arctic Council is more or less sheltered from major turbulence in world politics, largely because of its mandate excludes matters pertaining to traditional national security and military policy. This is mostly true also today. The most important and enduring element of the Council's work is the pragmatic scientific cooperation in its working groups, not high politics. More generally, a significant part of Arctic governance is carried out in various regional fora, where informal people-to-people relationships, rather than formal and political state-to-state relationships, are central to solving shared practical challenges.

However, the rapidly worsening relationship between the West and Russia after the outbreak of the crisis in Ukraine did have some effect on political cooperation in the Arctic, particularly in the context of the Arctic Council. First, the finalization of the EU's observer status in the AC—agreed upon in principle at the 2013 Kiruna ministerial meeting—was put on hold due to the crisis. Russia increasingly perceives the EU as an antagonistic geopolitical actor (see, e.g., Moshes 2017). With the crisis in Ukraine and deteriorated relations between Russia and the EU in particular, Russia quietly expressed that it will not agree to the EU becoming an observer (see, e.g., The Arctic Journal 2015; Garcés de los Fayos 2015). Russia's objection meant that the whole issue was not even raised in the 2015 and 2017 ministerial meetings, and the EU was to remain in a liminal state. However, the EU did continue its de facto participation in the AC and its working groups.

Furthermore, as part of their broader recalibration of relations with Russia (see, e.g., Carney 2014; Government of Canada: Foreign Affairs, Trade and Development Canada 2014), the US and Canada initially decided to boycott Arctic Council meetings that are organized in Russia or chaired by a Russian, such as the AC working group meeting on black carbon and methane in Moscow in April 2014. However, this policy of boycott was not continued systematically, and a widespread agreement emerged in the US, Canada, Europe, and Russia that it is important to continue the pragmatic co-operation in Arctic governance structures. For example, already in autumn 2014 the US Administration declared in a conciliatory tone that it "is keenly aware of the value of maintaining scientific cooperation on collaborative research projects, especially in the Arctic, and will assess our interactions consistent with that awareness" (quoted in Rosen 2014). This co-operative ethos has persisted, as indicated by the relatively smooth US Arctic council chairmanship period and the establishment of new forums of co-operation, including the Arctic Coast Guard Forum and the Arctic Offshore Regulators Forum. Furthermore, the work of the Arctic Economic Council has proceeded without extra-Arctic problems, and a new legally binding agreement on Arctic scientific cooperation has been achieved under the auspices of the Arctic Council.

However, if the crisis in Ukraine and the tension in the Russia–West relationship continue[3]—and especially if the inflamed situation were to worsen given further conflictual dynamics (such as Syria, election meddling)—it is not out of the question that conflictual political dynamics could start affecting the Arctic Council in a more serious manner. In other words, there is likely to be a "tipping point" in the resilience of Arctic Council co-operation (Koivurova 2016). As Fran Ulmer, the chair of the US Arctic Research Commission, acknowledged in 2014, while there is hope that the Arctic will remain mostly isolated from the crisis in Ukraine, "obviously, everyday decisions are being made in Moscow and Washington and other capitals that could set us back" (Ulmer quoted in Rosen 2014).

Fourthly, it is in the *economic sector* that the most significant effects of the crisis in Ukraine in the Arctic have been seen. Through the policy of sanctions, the Ukraine crisis has clearly spilled over to hinder the economic sphere of cooperation in the Arctic. The gradually tightened sanctions imposed on Russian Arctic offshore oil projects have been one of the West's primary tools for countering Russia's actions in Ukraine. In the third round of sanctions, the West decided to prohibit the exportation of Western goods, services and technology for the development of Russian Arctic offshore oil prospects. The US and the EU had already imposed financial sanctions that restricted the access of Russia's oil companies and their highly expensive Arctic megaprojects to Western capital. As a result, joint ventures in the Russian offshore Arctic, such as the exploratory drilling in the Kara Sea between Exxon-Mobil and Rosneft, have been halted (Mikkola and Käpylä 2014).

10.5 Russia, Growing Uncertainty, and the Arctic

Beyond the above-mentioned effects of the crisis in Ukraine, there are on-going adverse dynamics related primarily to Russia and its foreign and security policy that may further challenge the exceptional status of, and forms of co-operation in, the Arctic. While speculative, this kind of reflection on possible dynamics in the Arctic in an uncertain politico-strategic conjuncture in Europe is worth presenting given the current worrisome direction of Russia, both domestically and internationally (see Martikainen et al. 2016; Laine et al. 2015).

First, strategic considerations play a major role in the way different actors regard cooperation and international law in the region. This is especially the case with Russia, which is *the* most important player in the Arctic and which views the region as a strategically vital resource base for the country. For instance, up to one-fifth of Russia's GDP is produced north of the Arctic Circle, and 95% of the country's natural gas and 75% of its oil reserves are located in Arctic and sub-Arctic regions (Laruelle 2014, xxi). Consequently, the Arctic's role in Russia's economic equation is

[3]For an illustrative discussion, see, e.g., Rogin (2016).

substantial: domestic social programs, infrastructure investments, and military modernization are all critically dependent on revenues from natural resource exports, as is the rent-seeking system that extends its reach to all economic activity in Russia. From a foreign policy perspective, Arctic hydrocarbons also provide important leverage for Russia's external influence, especially in Europe (Martikainen and Vihma 2016). Given the critical role that Arctic energy plays in the country, Russia has considered it prudent to endorse co-operation and international law in the region, not only to gain access to new resources, but also to generate a stable and predictable investment and operating environment.

However, Russian Arctic economic ambitions are becoming increasingly difficult to realize. This is not only due to challenging operating conditions, but also to deteriorated national economy, adverse market conditions, and Western sanctions that combine to hinder the pace and scope of economic development, particularly in the offshore Arctic. If the Arctic economic potential does not materialize, and if the biggest stabilizing factor—that is, common economic interests—were eliminated from the equation, the region could still be utilized as a tool in domestic and international politics.

Russia has invested considerable international and domestic political capital in developing the Arctic and has utilized the region in nation-building and identity politics. In Russian identity, the development of Arctic mega-projects resembles the Soviet space program of the 1960–1970s, both as evidence of country's greatness and as a tool for general technological development (see Laruelle 2014, 27, 39–40). Subsequently, Russia has a lot at stake in the region and the Arctic may increasingly witness uses other than just the economic one. For example, it could be increasingly used in the construction of enemy images that incite nationalism, a sense of being a great power, and/or a siege mentality at home. Furthermore, as Russia's hard military capabilities are superior to other state actors in the Arctic, the region could even be constructed as a new hostile theater for domestically targeted foreign policy victories or shows of force that aim to secure regime stability in a situation where Russian domestic political and economic system is facing severe problems.[4]

Secondly, Russia's consistent commitment to international law can no longer be taken for granted under the current regime. The annexation of Crimea and the ongoing conflict in Ukraine highlight clearly the fact that, if necessary, Russia is prepared to dismiss foundational international norms and commitments it has previously endorsed. These include key principles—sovereign equality, non-use of force, inviolability of frontiers, and the territorial integrity of states—agreed upon in the 1975 Helsinki Final Act, as well as other international obligations such as the security assurances to Ukraine (in exchange for its nuclear non-proliferation) agreed upon in

[4]Recent media analysis indicates a potential for more non-cooperative framing and usage of the region. Before the crisis in Ukraine, Arctic-related reporting in mainstream Russian newspapers highlighted the commercial development of the Arctic. After the crisis, newspaper coverage has started to emphasize a more statist and militarized rendering of the region. In particular, the keyword "military" has become popular: in an analysis of keywords, it has come up from the 89th position before the crisis up to the 4th position in post-crisis media coverage in mainstream newspapers (Pilli-Sihvola et al. 2016, 53–54).

the 1994 Budapest Memorandum, conventional and nuclear arms limitation frameworks, and best practices of conducting military exercises. Subsequent actions in Syria—such as the targeting of civilians and medical facilities that suggests a systematic disregard for basic humanitarian principles in warfare (see, e.g., Medecins Sans Frontiers 2016)—continue this trend. Given all this, it has become more uncertain whether, or *under what conditions*, one should expect Russia to remain consistently committed to its legal and diplomatic obligations *also* in the Arctic, especially if it does not directly play into its national interests.

Third, the annexation of Crimea and the ongoing conflict in Eastern Ukraine indicate that Russia has chosen the path of a revisionist power in Europe. The current regime in Russia considers the collapse of the Soviet Union as a geopolitical catastrophe that not only diminished the status of Russia, but also shattered the perceived legitimate territorial integrity of the state (see Laruelle 2014, 9–11). The annexation of Crimea can be interpreted as an act to reclaim lost territory despite significant financial and reputational cost. In the international arena, Russia has shown its readiness to use military and other means for geopolitical ends to expand Russian territory and to challenge the European, if not the liberal political order more broadly.

But what are the implications of a revisionist Russia for territorial stability in the Arctic? In this respect, Russia has been known for sending mixed signals. In the late 2000s, at a time when Russia under President Vladimir Putin started to emphasize a more belligerent rhetoric towards the West and made what appeared a unilateral claim to the seabed of the North Pole (in 2007), it also endorsed UNCLOS (in 2008) and resolved a border dispute with Norway in the Barents Sea (in 2010). More recently, during the conflict in Ukraine, Russia's public endorsements of international law and co-operation have co-existed with bolder rhetoric about the territorial value of, and Russia's territorial designs in, the Arctic that equals "the historical reunification of Crimea" and "the development of the Arctic" (see The Barents Observer 2015).

Even if such statements were merely nationalistic rhetoric meant for domestic consumption, they are nevertheless public speech acts that are noticed internationally and unhelpful in reducing the uncertainty about Russia's territorial intentions. Today, in the light of the annexation of Crimea, it is not altogether unreasonable to at least ponder whether—and again, under what conditions and with what risks—Russia could decide to further "restore" its territorial integrity in the high north.

Fourth, the nationalistic comments are also worrisome in the context of Russia's military build-up in the Arctic. For security analysts, threat is typically understood as a combination of capability and harmful intent, or the perception of such intent. Due to Russia's changed foreign policy behavior, the latter part of the equation has turned for the worse in the West. At the same time, Russia has signaled a willingness to improve its Arctic capabilities by re-opening various military bases, establishing a new strategic military command, and conducting massive unannounced military drills in the region (see Conley and Rohloff 2015, 69–88). The securing of the Arctic was also recently highlighted in Russia's new 2014 military strategy (see Klimenko 2016).

The growing uncertainty about Russia's intentions has opened the door for interpretations that highlight the potential for aggressive behavior and/or militarization in the Arctic. For example, the establishment of military bases along the NSR can be read as de facto control of the maritime area with potential implications to navigation of freedom[5] and territorial stability, as military presence could act as a coercive back-up or backdrop to secure Russia's interest to extend its continental shelf northwards (especially if the on-going UNCLOS procedure does not produce the desired recommendation).[6] Of course, Russia's Arctic military assets, such as naval vessels of the Northern Fleet, have already been deployed to conflicts outside the Arctic, most notably to the war in Syria (see, e.g., The Guardian 2016a, b).

The concern about Russia's intentions has also grown with its adoption of so-called hybrid warfare or hybrid influencing (see, e.g., Giles 2016; Martikainen et al. 2016), which combines the threat or use of military capabilities with other forms of strategic influence to create desired strategic effects. Some of these methods, such as airspace incursions, cyber activity, ominous diplomatic communication, energy politics, and the organization of refugee flows, have taken place in the Nordic/Arctic region.[7]

Growing military (and other strategic) capabilities, especially in a time of uncertainty about Russia's intentions, may eventually reintroduce the classical security dilemma to the Arctic as militaries tailor their respective capabilities in relation to others based on their assessments. In the worst-case scenario, this could lead to a negative spiral in which states feel compelled to respond in kind; that is, to match growing capability of Russia with counter-capabilities of their own, and vice versa. This would be detrimental to the overall spirit of co-operation in the Northern region.

That said, the Western military response to a more unpredictable and militarily active Russia has so far remained moderate in the Arctic region. Illustrative of this, the United States continues to have difficulty finding adequate funding even to maintain its degrading ice-breaking capability.[8] However, there have been some developments detectable from public sources. These include a plan to redeploy US Air Force assets capable of maritime patrolling to the retrofitted Keflavik airport in Iceland (Pettersen 2016), plans to increase maritime surveillance by NATO countries, most notably Norway, in the Arctic (Nilsen 2016), and even a deployment of a small contingent of US Army to Norway (BBC 2016), akin to the battalion-size NATO reassurance deployments in the Baltic. The United States has also announced a plan to deploy in the future new advanced F-35 fighters in Alaska (U.S. Air Force 2016).

Two militarily non-aligned Nordic/Arctic countries, Finland and Sweden, have also responded to growing uncertainty stemming from a more active and aggressive Russia. Among other things, they have come to share an updated threat assessment in the Nordic-Baltic region, plan further military co-operation, and have plans to

---

[5]For example, the US has reservations about Russia's view of the NSR as internal waters as well as its willingness to control movement along the route through ice-breaker policy.

[6]That said, currently, Russia proceeds legitimately through the UN process.

[7]For more on Russian hybrid influence in the Nordic-Baltic region, see Pynnöniemi and Salonius-Pasternak (2016), Martikainen et al. (2016).

[8]See, for example, Koivurova (2016). For a discussion of icebreakers and their limited relevance for the US military capability in the Arctic, see Kuersten (2016).

upgrade both their military hardware and ability to respond swiftly to various kinds of security scenarios, including those that might not pass the threshold of traditional war. Both have also signed non-binding agreements of intent to improve military co-operation with the US and the UK (see YLE 2016; Finland Times 2016).

In sum, it is important to realize that the broader strategic situation has changed and that we no longer live in the "end of history" world where Arctic exceptionalism emerged in the post-Cold War era. While there is no need to be overly alarmist, it is crucial to realize that contemporary Russia has two overriding strategic goals—maintaining regime stability and achieving a great power status—that it pursues with all its available, and sometimes harsh, means. The second of these goals effectively also involves the pursuit of a new multipolar or post-hegemonic world order. Because Russia is the most important Arctic player, it plays an essential role when it comes to the future of Arctic exceptionalism. Russia has explicitly emphasized that the Arctic is a strategic area, not least as a future resource base and as a crucial site for its conventional and nuclear military capability. Hence, the Arctic is likely to be (increasingly) significant for Russia's domestic development and its foreign influence more broadly. As such, Russia's Arctic policy ought to be seen in the light of its broader strategic goals, wherein the Arctic policy serves ultimately a subordinate role.

If developments in Russia's Arctic policy are related to its broader strategic calculus, the key question becomes whether cooperation or conflict in the Arctic serves Russia better. Currently, Russia seems determined to develop the region in the long term, and for this it needs international cooperation and regional stability and predictability. Thus, there are still reasons to believe that the Arctic remains largely a zone of peaceful co-operation—at least for the time being.

However, this can no longer be taken as self-evident. The implications of the Western sanctions regime, intensified perceptions of external threats, geopolitical isolation, and the growing influence of the security apparatus in the Russian domestic political system may all indicate increased self-sufficiency and a stronger security focus also on Arctic matters in Russia. If these adverse dynamics became even more widely entrenched at a time when Russia perceives itself in a protracted strategic conflict with the West, it could end up producing non-cooperative developments in the Arctic, and by doing so, also damage the very idea of Arctic exceptionalism.

10.6 The US, Novel Uncertainty, and the Arctic

Russia is not the only acute source of concern with regard to the future of Arctic exceptionalism. A new and additional layer of uncertainty has emerged from the United States. The US has traditionally been a "reluctant Arctic power" (Huebert 2009) that has paid a limited amount of policy attention to the region. While the Arctic continues to be a relatively minor topic on the overall US foreign policy agenda even today, the US has started to pay closer attention to the region with the publication of key strategic documents and high-profile participation in Arctic affairs. In short, the Arctic has gradually emerged as a "new" foreign policy frontier in the

US (Conley 2013). Particularly during the Obama era, the US approach to the Arctic tended to focus international co-operation, environmental protection (especially climate change), and scientific research.

The election of Donald Trump as US president has raised a number of critical concerns about the direction of US Arctic policy, and more broadly about America's commitment to co-operation in the Arctic (see, e.g., Huebert 2016). First, co-operation on environmental protection and especially climate change may be in jeopardy. During his presidential campaign, Trump was infamously dismissive of global warming and highly critical of the Paris Climate Agreement. As president, Trump has decided to withdraw the US from the Paris Agreement as the international accord apparently represented an insurmountable challenge to US sovereignty and its ability to tailor domestic environmental laws to benefit the US economy (Shear 2017). This contradicts the US endorsement of the Declaration of Foreign Ministers in the recent 2017 Arctic Council ministerial meeting in which, albeit in a revised form, the US acknowledged the existence of climate change and a need to tackle it (Arctic Council 2017). Trump's administration has also sought to downsize the Environmental Protection Agency and to slash international funding to tackle climate change and pursue sustainable development, two broad frameworks that Finland has decided to advance as the chair of the Arctic Council from 2017 to 2019. The Trump Administration's critical views on conducting and funding climate research are also likely to affect negatively scientific work done in the US, with potential implications also to the Arctic Council (Harvey 2017; Milman 2017). In the long run, it might even affect the US's status as an "Arctic science power" (Conley and Rohloff 2015: xiv).

Secondly, the Trump Administration has also pursued a climate-unfriendly energy policy. It has decided to re-open the Arctic for offshore hydrocarbon production by reversing President Obama's key decisions. These include the exclusion of US Arctic waters from the next offshore oil and gas leasing strategy, and the subsequent decision to indefinitely ban future hydrocarbon production in the majority of US offshore Arctic. The Italian energy company Eni has received permission to start drilling in Arctic waters in December of 2017 (Cama 2017).

Thirdly, President Trump is likely to emphasize, as much as possible, a transactionalist approach to foreign policy, as opposed to a multilateral approach. Transactionalist foreign policy emphasizes bilateral relations and the sovereign ability to make the best possible deals that advance America's interests as defined by, in this case, President Trump himself. As such, it entails "the implicit rejection of the US role in the international system that has underpinned the west" (Stephens 2017). While Secretary of State Rex Tillerson has stated that the US "will continue to be an active member in [the Arctic] Council" (quoted in Quinn 2017), the overall transactionalist emphasis on foreign policy and US engagement in the world may result in lesser emphasis on co-operation in multilateral governance fora in the Arctic, including the Arctic Council.

Finally, President Trump has also made it clear that he wishes to improve the relationship between the US and Russia. As his Secretary of State Tillerson has argued, the "world's two foremost nuclear powers cannot have this kind of [poor] relationship" (The Economist 2017). These kinds of comments have raised the possibility that president Trump might singlehandedly relax or undo parts of, or all, sanctions that the US has placed on Russia and Russian entities as a result of the crisis in Ukraine and, subsequently Russia's interference in the 2016 US presidential elections. The cancellation of sanctions would both effectively legitimize Russia's recent revisionist activities and make it possible to resume offshore energy co-operation in the Kara Sea among Exxon Mobil, Rosneft, and others. The latter development, in turn, would be detrimental to efforts to combat climate change as all the hard-to-recover resources in the Arctic ought to be left untouched in order to successfully limit global warming in the next decades (Koivurova 2018, 23). However, recent legislation that codifies the above-mentioned Russia sanctions—passed almost unanimously by the Congress in July 2017 and reluctantly signed into law by the president in August 2017—makes any unanimous decision to undo them by the executive branch impossible.

In sum, recent developments in the US have a real potential to affect negatively international co-operation in the Arctic. The first indications of this were seen in the recent ministerial meeting of the Arctic Council in Fairbanks, where last-minute uncertainty over climate change references due to US reservations put the unanimous acceptance of the final declaration exceptionally in jeopardy. As glimpsed in Fairbanks, in addition to affecting concrete co-operative work done in the Council, uncertainty over US policy has the potential to create a new line of division within the Arctic between the US and other Western nations.

10.7 Conclusion

This chapter has discussed Arctic exceptionalism in the post-Cold War era; specifically, the historical background to the idea of a peaceful and cooperative region, factors which continue to support the idea in practice, developments that have challenged some of its aspects in recent years, and potentialities that might undermine it in the future. In particular, we have argued that there is a close but often downplayed connection between the Arctic and global political dynamics, including those dynamics related to international conflicts that involve Arctic states as parties. The crisis in Ukraine was discussed as a case in point. This crisis has pitted Russia and seven other Arctic states against each other. As a result, extra-Arctic political and conflict dynamics have spilled over to the region in various sectors—albeit in a limited manner.

In addition to the worrisome changes in the Russian domestic and foreign policy, recent acute developments in the US political landscape have increased uncertainty in world politics in general. The US-Russo relationship, in particular, remains at a post-Cold War low, all the while Trump administration's policy choices and regard for international co-operation and partners have raised concern. As the Arctic is not shielded from these dynamics, the future of Arctic exceptionalism has become increasingly difficult to predict—and to realize.

Fortunately, despite the crisis in Ukraine, the foreign and security policy concerns in Europe (and beyond) and the uncertainty over US foreign policy, co-operation in the Arctic has shown continuing resilience in a difficult international situation. This is the case especially, though not exclusively, with the Arctic Council. Arctic states share many common interests and challenges, primarily related to safe and sustainable (at least according to official discourse) socio-economic activities in the Arctic,—including hydrocarbon exploitation, maritime logistics, (maritime) tourism, commercial fishing, infrastructure building—that continue to create incentives to seek common ground and shared practical solutions. However, in the current environment, predicting future Arctic trajectories has become increasingly difficult due to heightened uncertainty stemming from various, global sources.

Bibliography

Arctic Council (2013) The Kiruna declaration, 15 May 2013. https://oaarchive.arctic-council.org/handle/11374/93
Arctic Council (2015) Iqaluit declaration, 24 Apr 2015. https://oaarchive.arctic-council.org/handle/11374/662
Arctic Council (2017) Fairbanks declaration, 11 May. https://oaarchive.arctic-council.org/handle/11374/1910
Åtland K (2008) Mikhail Gorbachev, the Murmansk initiative, and the desecuritization of interstate relations in the Arctic. Cooperation and Conflict 43(3):289–311
Baev PK (2015) Russia's Arctic aspirations. In: Jokela J (ed) Arctic security matters, EUISS Report 24, June 2015. EU Institute for Security Studies. http://publications.europa.eu/resource/cellar/6fe2f868-6856-11e5-9317-01aa75ed71a1.0001.04/DOC_1
BBC (2016) US troops to be stationed in Norway in break with tradition, 25 Oct 2016. http://www.bbc.com/news/world-europe-37761376
Borgerson SG (2008) Arctic meltdown: the economic and security policy implications of global warming. Foreign Affairs, March/April 2008. http://www.foreignaffairs.com/articles/63222/scott-g-borgerson/arctic-meltdown
Byers M (2013) International law and the Arctic. Cambridge University Press, Cambridge
Cama T (2017) Trump admin approves oil company's Arctic drilling plan. The Hill, 13 July 2017. http://thehill.com/policy/energy-environment/341833-trump-admin-approves-oil-companys-arctic-drilling-plan
Carney J (2014) Ukraine in the White House Daily Press briefing. The Embassy of the United States, Kiev, Ukraine, 3 Apr 2014. http://ukraine.usembassy.gov/statements/wh-ukraine-04032014.html
Chivers CJ (2007) Russians plant flag on the Arctic seabed. New York Times, 3 Aug 2007. http://www.nytimes.com/2007/08/03/world/europe/03arctic.html?_r=0
Claes DH, Moe A (2014) Arctic petroleum resources in a regional and global perspective. In: Tamnes R, Offerdal K (eds) Geopolitics and security in the Arctic: regional dynamics in a global world. Routledge, New York
Conley H (2013) The new foreign policy Frontier: U.S. interests and actors in the Arctic. Center for Strategic & International Studies, Washington, D.C. https://csis-prod.s3.amazonaws.com/s3fs-public/legacy_files/files/publication/130307_Conley_NewForeignPolFrontier_Web_0.pdf
Conley H (2014) A role for NATO in the Arctic? In: Dahl AS, Järvenpää P (eds) Northern security and global politics: Nordic-Baltic strategic influence in a post-unipolar world. Routledge, New York
Conley H, Rohloff C (2015) The new ice Curtain: Russia's strategic reach to the Arctic. Center for Strategic & International Studies. https://csis-prod.s3.amazonaws.com/s3fs-public/legacy_files/files/publication/150826_Conley_NewIceCurtain_Web.pdf

Finland Times (2016) Finland, USA sign defence cooperation deal, 8 Oct 2016. http://www.finlandtimes.fi/national/2016/10/08/30690/Finland,-USA-sign-defence-cooperation-deal
Garcés de los Fayos F (2015) The outcome of the ninth Arctic Council ministerial meeting. European Parliament, Directorate-General for External Policies, Policy Department, DG EXPO/B/PolDep/Note/2015_17, May 2015—PE 549.036. http://www.europarl.europa.eu/RegData/etudes/ATAG/2015/549036/EXPO_ATA(2015)549036_EN.pdf
Giles K (2016) Russia's 'New' tools for confronting the west: continuity and innovation in Moscow's exercise of power. Chatham House. https://www.chathamhouse.org/sites/files/chathamhouse/publications/research/2016-03-21-russias-new-tools-giles.pdf
Gorbachev M (1987/2012) Mikhail Gorbachev's speech in Murmansk at the ceremonial meeting on the occasion of the presentation of the order of Lenin and the gold star to the city of Murmansk. http://www.lecerclepolaire.com/en/documentation-uk/news-a-views-uk/454-meeting-with-the-former-president-mikhail-gorbachev
Government of Canada: Foreign Affairs, Trade and Development Canada (2014) Canada takes principled stand on Arctic council meetings, 15 Apr 2014. http://www.international.gc.ca/media/arctic-arctique/news-communiques/2014/04/15a.aspx?lang=eng
Harvey C (2017) Rex Tillerson signed a statement supporting Arctic science. Trump's budget would cut it. The Washington Post, 26 May 2017. https://www.washingtonpost.com/news/energy-environment/wp/2017/05/26/rex-tillerson-signed-a-statement-supporting-arctic-science-trumps-budget-would-cut-it/?utm_term=.90bb07f89f06
Heininen L, Exner-Pirot H, Plouffe J (2013) From regional transition to global change: dualism in the Arctic. In: Heininen L (ed) Arctic yearbook 2013. Northern Research Forum, Akureyri, Iceland
Hilde PS (2014) Armed forces and security challenges in the Arctic. In: Tamnes R, Offerdal K (eds) Geopolitics and security in the Arctic: regional dynamics in a global world. Routledge, New York
Huebert R (2009) The United States Arctic policy: the reluctant Arctic power. SPP Briefing Pap 2(2), May 2009
Huebert R (2016) What Donald Trump's Presidency might mean for the Arctic. Arctic Deeply, 14 Dec 2016. https://www.newsdeeply.com/arctic/community/2016/12/14/what-donald-trumps-presidency-might-mean-for-the-arctic
Humrich C, Wolf KD (2012) From meltdown to showdown? Challenges and options for governance in the Arctic. PRIF Report No. 113, 2012
ITAR-TASS (2014) No need for NATO presence in Arctic—Lavrov. 20 Oct 2014. http://en.itar-tass.com/russia/755311
Keskitalo C (2007) International region-building: development of the Arctic as an international region. Cooperation and Conflict: J Nord Int Stud Assoc 42(2):187–205
Klimenko E (2016) Russia's Arctic security policy: still quiet in the high north? SIPRI Policy Paper 45. http://books.sipri.org/files/PP/SIPRIPP45.pdf
Koivurova T (2010) Limits and possibilities of the Arctic council in a rapidly changing scene of Arctic governance. Polar Rec 46(2):146–156
Koivurova T, Käpylä J, Mikkola H (2015) Continental shelf claims in the Arctic: will legal procedure survive the growing uncertainty? FIIA Briefing Paper 178. The Finnish Institute of International Affairs. http://www.fiia.fi/en/publication/516/continental_shelf_claims_in_the_arctic/
Koivurova T (2016) The Arctic conflict—truth, fantasy or a little bit of both? High North News, 18 Nov. http://www.highnorthnews.com/analysis-the-arctic-conflict-truth-fantasy-or-a-little-bit-of-both/
Koivurova T (2018) Framing the problem in Arctic offshore hydrocarbon exploitation. In: Pelaudeix C, Basse EM (eds) Governance of Arctic offshore oil and gas. Routledge, New York
Kuersten A (2016) Icebreakers and U.S. power: separating fact from fiction. War on the Rocks, 11 Oct. http://warontherocks.com/2016/10/icebreakers-and-u-s-power-separating-fact-from-fiction/

Käpylä J, Mikkola H (2013) Arctic conflict potential: towards an extra-Arctic perspective. FIIA Briefing Paper 138. The Finnish Institute of International Affairs. http://www.fiia.fi/en/publication/361/arctic_conflict_potential/
Käpylä J, Mikkola H (2015) On Arctic exceptionalism: critical reflections in the light of the Arctic sunrise case and the crisis in Ukraine. FIIA Working Paper 85. The Finnish Institute of International Affairs. http://www.fiia.fi/en/publication/502/on_arctic_exceptionalism/
Käpylä J, Mikkola H, Martikainen T (2016) Moscow's Arctic dreams turned sour? Analysing Russia's policies in the Arctic. FIIA Briefing Paper 196. The Finnish Institute of International Affairs. http://www.fiia.fi/en/publication/575/moscow_s_arctic_dreams_turned_sour/
Laine V, Martikainen T, Pynnöniemi K, Saari S (2015) Zugzwang in slow motion? The implications of Russia's system-level crisis. FIIA Analysis 6, Dec 2015. The Finnish Institute of International Affairs. http://www.fiia.fi/fi/publication/554/zugzwang_in_slow_motion/
Laruelle M (2014) Russia's Arctic strategies and the future of the far north. New York, M.E, Sharpe
Lasserre F et al (2012) Is there an arms race in the Arctic? J Mili Strateg Stud 14(3&4)
Le Miére C (2014) Ukraine brings cool wind to the Arctic. IISS voices, 20 Sept. http://www.iiss.org/en/iiss%20voices/blogsections/iiss-voices-2014-b4d9/september-2f2a/ukraine-arctic-9239
Martikainen T, Vihma A (2016) Dividing the EU with energy? Unpacking Russia's energy geoeconomics. FIIA Briefing Paper 191. The Finnish Institute of International Affairs. http://www.fiia.fi/en/publication/571/dividing_the_eu_with_energy/
Martikainen T, Pynnöniemi K, Saari S (2016) Neighbouring an unpredictable Russia: implications for Finland. Finnish Foreign Policy Papers 05, November. The Finnish Institute of International Affairs. http://www.fiia.fi/fi/publication/629/neighbouring_an_unpredictable_russia/
Medecins Sans Frontiers (2016) Eastern Aleppo: 23 attacks on hospitals since July; No hospital left undamaged. MSF press release, 7 Oct. http://www.doctorswithoutborders.org/article/eastern-aleppo-23-attacks-hospitals-july-no-hospital-left-undamaged
Mikkola H, Käpylä J (2013) Arctic economic potential: the need for a comprehensive and risk-aware understanding of Arctic dynamics. FIIA Briefing Paper 127. The Finnish Institute of International Affairs. http://www.fiia.fi/en/publication/337/arctic_economic_potential/
Mikkola H, Käpylä J (2014) Russian Arctic sanctioned: western measures against Arctic oil projects do not signify the geoeconomic importance of the region. FIIA Comment 16/2014. The Finnish Institute of International Affairs. http://www.fiia.fi/en/publication/447/russian_arctic_sanctioned/
Milman O (2017) Trump budget would gut EPA programs tackling climate change and pollution. The Guardian, 16 Mar. https://www.theguardian.com/environment/2017/mar/16/trump-budget-cuts-climate-change-clean-up-programs-epa
Moshes A (2017) EU-Russia relations: Quo Vadis? Muddling, normalization, or deterioration. PONARS Eurasia Policy Memo No. 454, January. http://www.ponarseurasia.org/sites/default/files/policy-memos-pdf/Pepm454_Moshes_Jan2017.pdf
Nilsen T (2016) Norway acquires five new maritime patrol aircraft for Arctic waters. The Independent Barents Observer, 26 Nov. https://thebarentsobserver.com/en/security/2016/11/norway-acquires-five-new-maritime-patrol-aircrafts-arctic-waters
Pedersen T (2012) Debates over the role of the Arctic council. Ocean Dev Int Law 43(2):146–156
Pettersen T (2014) USA cancels joint exercises with Russia. Barents Observer, 5 Mar. http://barentsobserver.com/en/security/2014/03/usa-cancels-joint-exercises-russia-05-03. Accessed 19 Dec 2016
Pettersen T (2016) U.S. military returns to Iceland. The Independent Barents Observer. 10 Feb. https://thebarentsobserver.com/en/security/2016/02/us-military-returns-iceland. Accessed 11 Nov 2016
Pilli-Sihvola K et al (2016) Suomi arktisen alueen vastuulliseksi edelläkävijäksi: toimenpide-ehdotuksia yleisen kehityksen, meriklusterin ja matkailun edistämiseksi vuoteen 2035. Valtioneuvoston selvitys- ja tutkimustoiminnan julkaisusarja 10/2016, maaliskuu 2016. http://tietokayttoon.fi/documents/10616/2009122/10_Suomi+arktisen+alueen+vastuulliseksi+edell%C3%A4k%C3%A4vijaksi.pdf/95f162ea-593c-4a49-a842-5db79ba0165d?version=1.0

Pynnöniemi K, Salonius-Pasternak C (2016) Security in the Baltic Sea region: activation of risk potential. FIIA Briefing Paper 196. The Finnish Institute of International Affairs. http://www.fiia.fi/en/publication/597/security_in_the_baltic_sea_region/
Quinn E (2017) Arctic council presents united front as Finland takes over from U.S. Eye on the Arctic, 9 Aug. http://www.rcinet.ca/eye-on-the-arctic/2017/05/11/arctic-council-presents-united-front-as-finland-takes-over-from-u-s/
Rogin J (2016) Congress may spoil Trump's Russian reset. The Washington Post, 20 Nov. https://www.washingtonpost.com/opinions/global-opinions/congress-may-spoil-trumps-russian-reset/2016/11/20/6b679a74-adc4-11e6-8b45-f8e493f06fcd_story.html?utm_term=.5fefddb830ef
Rosen Y (2014) US–Russia tensions create worries for Arctic scientists. Alaska Dispatch News, 9 May. http://www.adn.com/article/20140509/us-russia-tensions-create-worries-arctic-scientists
Shear MD (2017) Trump will withdraw U.S. From Paris climate agreement. The New York Times, 1 June. https://www.nytimes.com/2017/06/01/climate/trump-paris-climate-agreement.html
Stephens P (2017) Trump presidency: America first or America alone? Financial Times, 9 Jan. https://www.ft.com/content/ae092214-d36f-11e6-b06b-680c49b4b4c0
The Arctic Journal (2015) Council embraces Russia but puts new observers on hold. http://arcticjournal.com/politics/1523/council-embraces-russia-puts-new-observers-hold
The Barents Observer (2015) Expansionist Rogozin looks to Arctic. http://barentsobserver.com/en/arctic/2015/04/expansionist-rogozin-looks-arctic-21-04
The Guardian (2016a) Royal Navy sends ships to shadow Russian fleet passing Britain. 19 Oct. https://www.theguardian.com/world/2016/oct/19/convoy-of-russian-warships-on-route-past-british-isles
The Guardian (2016b) Ominous news for Aleppo as Russian frigate reaches Syrian coast. 4 Nov. https://www.theguardian.com/world/2016/nov/04/russia-further-bolsters-naval-presence-in-mediterranean
The Economist (2017) Relations between America and Russia are at a low point. 13 Apr. https://www.economist.com/blogs/democracyinamerica/2017/04/sober-moscow
The Ilullissat Declaration (2008) http://www.oceanlaw.org/downloads/arctic/Ilulissat_Declaration.pdf
U.S. Air Force (2016) Eielson selected to receive operational F-35A aircraft. 4 Apr. http://www.af.mil/News/ArticleDisplay/tabid/223/Article/712879/eielson-selected-to-receive-operational-f-35a-aircraft.aspx
United Nations Convention on the Law of the Sea (1982) http://www.un.org/depts/los/convention_agreements/texts/unclos/unclos_e.pdf
US Geological Survey (2008) Assessment of undiscovered oil and gas in the Arctic. http://www.usgs.gov/newsroom/article.asp?ID=1980&from=rss_home
Wezeman ST (2012) Military capabilities in the Arctic. SIPRI Background Paper, March
YLE (2016) Daily: Finland to sign defence cooperation pact with UK. 1 July. http://yle.fi/uutiset/osasto/news/daily_finland_to_sign_defence_cooperation_pact_with_uk/8996698
Young OR (1992) Arctic politics: conflict and cooperation in the circumpolar north. University Press of New England, Hanover
Young OR (2009) Whither the Arctic? Conflict or cooperation in the circumpolar north. Polar Rec 45(1):73–82
Zysk K (2011) Military aspects of Russia's Arctic policy: hard power and natural resources. In: Security Arctic, Kraska J (eds) Arctic security in an age of climate change. Cambridge University Press, Cambridge

Chapter 11
The Changing Role of Military Power in the Arctic

Valery Konyshev and Alexander Sergunin

11.1 Introduction

In the Cold War era, military power was a coercive instrument in a global confrontation between two superpowers and capitalist and socialist systems. The Arctic region was part of this global confrontation; it was a home for strategic nuclear forces (especially Soviet ones) and an important area for significant military activities. Both the United States and the Soviet Union pursued containment strategies, with mutually assured destruction (MAD) doctrine at their cores.

In the post-Cold-War world, however, the roles of military power and the nature of military strategy have been radically transformed because of the global geopolitical changes and revolution in military affairs (RMA).

The recent international developments have repeatedly cast doubt on the efficacy of military force and the ability to achieve political objectives by military methods. The decision to use force has frequently been based on incorrect calculations or on ideological arguments (Afghanistan, Iraq, Libya, Syria, Yemen, and the former Yugoslavia), which had nothing to do with the real national interests of the countries involved. There is a new phenomenon in international relations: present-day wars no longer aim to acquiring enemy's territory and wealth. Now we see wars unleashed with the aim of changing political regimes or under the banner of "human rights" protection (the doctrines of "humanitarian interventions" and "responsibility

This research was conducted within the framework of the GlobalArctic project run by the Northern Research Forum and CANARCT project no. 257638 funded by the Research Council of Norway, 04-2016–03-2019.

V. Konyshev (✉) · A. Sergunin
Saint Petersburg State University, Saint Petersburg, Russia
e-mail: valery.konyshev@spbu.ru

A. Sergunin
e-mail: sergunin60@mail.ru

M. Finger and L. Heininen (eds.), *The GlobalArctic Handbook*,
https://doi.org/10.1007/978-3-319-91995-9_11

to protect"). Armed forces have new, non-traditional roles, such as fighting terrorists, piracy, policing conflict zones, protecting a country's economic interests, conducting search and rescue (SAR) operations, and coping with natural and man-made catastrophes.

The RMA has changed the nature of war as well. The use of precision weapons, unmanned aerial vehicles, hybrid tactics, and strikes against information infrastructure are now the most popular methods for waging wars. The supremacy in military technologies became a crucial factor in achieving regional or world military hegemony. For this reason, the competition between major powers has moved to the technological sphere and to equipping armed forces with advanced weaponry.

The RMA has also changed the nature of war, in the sense that it made the concept of territorial defense obsolete. In the age of non-contact and network-centric warfare, there is no need to deploy massive land forces close to the national borders in order to prevent a potential enemy's invasion. The world's major powers' current focus regarding military reform is on making armed forces more compact, mobile, and better armed and trained for multipurpose missions.

To what extent have these dramatic changes affected the military situation in the High North? How do the Arctic states perceive the role of military power in their regional strategies? Is there an arms race or militarization in the High North, as some journalists and politicians believe? These are questions that form the academic community's research agenda and should be attacked by students of Arctic politics.

In this chapter, we aim to examine the paradigmatic shifts in the nature of military power in the Arctic. More specifically, we focus on the interplay of "old" and "new" components of military force in the region and how it affects the military-strategic situation in the Far North.

11.2 Changing Traditional Roles

Despite its new roles, which will be described below, the military power in the Arctic still has some traditional functions, such as defense of a country's sovereignty and territory from potential aggression, power projection, deterrence, containment, etc.

Along with the significant economic, environmental and humanitarian interests the Arctic states' perceptions of the region are still largely based on hard security considerations. For example, the Kola Peninsula and the adjacent area are still considered a region of special strategic importance to Russia's national security. The direct access to the Arctic and Atlantic oceans, close proximity to potential US/NATO targets, and a relatively developed military infrastructure make this region well-suited for strategic naval operations (Khramchikhin 2011, 2013). Above all, the strategic importance of the Kola Peninsula is explained by the fact that it hosts two-thirds of Russia's sea-based nuclear forces.

Russian military analysts believe that the Archangelsk Air Defense Sector is still crucial for the prevention of surprise attack over the North Pole. The Norwegian and Barents Seas can still serve as the main launching areas for Western seaborne attack;

therefore, these analysts maintain, the Russian Navy should still be concerned about the readiness of its anti-submarine forces in the Arctic (Khramchikhin 2011, 2013; Konyshev and Sergunin 2014a, b). "There are [US] submarines there and they carry missiles," President Putin told students at a meeting at Moscow State University. "It only takes 15–16 min for US missiles to reach Moscow from the Barents Sea. So should we give away the Arctic? We should, on the contrary, explore it" (Anishchuk 2013).

The US strategic perceptions of the Arctic mirror the Russian ones. On the one hand, according to the recent US nuclear doctrines, the High North is a region where the US can make a "prompt global strike" (including its both nuclear and conventional components) (Woolf 2012).

On the other hand, the Arctic is crucial for US and Canadian strategic defense in terms of protecting those countries from a potential Russian nuclear strike (either preemptive or retaliatory). The Arctic and contiguous regions of Canada and the US are a home for the North American Aerospace Defense Command (NORAD), which is a combined organization of two countries that provides aerospace warning, air sovereignty, and protection for all of North America.

In addition to the continental part of the NORAD is the Thule Air Force Base in Greenland, which is the northernmost military facility operated by the US Air Force (USAF). The base hosts the USAF 821st Air Base Group and the 12th Space Warning Squadron under the USAF Space Command, which operates an early warning radar with a role in the US and NORAD ballistic missile defense (BMD) system. The Pentagon still views Thule Air Force Base as a strategic asset because it is part of a larger effort to protect the US from intercontinental ballistic missile (ICBM) attack through its Upgraded Early Warning Radar (UEWR) system (Bender 2014; Conley and Kraut 2010: 21).

The only US Arctic state—Alaska—is home to the United States' national BMD system, the Ground-based Midcourse Defense (GBMD), which currently protects against the threat of a limited ICBM attack. Officially, this system is designed to counter an alleged North Korean ICBM threat but it is actually directed against the Russian and Chinese ballistic missile potentials because only they are currently capable of reaching the US territory. The GBMD system relies on 26 ground-based interceptors (GBIs) at Fort Greely (in addition to four interceptors at the Vandenberg Air Force Base, California). To enable these ground-based systems to successfully intercept attacking missiles in the midcourse part of their trajectory, the US employs early warning radars in Alaska, California, Greenland (Thule), and the United Kingdom; afloat radar systems (Aegis destroyers, Aegis cruisers, and Sea-Based X-band radar [SBX]); and a sophisticated command and control infrastructure (U.S. Department of Defense 2010: 15). The US Missile Defense Agency plans to add 14 GBIs to the operational fleet at Fort Greely by 2017 (Syring 2013).

It should be noted that the nuclear deterrent and MAD doctrine not only still remain key elements of the US and Russian military strategies, but also serve as symbols and guarantees of their superpower/great power status (Conley 2013; Conley and Kraut 2010; Klimenko 2016; Zysk 2008: 81). Therefore, maintaining strategic nuclear capabilities and modernizing strategic nuclear forces are the highest priorities of Moscow's and Washington's military policies, both in the High North and globally. The two countries' modernization programs aim to make strategic forces more compact and efficient.

As far as the Russian fleet of strategic nuclear submarines (SSBNs—Ship Submarine Ballistic Nuclear) is concerned, only Delta IV-class submarines undergo the process of modernization. They will be provided with a new sonar system and the new submarine-launched ballistic missiles (SLBMs) *Sineva* (Skiff SSN-23), which entered service in 2007. *Sineva* is a third-generation liquid-propelled SLBMs that can cover a distance up to 8300 km and carry either four or 10 nuclear warheads (http://www.armsexpo.ru/049055051051124052049049.html). Russia is planning to equip its Delta IV-class submarines with at least 100 *Sineva* missiles, which will remain on alert status until 2030. The *Sineva* missiles can be launched from under the ice while remaining invisible to enemy's satellites until the last moment (Laruelle 2014: 122).

Another class of the Russian strategic submarines—the *Typhoons*, which are considered the world's largest—will be reequipped with long-range cruise missiles. For the time being, only one *Typhoon*-class strategic submarine, the *Dmitri Donskoy*, has been modernized and deployed to the Northern Fleet. It serves to conduct test firing for the *Bulava* system, a new-generation solid-fuel SLBM that is designed to avoid possible future US BMD weapons and can cover a distance of more than 9000 km (http://www.arms-expo.ru/049057054048124050052056054051.html).

In the future, it is planned that the *Typhoon*-class submarines will replaced with the new *Borey*-class fourth-generation nuclear-powered strategic submarines. The first *Borey*-class submarine, the *Yuri Dolgoruky*—which was the first strategic submarine to be built in Russia after the collapse of the Soviet Union—has been in operation by the Northern Fleet since January 2013. Two other *Borey*-class submarines, the *Alexander Nevsky* and the *Vladimir Monomakh*, run the sea trials, while the fourth one, *Prince Vladimir*, is under construction at the Severodvinsk shipyard (http://bastion-karpenko.narod.ru/955_more_01.html). These three submarines will be placed with the Pacific Fleet. The *Borey*-class submarines, which are to be deployed to the Northern Fleet, will be based at the Gadzhievo Naval Base (approximately 100 km from the Norwegian border), where new infrastructure is being built to host them.

This new generation of the Russian strategic submarines is almost invisible at deep ocean depths and, having several types of cruise missiles and torpedoes, will be able to carry out multipurpose missions, including attacks on enemy aircraft carriers and missile strikes on coastal targets. According to the Defense Ministry's plans, the building of eight *Borey*-class submarines (four for the Northern Fleet and four for the Pacific Fleet) are scheduled to be completed by 2020, although this seems quite ambitious and unlikely in the context of budget constraints caused by the ongoing economic crisis.

The US SSBN fleet consists of 14 Trident (Ohio-class) submarines, each equipped to carry 24 Trident missiles. They are deployed to two naval bases at Bangor, Washington, and Kings Bay, Georgia. In the 2000s, the US Navy converted four Trident submarines to carry conventional cruise missiles and other conventional weapons and designated them "guided missile" submarines (SSGNs) (Woolf 2016: 19).

Similar to the Russian SSBNs the US ones are specifically designed for extended deterrent patrols. On average, the submarines spend 77 days at sea, followed by 35 days in-port for maintenance (U.S. Navy Navy 2016a). Each SSBN has two crews, Blue and Gold, which alternate manning the submarines and taking them on patrol. This maximizes the SSBN's strategic availability, reduces the number of submarines required to meet strategic requirements, and allows for proper crew training, readiness, and morale.

Since 1958, the US Navy has organized ice exercises (ICEX); in these, submarines conduct Arctic transits in which they surface and break the ice (usually 60–90 cm thick), collect data, and run other training exercise to gain experience working in this region. According to one of the leaders of the ICEX-2016, the importance of this exercise is that "submarine operations as part of ICEX provide the necessary training to maintain a working knowledge of an extremely challenging region that is very different than any other ocean in the world … Navigating, communicating, and maneuvering are all different in an Arctic environment as there are surfaces both above and below a submarine" (U.S. Navy 2016b http://www.navy.mil/submit/display.asp?story_id=93648). Other than collecting data and training in this region, the ICEX exercise also displays the US Navy's Arctic defense capabilities and readiness for roles in this region, increases the experience of sailing and working in the area, and gathers broader knowledge about this region.

As far as the modernization plans are concerned, the US Navy initially planned to keep Trident submarines in service for 30 years, but then extended that time period to 42 years, with two 20-year operating cycles separated by a two-year refueling overhaul. With this schedule, the submarines will begin to retire from the fleet in 2027. The Navy has also pursued a number of programs to ensure that it has enough missiles to support this extended life for the submarines.

The US Navy is currently conducting development and design work on a new class of ballistic missile submarines, originally known as the SSBN(X) program and the Ohio Replacement Program. The Navy has recently announced that these submarines will be known as the Columbia class and they will replace the Ohio-class Trident submarines as they reach the end of their service lives. The Trident submarines will begin to retire in 2027 and the Navy initially indicated that it would need the new submarines to begin to enter the fleet by 2029, before the number of Trident submarines falls below 12. However, with the FY2013 budget request, the Navy delayed the procurement of the new class of submarines by two years. As a result, the first new submarine will enter the fleet in 2031 and the number of SSBNs in the fleet is expected to decline to 10 for most of the 2030s (Woolf 2016: 24).

The UK Royal Navy has four Vanguard-class SSBNs armed with American Trident II D5 missiles planned for decommissioning in the late 2020s and early 2030s. According to a UK Defense Ministry review of alternative scenarios for upgrading the strategic nuclear forces, the optimal path lies in building the new-generation SSBNs carrying Trident missiles. Plans provide for the construction of four submarines under the Successor program, with construction of the British leading missile submarine due to start in 2016 and its commissioning envisaged in 2028 (Tebin 2015).

It should be noted that some climate change implications, such as the Northern pole ice cap's meltdown necessitate some serious changes in the Arctic states' military strategies. On one hand, as the recent US Navy's document argues, the extension of the ice-free season can result in a significant expansion of surface naval activities in the Arctic (The United States Navy Arctic Roadmap for 2014 to 2030, 2014: 8, 16–19). On the other hand, the shrinking ice cap provides less protection to submarines, making them more visible to enemy's satellites and aircraft.

Given the ice-free Arctic in the foreseeable future (at least for part of the year), Russian military analysts have not excluded the possibility that the US could permanently deploy a nuclear submarine fleet and sea-based BMD systems in the Arctic Ocean (Khramchikhin 2013; Konyshev and Sergunin 2014a, b). In this case, the US could create capabilities for intercepting Russian ICBM launches at the initial (boost) phase and making a preventive/'disarming' strike by ICBMs, SLBMs, and cruise missiles, regardless of whether they are nuclear or non-nuclear. In turn, this way of American strategic thinking can provoke Russia's continuing efforts to regularly modernize its strategic nuclear forces, with the aim of having sufficient potential to overcome the US BMD system.

It should be noted that there are differences between Russian and American visions of the strategic and operative-tactical forces' roles in the region. For the Russian strategic forces, the Arctic, North Atlantic, and North Pacific create a single operation zone or military theater in which they confront the US strategic forces. For the Russian conventional forces, the Arctic is an area where they should mainly protect Russia's economic interests and state borders (land, maritime, and air). From an operative-tactical point of view, the Arctic is split into several sectors that represent various zones of responsibility. In the Western sector, the Russian land and air forces confront the NATO (Norwegian) troops, while the conventional component of the Northern Fleet mainly protects Russia's economic interests in the Barents Sea and provides nuclear forces with support/auxiliary services. The Northern Fleet and the Murmansk Command of the Border Guards are responsible for the protection of the NSR and the Arctic Ocean's coastline, while the Pacific Fleet and the Petropavlovsk-Kamchatsky Command of the Border Guards control the Bering Sea, Bering Strait, and access to the Chukchi Sea (Konyshev and Sergunin 2014a, b).

Given that the Soviet-era military machine degenerated significantly in the 1990s and early 2000s, the Russian conventional forces badly needed modernization in order to effectively meet new challenges and threats. The main idea behind the modernization plans is to make the Russian armed forces in the Arctic more compact and better equipped and trained (Konyshev and Sergunin 2014a, b). To reorganize in a more efficient way, the Russian land forces in the Western part of the AZRF planned to transform the motorized infantry and marine brigades located near Pechenga (Murmansk region) to the Arctic special force unit, with soldiers trained in a special program and equipped with modern personal equipment for military operations in the Arctic. It was initially planned that the Arctic brigade should be operational by 2016. There were also plans to create another Arctic brigade somewhere in the Arkhangelsk region. All conventional forces in the AZRF should form an Arctic

Group of Forces (AGF) to be led by the joint Arctic command (to be established in 2017).

However, the Ukrainian crisis has forced adjustments to Russia's military planning. While two Pechenga-based brigades were left in place, the Arctic brigade was surprisingly created ahead of schedule (in January 2015) and deployed in Alakurtti, which is close to the Finnish-Russian border. Another surprise was that, given an "increased NATO military threat" in the North, President Putin decided to accelerate the creation of a new strategic command, "North," which was established in December 2014 (three years ahead of the schedule). It was also announced that the second Arctic brigade will be formed soon and will be stationed in the Yamal-Nenets autonomous district (east of the Ural Mountains in the Arctic Circle).

Other measures included the deployment of the Pantsir S-1 short-range air defense system on the Kola Peninsula, plans to replace S-300 long-range air defense system with a more advanced S-400 'Growler' system, tactical training for fighter jet pilots in Arctic conditions, plans to establish 16 deep-water ports, 10 air defense radar stations, and 13 airfields along its Arctic periphery.

On one hand, the US strategic forces have, like Moscow, a vision of the Arctic as a separate military theater in global containment of Russia. On the other hand, North Atlantic is seen as a part of the European military theater, while the North Pacific's "second identity" is the Asia-Pacific military theater, with an emphasis on preventing supposed North Korean and Chinese threats.

The US conventional/Army forces in Alaska (USARAK) include the 1st Stryker and 4th Airborne Brigade Combat Teams, the 17th Combat Sustainment Support Battalion, Army Field Support Battalion-Alaska, the 59th Signal Battalion, the US Army's Northern Warfare Training Center (NWTC), a Noncommissioned Officer Academy, the Army National Guard's 297th Battlefield Surveillance Brigade and 1st Battalion, the 207th Aviation Regiment, and Army Reserve Component units (U.S. Army 2016a, b).

Most of these units are oriented to the Pacific theater rather than the Arctic military theater. Only the NWTC is designed for the Arctic missions. It provides Cold Weather and Mountain Warfare training to US military and designated personnel in order to enhance war-fighting capabilities of US and coalition partners. On order, the Center assists in SAR operations worldwide. It also conducts missions in support of a memorandum of agreement between US Department of Interior and USARAK for high altitude rescue vicinity Denali National Park in Alaska (U.S. Army 2016b).

Canada's military modernization programs are also focused on making the armed forces capable of fulfilling functions such as ascertaining Canada's sovereignty over its Arctic sector and protecting Ottawa's economic interests in the region. Canada's Arctic land forces are mostly formed by the Rangers, a lightly armed paramilitary force with a patrol and reconnaissance role in northern Canada. Its size was increased from 4100 personnel in 2008 to 5000 by 2012, and it received new equipment and weapons.

Other Canadian land forces receive basic cold weather training and have cold weather personal equipment. Since the 1950s a small military base has been located at Alert on Ellesmere Island, Nunavut, in the extreme north of Canada, facing Greenland. To improve Arctic training, a special Arctic training base was set up at Resolute Bay, Nunavut, in 2007.

As for the air force, Ottawa has plans to replace its 80 F/A-18 combat aircraft, which are stationed in south-east and central Canada and regularly deployed in the Arctic region, with 65 F-35 Joint Strike Fighters (JSFs) from 2020. Their purchase has repeatedly been linked by the government to Russian long-range bomber aircraft operations over the Arctic (Wezeman 2012: 3).

The Canadian Navy has 15 major surface warships and four conventional submarines, which have enough range to operate in the Arctic Ocean but are not designed to ply Arctic waters. The only ships that can navigate in thin ice are the two old (1969) *Protecteur*-class Auxiliary Oil Replenishment and the Kingston-class Maritime Coastal Defense Vessel (Lasserre et al. 2012: 46).

The National Shipbuilding Procurement Strategy was launched in 2010. It was developed in an effort to renew the fleets of the Canadian Navy and the Coast Guard. The strategy was broken into three sections: the combat package, the non-combat package and the smaller vessel package. The combat package (worth $30 billion) includes five or six vessels from the Arctic Patrol Ship Project and up to 15 vessels from the Single-Class Surface Combatant Project (Government of Canada 2016). In September 2015, the Irving Shipbuilding Company started production of Canada's first Arctic Offshore Patrol Ship (OPV), which should be delivered in 2018 (http://armscom.net/news/irving_shipbuilding_begins_construction_of_first_arctic_offshore_patrol_ship_aops_for_canada).

To develop naval infrastructure, the Nanisivik Naval Facility is being constructed on Baffin Island, Nunavut. The station will be built at the former lead-zinc mine site near the former company town of Nanisivik. It is expected that the port will be operational by 2018 (De Souza 2015).

The Nordic nations also emphasize the development of land, air, and naval forces capable of protecting their sovereignty over and economic interests in the High North. For example, Denmark has a small Frømandskorps (frogman corps) special forces unit and a small military (sledge) patrol force (Sirius) on Greenland (Wezeman 2012: 6). The Arctic Military Command, with headquarters in Nuuk, Greenland, was established in 2014 to merge the Greenland and Faroe Islands commands.

Similar to Russia, Oslo decided to create a special Arctic unit on the basis of the mechanized infantry battalion located in Skjold, Trøms County (Norwegian Ministry of Defense 2012). Over the last decade, has Norway moved major military command structures to the North: the armed forces headquarters—to Reitan, the land forces headquarters—to Bardufoss, and the Coast Guard headquarters—to Sortland (http://mil.no/organisation/about/norwegianjointheadquarters/; Wezeman 2012: 7).

Plans to modernize the Royal Danish air force include a potential deployment of F-16 combat aircraft to Greenland. Norway has even more ambitious plans: in 2008 Oslo decided to buy 52 brand new F-35 JSFs (a program worth $10.4 billion) to replace its 60 F-16th (Norwegian Ministry of Defense 2012, 2014). The aircraft should be delivered in 2017–2024. However, experts are skeptical about the feasibility of the project (it is being implemented with numerous delays) and about the ability of F-35 fighters to efficiently operate in the Arctic conditions (Wezeman 2012: 7).

The real Arctic capability lies with the six P-3 long-range maritime patrol aircraft. However, these planes are now over 20 years old and, while they are to be modernized, no plans have yet been announced for a replacement.

Both Copenhagen and Oslo are modernizing their navies and coast guards to navigate icy waters. Two Danish ice-strengthened *Knud Rasmussen* class OPVs, which were commissioned in 2008–2009, are dedicated for patrols off Greenland and a third one is planned for 2017 (McGwin 2013).

The Norwegian Navy replaced its five small frigates with five much larger and more capable *Fridtjof Nansen*-class frigates in 2011. Because of their size and equipment, the new frigates are much more able to operate in Arctic waters, as are Norway's six *Ula*-class submarines (which should be modernized in the near future). Six *Skjöld*-class patrol boats, which are new, large and very fast (maximum speed of 60 knots) stealth missile crafts, are now in active service. The *Svalbard*, an Arctic Offshore patrol vessel that was built specifically for Arctic operations, was commissioned in 2002 (Norwegian Ministry of Defense 2012; Lasserre et al. 2012: 41–42). Norway also operates a large 'research ship' with electronic and signals intelligence equipment, which is capable of operations in thin ice. Currently, the fourth generation of the vessel, called *Marjata*, which was commissioned in 2016, is in active service (https://newinform.com/9990-maryata-morskoy-razvedchik-nato).

A controversy emerged regarding the new OPVs launched by Denmark and Norway. Some experts have questioned why, if their main tasks were constabulary, they were armed like warships and so well equipped (Lasserre et al. 2012: 42–43). For example, *Fridtjof Nansen*-class frigates are equipped with the *Aegis* multi-target air tracking system (with BMD capabilities). The *Svalbard* is NBC-proof (nuclear, bacteriological, and chemical), which appears unnecessary unless it was designed to operate in a hostile maritime environment, but the ship is also is equipped for fire-fighting and environmental protection.

According to the strategists from the Arctic states, these military modernization programs aim only to update their armed forces and make them better equipped to cope with new challenges in the High North, rather than to provide them with additional offensive capabilities.

11.3 Changing Roles of Military Power in the High North

Along with the above traditional functions, the modern military has new roles in the High North. Among the relatively new roles that the armed forces acquired in the post-Cold war era, the primary one was the mission to ascertain coastal states' *sovereignty* over their exclusive economic zones (EEZs) and continental shelves in the Arctic Ocean. This mission emerged only in the post-Cold-War era when the major polar players have signed and ratified the UN Convention on the Law of the

Sea (UNCLOS) of 1982.[1] All littoral states have special sections on this issue in their Arctic strategies.

For example, the Norwegian Arctic strategy says: "Importance is also attached to maintaining a visible military presence that has the relevant capacity to exercise sovereignty, safeguard our sovereign rights in our 200-mile zones, exercise authority, and carry out surveillance, intelligence and crisis management" (Norwegian Ministry of Foreign Affairs 2009).

The Canadian Arctic doctrine echoes the Norwegian one: "The Government of Canada is firmly asserting its presence in the North, ensuring we have the capability and capacity to protect and patrol the land, sea and sky in our sovereign Arctic territory. We are putting more boots on the Arctic tundra, more ships in the icy water and a better eye-in-the-sky" (Minister of Public Works and Government Services Canada 2009).

Former Canadian Prime Minister Stephen Harper repeatedly noted that the first rule of Arctic sovereignty is "use it or lose it" and his government "intends to use it."[2] To this end, Canada started annual summer military exercises entitled Operation Nunalivut in its northern territories, which are explicitly designed to project Canadian sovereignty in the High Arctic.

The US national strategy in the Arctic region has the same postulate: "Our highest priority is to protect the American people, our sovereign territory and rights, natural resources, and interests of the United States" (President of the United States 2013).

Unsurprisingly, the Kremlin aims "…to ensure the sovereign rights of Russia's Arctic and features the smooth implementation of all of its activities, including the exclusive economic zone and the continental shelf of the Russian Federation in the Arctic…" (Putin 2013).

Denmark has a specific situation in terms of enforcing its sovereignty rights in the High North. Being represented by Greenland in the Arctic, the Kingdom of Denmark still retains its prerogatives in the defense area, regardless of the self-rule status of the island. According to the Danish Arctic strategy, "Sovereignty enforcement is the primary task of the Danish Armed Forces in the Arctic and the level of presence in the area is determined accordingly. Units from the army, navy and air force carry out tasks in the Arctic. They undertake surveillance and enforcement of sovereignty of Greenland and Faroese territorial waters and air space, as well as the Greenland exclusive economic zone and the fishing zones to ensure that no systematic violations of territory can take place. Likewise, the Sirius Patrol oversees the National Park in Northeast Greenland and enforces sovereignty there" (Danish Ministry of Foreign Affairs 2011: 21).

In contrast to the coastal states, Finland, Iceland, and Sweden do not push forward sovereignty-related issues in their Arctic strategies and do not mention their armed forces in this context. They prefer to focus on issues, such as climate change's implications, environment, indigenous peoples, and the prospects for multilateral

[1]The US has signed but not ratified the UNCLOS, although Washington de facto observes most of the Convention's rules.

[2]http://www.canada.com/topics/news/story.html?id=7ca93d97-3b26-4dd1-8d92-8568f9b7cc2a.

cooperation in the Arctic. The Finnish Arctic strategy of 2013 barely mentions the need to monitor security developments in the High North, maintain the Finnish armed forces' preparedness at high level, or Nordic defense cooperation in the NORDEFCO framework (Prime Minister's Office 2013: 40–41).

Another "new" mission of the armed forces is to protect the Arctic countries' *economic interests* in the High North, including mineral and bio-resources, fighting smuggling and poaching. The melting of the northern polar ice has dramatically altered this once-static geographic and oceanic region and is responsible for the new-found profitability and geostrategic/geoeconomic relevance of the region. Access to oil, gas, minerals, fish, and transportation routes, formerly locked in by thick ice, are for the first time becoming accessible and viable sources of profit. These resources include an estimated 13% of the world's undiscovered oil, 30 percent of its undiscovered gas, and around a trillion dollars' worth of minerals, including gold, zinc, palladium, nickel, platinum, lead, rare-earth minerals, and gem-quality diamonds (U.S. Coast Guard 2013: 7).

In addition to mineral reserves, the Arctic possesses abundant bio-resources. More than 150 fish species can be found in Arctic waters, including important varieties for international commercial fishing, such as herring, cod, butterfish, haddock and flatfish. The Arctic Zone of the Russian Federation (AZRF) produces 15% of Russia's seafood (Kochemasov et al. 2009). The Bering Sea and adjacent waters provide for a half of the US fish catch (Conley 2013: 61; US Coast Guard 2013: 7). The region is also populated by some unique animal species such as the polar bear, narwhal, walrus, and white whale.

With greater accessibility to the Arctic region and its abundant resources come both new opportunities for multilateral cooperation and the potential for regional competition and dispute, particularly conflicting territorial claims and managing maritime resources. Protracted disagreement among the Arctic littoral states could cause individual Arctic nations to become increasingly assertive in their resource and territorial claims, which has the potential to lead to the militarization of the Arctic. Although this scenario would appear unlikely, many coastal states believe that it is critical to articulate their strategic interests in the Arctic region and develop a sufficient military potential and plans of action to ensure their leadership in this evolving region to both anticipate challenges and offer multilateral and transparent resolution to these challenges (Conley and Kraut 2010: 3).

Some Arctic states have elevated the armed forces' mission to protect their economic interests in the High North to the level of their national doctrines. For example, both Russian Arctic strategies of 2008 and 2013 set a goal to make the AZRF a strategic resource basis that should be protected from foreign powers' encroachments (Medvedev 2008; Putin 2013). The Russian Security Council's Secretary Nikolay Patrushev explained the need to tighten control over the AZRF and its external borders by other polar players' increased activities in the region.[3]

The volume of illegal, unreported, and unregulated (IUU) fishing is concerned has reached a significant scale in the region; it amounted to approximately 1.3 million tons

[3]http://www.dni.ru/economy/2008/9/12/149006.html.

a year in the Bering Sea enclave (Zilanov 2016: 48). It is estimated that the fish caught in Russian waters exceeds the official quota by at least 150% (The International Bering Sea Forum 2006). Overfishing creates numerous ecological problems in the region. According to some accounts, intensive trawling has led to species such as crab and perch being in serious decline in the entire Bering Sea, while the stocks of pollock fluctuate in an unpredictable manner from year to year. The once-plentiful pollock have had especially dramatic declines on the Western (Russian) side of the Bering Sea because of illegal fishing. In the Eastern (US) Bering Sea, harvests of snow crab have declined by 85% since 1999 (The International Bering Sea Forum 2006). This is because poaching is rampant and the Russian organized crime is heavily involved in the fish trade. The Russian "fish, crab, and caviar mafias" not only aim to expand their commercial activities and sideline their foreign rivals, but also to establish control over the regional governments and federal agencies in the Russian Far North and East.

Illegal fishing in the Barents Sea constituted at least equal threat to fish stocks, although the scale of IUU fishing there was lower than that in the Bering Sea. Norway continues to object to Russian fishing around Spitsbergen. Since Norway introduced a 200-mile economic zone around the archipelago, it has regarded such fishing as poaching. Forcible arrests of Russian trawlers by the Norwegian navy have become more frequent. As Russia does not recognize the aforementioned decision by Norway and considers this area open to international economic activity, in 2004 Russia's Northern Fleet started regular patrols of the waters around Spitsbergen. Norway particularly objected to this move, viewing it as a sign of Russian imperial ambitions and of Moscow's unwillingness to cooperate with Oslo to settle maritime and economic disputes.

Given the continuation of ice melting in the High North and the opening of the Arctic maritime routes for navigation for several months a year, Canada and Russia are concerned about the possibility of the growth of smuggling activities along their Arctic Ocean's coastline and Northern Sea Route (NSR) and Northwest Passage (NWP). Both Canada and Russia have plans to further develop their border and coast guards in the region.

The Arctic states are concerned about smuggling, not only from outside the region (a threat that remains hypothetical for the time being), but also between the Arctic nations themselves, which is already a reality. For example, in 2012 a narwhal smuggling ring was disclosed by the Canadian and US law enforcement agencies. Between 2000 and 2010, an American family purchased the tusks legally in northern Canada and then used the Internet to find buyers in the US. This family is estimated to have sold between $400,000 and $1 million worth of tusks to as many as 150 buyers (McGwin 2015). It should be noted that Narwhals are protected under various national and international treaties. The Convention in International Trade in Endangered Species of Wild Fauna and Flora bans their hunting in Canada and Greenland by anyone other than the Inuit. Canadian tusks may be sold domestically or shipped abroad to countries where their sale is legal. Imports of tusks to the US are banned under the federal Marine Mammal Protection Act.

To prevent or reduce poaching, overfishing, and smuggling in the region, an Arctic Coast Guard Forum was established in October 2015. A joint statement formally established the operationally focused, consensus-based organization with the purpose of leveraging collective resources to foster safe, secure, and environmentally-responsible maritime activity in the Arctic (Fonseca 2015).

On July 16, 2015 the so-called "Arctic five" countries (Canada, Denmark, Norway, Russia and the US) signed a "Declaration concerning the prevention of unregulated high seas fishing in the Central Arctic Ocean" (2015) in Oslo. The agreement came at a time when there was no commercial fishing in international waters in the Central Arctic Ocean (CAO) and was presented as a precautionary measure. The idea behind the agreement, including its focus on need for further scientific research and its application of international law, is in line with what most Arctic fisheries stakeholders agree on. Negotiations regarding a mandatory agreement are underway and some other countries with global fishing interests and capabilities (China, Iceland, Japan and South Korea) have been invited. Again, Arctic nations' coast guards are in charge with enforcing the IUU fishing ban regime in the CAO.

Illegal migration is one more security challenge for the Arctic states. Over the last decade, Canada registered several cases of illegal migration. For example, Romanian citizens traveled from Greenland to Canada's Ellesmere Island by motor boat before trying to fly to Toronto. Several Turkish sailors have illegally left their ship in the Canadian port of Churchill in attempt to travel by train to Winnipeg (Gudev 2014).

The case of Arab refugees traveling to the Nordic countries via the polar routes is a much more serious security threat. Since the beginning of 2015, at least 29,000 people, mostly from Syria, have used various routes to seek asylum in Norway. The number of asylum seekers arriving plummeted by 95 per cent in 2016.[4] The refugees being bussed to Russia had taken the so-called "Arctic Route" through Russia, crossing the Norwegian border by bicycle as Russia does not allow anyone to cross by foot. In November 2015, Oslo announced it would deport people who had arrived from a safe country. The Norwegian government considers Russia safe, but has not given the refugees opportunity to appeal the decision. The Norwegian authorities have started sending the first of approximately 5500 mainly Syrian refugees, who have been housed in a transit camp in the north of the country, back to the Russian border they crossed in 2015.[5] Critics of the government have said the attempts to return refugees to Russia put them at risk and contravenes European human rights. Although Norway is not an EU member, it is in the border-free Schengen zone.

In addition to deportation, in 2016 Norway started to build a steel fence at its border with Russia to prevent a further influx of refugees.

As far as Finland is concerned, approximately 32,500 refugees, mostly natives of Iraq, came to the country in 2015.[6] Most of them arrived through the Swedish-Finnish border in the north of the country, but many others chose other routes, such

[4]http://www.independent.co.uk/news/world/europe/norway-border-fence-russia-refugees-refugee-crisis-schengen-syria-war-a7208806.html.

[5]http://www.euronews.com/2016/01/20/norway-sends-syrian-refugees-back-to-russia.

[6]http://tass.com/politics/852479.

as traveling by ferry from Germany and across the border with Russia in northern Lapland. The Russian-Finish border has become one of the main routes that refugees use to get to Finland. The influx of refugees from Russia to Finland increased after Norway tightened security measures on its borders and made the procedure of asylum application more difficult in late 2015, prompting migrants to seek alternative paths to enter the EU's borderless Schengen area and to get asylum there. In January 2016, Finland registered 500 asylum seekers arrivals from Russia against 700 border crossings throughout 2015.[7]

In March 2016, Russia and Finland agreed to introduce temporary restrictions at two checkpoints on their border, Salla and Raja-Jooseppi, for citizens of third countries. Similar to Norway, the asylum seekers were sent back to Russia, where they have valid residence permits. The Finnish Ministry of Interior said that the measures have aimed to curb undocumented migration and related threats and enhance the effectiveness of measures taken by both Helsinki and Moscow to combat illegal migration.

A threat of *international terrorism,* including the nuclear one, is seen by the Arctic states as a real danger. For example, one of the US Coast Guard's Arctic strategy's aims is to prevent terrorism, although it does not specify the source and nature of a terrorist threat (US Coast Guard 2013: 34).

Some experts believe that a temporarily ice-free Northwest Passage during the summer will expose Canada to new vulnerabilities. Increased commercial and tourist traffic may also increase illicit transportation of drugs and terrorists, requiring robust patrolling, monitoring, and emergency response capabilities. Many Canadian experts recommend establishing new rules to require all ships to register their presence in northern Canadian waters. "We're saying that all vessels—no matter what size and what they carry—should have to report to Canadian authorities," Liberal senator and committee chair Bill Rompkey. "The threat is not just oil spills and not just commercial vessels moving through. The threat is drugs and the threat is terrorism. And we've got to counteract that" (cited in: Conley and Kraut 2010: 17–18).

The Canadian government has made significant commitments to acquire urgently needed equipment to prepare the Coast Guard for any threats that may arise. The Canadian government has promised to spend billions of dollars building six to eight offshore patrol vessels capable of breaking up first-year ice, plus the construction of a new icebreaker, the CAD720 million (USD675 million) *John G. Diefenbaker*, which was scheduled to replace the aging *Louis St. Laurent* in 2017; a new Canadian Forces winter fighting school at Resolute Bay in the Northwest Passage; and an estimated $100 million to build a new naval base at the existing deep-water port Nanisivik on Baffin Island (Government of Canada 2010: 8–11). *Canada First Defense Strategy* proposed the acquisition of 10 to 12 maritime patrol aircraft to replace the Aurora fleet starting in 2020 to become part of a broader surveillance system, which will include sensors, unmanned aerial vehicles, and satellites (National Defense and Canadian Forces 2008: 17). However, critics have argued that neither the pledged icebreaker

[7]https://sputniknews.com/world/201604081037708246-finland-russia-border-restrictions/.

nor the promised Arctic patrol ships have progressed much beyond the announcement phase (Conley and Kraut 2010: 17–18; Lasserre et al. 2012: 46).

Russia shares the same security concerns with other Arctic nations. According to a statement in 2010 by the head of the FSB's Border Service, Vladimir Pronichev, the main challenges for the Russian Border Service were the unauthorized presence of foreign ships and research vessels in Russian Arctic waters, illegal migration, drug smuggling and poaching.[8] Terrorist attacks against oil platforms were also seen as a potential threat to security in the Arctic (Vasiliev 2012: 14). Based on these perceived security risks, Russia again began to prioritize the protection of Arctic borders and the strengthening of the Border Service in the region. This return to a focus on Arctic border protection has been reiterated by Presidents Medvedev and Putin on a number of occasions.[9]

An Arctic border guards unit was created as early as in 1994 with the aim of monitoring the circulation of ships and poaching at sea. The unit was reorganized in 2004–2005. In 2009, it was announced that new Arctic units had been established in border guard stations in Arkhangelsk and Murmansk. Furthermore, the FSS has established two new border guard commands: one in Murmansk for the western AZRF regions, and one in Petropavlovsk-Kamchatsky for the eastern Arctic regions. Now the border guards are assigned with the task of dealing with the new (soft security) threats and challenges such as the establishment of reliable border control systems, the introduction of special visa regulations to certain regions, and the implementation of technological controls over fluvial zones and sites along the NSR. The latter is currently controlled from the air by border guard aircrafts, and on the land and sea by the North-Eastern Border Guard Agency; the Russian border guards further plan to establish a global monitoring network from Murmansk to Wrangel Island. All in all, Moscow plans to build 20 border guard stations along the Arctic Ocean's coastline (Klimenko 2016: 14–15; Zagorsky 2013).

Another interesting structural change is an ongoing reorganization of the Russian Coast Guard (part of the Border Service). Now the Coast Guard has a wide focus in the Arctic: in addition to the traditional protection of biological resources in the Arctic Ocean, oil and gas installations and shipping along the Northern Sea Route are among the agency's new top priorities. There are plans to equip the Coast Guard in the AZRF with the brand new vessels of project 22100. The Ocean-class ice-going patrol ship, the Polyarnaya Zvezda (Polar Star), is currently undergoing sea trials in the Baltic Sea. Vessels of this class can break up to 31.4 inch-thick ice. They have an endurance of 60 days and a range of 12,000 nautical miles at 20 knots. They are equipped with a Ka-27 helicopter and can be supplied with Gorizont UAVs (unmanned aerial vehicles).

Some Arctic nations (especially Russia) are seriously concerned about the threat of nuclear terrorism. Moscow is afraid that not only the industrial infrastructure or oil platforms, but also nuclear power plants and nuclear waste storages, could be potential targets for terrorists. There are two nuclear plants—Kola and Bilibin—in the AZRF.

---

[8]http://www.rg.ru/2010/06/02/pronichev.html.

[9]http://www.ng.ru/economics/2011-08-31/4_arctic.html.

Most notably, more than 200 decommissioned nuclear reactors from submarines and icebreakers from the Soviet period are stored on the Kola Peninsula from the Soviet period. In 2016, Russia launched a large-scale program for removing nuclear waste from the former Soviet submarine base in Andreev Bay in the Murmansk region. A total of 22,000 containers of spent fuel from nuclear submarines and icebreakers were stored in three storage tanks. There were also approximately 18,000 cubic meters of solid waste and 3400 cubic meters of liquid radioactive waste, which, according to Norwegian sources, are collectively as radioactive as 5000 Hiroshima bombs.[10] These nuclear facilities must be reliably protected to prevent potential terrorist attacks.

Another new trend is the development of the *dual-use potential* of the military, including SAR operations, monitoring air and maritime spaces, providing navigation safety, mitigating natural and man-made catastrophes (such as responses to oil spills), etc.

For instance, American experts acknowledge that building US military capacity in the Arctic extends beyond pure acquisition and procurement. The US Navy and US Coast Guard have recognized certain capability gaps that must be filled, chief among which is search and rescue. The sudden and substantial increase in Arctic commercial shipping and tourism poses significant challenges to the existing SAR infrastructure. Given the location of current US Coast Guard operating bases, it could take coast guard aircraft several hours, and coast guard cutters a few or several days, to reach a ship in distress in Arctic waters. To enable specialized training for enhanced SAR capabilities, the US Coast Guard would need to improve or create new operating bases in the region; procure additional Arctic-capable aircraft, cutters, and rescue boats; and add systems to improve Arctic maritime communications, navigation, and domain awareness (Conley and Kraut 2010: 10).

To increase SAR and monitoring capabilities, Canada decided to go ahead with Radarsat Constellation Mission (RCM), which will see the launch of at least three satellites by 2018. All three satellites will be designed to gather radar-imaging data. At present, the Royal Canadian Air Force operates a single radar-imaging satellite, Radarsat-2, which provides certain maritime surveillance data, but this is insufficient to give a complete picture in the region. According to Colonel Andre Dupuis, the Department of National Defense's director of space requirements, "Three satellites will give us a complete picture every single day of every ship in our area of responsibility, all the way out to about 2000 nautical miles."[11]

Canada's Department of National Defense is also funding an Automatic Identification System (AIS) package for installation on the RCM. For navigational safety, the International Maritime Organization (IMO) requires ships larger than 300 tons to carry an AIS beacon, which enables other ships or land-based receivers to track a vessel's identity, speed, and course. The Department of National Defense is funding the AIS sensor design and its integration into RCM, currently estimated at $55 million.

[10]http://sputniknews.com/environment/20160610/1041126139/russia-norway-arctic-nuclear-waste.html.

[11]http://www.spacedaily.com/reports/Canada_builds_up_arctic_maritime_surveillance_999.html.

The five Nordic nations decided to create a joint maritime monitoring system in the NORDEFCO (Nordic Defense Cooperation) framework. As the 2009 Stoltenberg reports noted, a Nordic system should be established for monitoring and early warning in the Nordic sea areas (Stoltenberg 2009: 12–14). In principle, the system should be civilian and designed for tasks such as monitoring the marine environment and pol-lution and monitoring of civilian traffic. The existing military surveillance systems are not specifically designed to carry out these tasks. Nevertheless, the military systems collect a great deal of information that is relevant for civilian maritime monitoring; therefore, an overall Nordic system will only be truly effective if it is coordinated with and can exchange data with military systems.

According to the Stoltenberg report, once a Nordic maritime monitoring system is in place, a Nordic maritime response force should be established, consisting of elements from the Nordic countries' coast guards and rescue services. This force should patrol regularly in the Nordic seas, and one of its main responsibilities should be SAR operations (Stoltenberg 2009: 15–16).

By 2020, a Nordic polar orbit satellite system (with no fewer than three satellites) should be established in connection with the development of a Nordic maritime monitoring system. Such a satellite system could provide frequently updated real-time images of the situation at sea, which is essential for effective maritime monitoring and crisis management (Stoltenberg 2009: 17–18).

Russia believes that by improving NSR infrastructure and safety, this maritime route will be attractive not only for Russian business but also for foreign shipping companies. The construction of 10 search and rescue centers along the NSR by 2018 (in addition to three already operation SAR centers) will be helpful in promoting this route internationally. Also, as the Yamal LNG (liquefied natural gas) plant becomes operational in 2017, LNG shipments from Sabetta to East Asia (and potentially to Europe and North America) will be facilitated.

It should be also noted that Russia's modernized military infrastructure in the Arctic, including the Soviet air and naval bases that have been reopened over the last years, is of dual-use nature. Such an infrastructure can be used not only for military but also for civilian purposes, including SAR operations.

In general, all the power structures of the Arctic nations (army, navy, border and coast guards, and agencies dealing with emergency situations) are charged with implementing the Arctic Council's (AC) agreement of 2011 on the creation of a Maritime and Aeronautical Sea and Rescue System. Each country is responsible for its own sector of the Arctic and Russia has the biggest one. The SAR agreement's signatories undertake joint exercises on the regular basis. As many experts believe, the SAR activities are a clear sign of the shift from the armed forces' purely military functions to the soft security missions.

Arctic research has become one of the important missions of the military as well. For example, the Russian Navy was very helpful in preparing Moscow's second submission to the UN Commission on the Limits of the Continental Shelf (UNCLCS) in 2015. Over the last decade, the Russian Navy has sent several expeditions to Franz Josef Land, Severnaya Zemlya, the Novosibirsk Islands archipelago, and Wrangel Island. For example, the objective of the Russian Navy's mission within the framework of the expedition *Arktika-2012* was to prove that its landmass extends to the North Pole by drilling into the sea floor (2.5–3.0 km depth) to collect rock samples for scientific analysis. In September 2012, the *Kalitka*, a *Losharik* class nuclear-powered auxiliary submarine, was used to guide the *Kapitan Dranitsyn* and *Dickson* ice breakers in drilling three boreholes at two different sites on the Mendeleev ridge, collecting over 500 kg of rock samples (Mikhailov and Voloshin 2012). The Navy has also shared the bathymetric data with civilian scholars to substantiate the Russian submission to the UNCLCS.

The declassified bathymetric soundings acquired by US Navy submarines from cruises between 1993 and 2005 helped create an International Bathymetric Chart of the Arctic Ocean (Jakobsson et al. 2012: 3). This data was later used by the Geological Survey of Denmark and Greenland to prepare the 2014 Danish submission to the UNCLCS because Copenhagen lacked technical capabilities of its own to make bathymetrical research for this purposes.

The military power carries out some *symbolic functions* for the Arctic nations. For example, for the five Nordic countries the above-mentioned NORDEFCO project symbolizes their Nordic solidarity. To demonstrate such solidarity, these countries decided to take the responsibility for air surveillance and patrolling over Iceland after the US withdrawal from Keflavik airbase in 2006.

Thorvald Stoltenberg even suggested that the Nordic governments should issue a mutual declaration of solidarity in which they commit to clarifying how they would respond if a Nordic country were subject to external attack or undue pressure. In other words, such a declaration should contain a security policy guarantee (Stoltenberg 2009: 34). However, this initiative was not supported by non-NATO and non-aligned Nordic countries, such as Finland and Sweden, which did not want to put themselves in an awkward position between NATO and Russia.

For Sweden, its armed forces and a rather developed military-industrial complex are symbols of self-sufficiency and self-reliance in security affairs, the guarantee of its non-aligned status.

For Russia, deployment of significant forces in the region and development of the military infrastructure in the High North shows that the country retains its great power status and still has world-class military capabilities.

Some nationalistic authors have put forward a more spiritual view of the role of the High North in the construction of Russian identity and the pursuit of its traditional messianism. For instance, in his *The Mysteries of Eurasia*, Dugin (1991) elaborated a cosmogony of the world in order to make Siberia, the last "empire of paradise" after Thule, the instrument of his geopolitical desire for a domination of the world, justified by Russia's "cosmic destiny". This group of theorists claims that the North is not only Russia's strategic resource base (as stated by the Kremlin), but also its territory of the spirit, of heroism, and of overcoming, a symbolic resource of central importance for the future of the country (Laruelle 2014, 39–43). The Arctic is presented as Russia's "last chance," and as a possible way to take "revenge on history." The Arctic is portrayed as rightful compensation for the hegemony that was lost with the disappearance of the Soviet Union.

11.4 Cooperative Agenda

Given that the 'hard' security situa-tion in the Arctic is relatively benign, serious threats and chal-lenges such as WMD (weapons of mass destruction) prolifera-tion, large-scale ter-rorist attacks, or military conflicts are hardly prob-able in the region—at least in the foreseeable future.

However, it should be noted that the Arctic lacks a special arms control regime. Only two international arms con-trol regimes have been applicable to the area. The first was a system of the US-Soviet/Russian strategic arms control and reduction agreements that regulated a number of launchers and nuclear warheads on the Russian strategic submarines based on the Kola Peninsula. The second regime was the Conventional Forces in Europe (CFE) treaty, which was concluded between NATO and the Warsaw Pact in 1990 and adapted in 1999 under the aegis of the OSCE. However, the Baltic States refused to abide by the treaty because it was concluded when they were still part of the USSR. Finland and Sweden have also refused to sign the treaty referring to their neutral (now non-aligned) status. In addition, none of the Western signatories of the 1999 Adaptation Treaty ratified it. As a result, Russia suspended its participation in the treaty in 2007.

However, Moscow hopes that the CFE process can be reanimated in the foreseeable future. Drawing lessons from past negative experiences, Russia believes that there are two preconditions for resumption and successful continuation of the CFE process. Firstly, a new treaty should be fully ratified by all signatories. Secondly, all countries of the Arctic region should partake in this arms control regime.

It should also be noted that the CFE treaty was applicable only to land forces. Naval armaments were (and are) mainly excluded from any arms control regime. Unilateral measures were taken by some countries (including Russia) in the 1990s for the reduction of naval armaments and naval activities, but they related only to obsolete weapons and cannot be a substitute for a real arms control regime. According to some assessments, the basic hesitancy of the EU and NATO nations regarding naval arma-ment limitations in the High North seems to be that if you initiate naval arms control in one of the seas within their zone of responsibility, this could also lead to restrictions on maritime flexibility in other seas as well. However, these par-ties should initiate negotiations on naval arms con-trol if they are serious about further improvement of the security environment in the region.

It should be noted, with concern, that the Arctic region currently has no confidence- and security-building measures regime; this gap should be filled with great urgency because CSBMs development is a very impor-tant element of any regional security system. The regional CSBMs could be based primarily on the 1994 OSCE Vienna Document, which proved to be efficient in Europe. In addition, the following measures could be suggested:

- Given the specifics of the region, CSMBs should cover not only land, but also naval military activities.
- Along with spatial limitations, temporal limi-tations on Russian, NATO, and EU military activi-ties in the region could also be established.

- Military-to-military contacts, joint exercises, exchanges and visits should be further encour-aged. Since the eruption of the Ukrainian crisis in 2014, however, Russian representatives have no longer been invited to meetings of the Arctic Security Forces Roundtable. Joint military exercises such as Northern Eagle—an opera-tion involving US, Russian, and Norwegian forces—have been cancelled. Cooperation in military affairs has been disrupted almost entirely.
- The countries of the region should intensify exchange information on their military doc-trines, defense budgets as well as on major arms export/import programs.
- Not only regional but also bilateral CSBMs should be further encouraged.
- The establishment of a limited nuclear weapon-free zone in the Arctic (say, in Central Arctic) could be discussed. For example, Russia and US could consider Canada's initiative to ban nuclear weapons in the region. Russia has been positive about such ideas in the past (Moscow raised a similar idea under Mikhail Gorbachev), but has questions about the geographical scope of such a zone. Russia supports making the Arctic a nuclear weapon-free zone, provided this would not affect the Kola Peninsula, which is a home to two-thirds of Russia's strategic nuclear submarines.

Moscow also considers the field of civil protection as a promising venue for the Arctic regional cooperation. For example, according to the EU–Russia 2005 roadmap to the Common Space on External Security, one of the strate-gic objectives of Brussels–Moscow cooperation is to strengthen EU–Russia dialogue on promoting common ability to respond to disasters and emer-gencies, specifically including crisis management situations (Commission of the European Communities 2005). The positive experience accumulated in this area could be replicated to the Arctic regional cooperation. The priority areas for such cooperation could be as follows:

- Strengthening coordination of the Arctic states' agencies responsible for civil protection. This requires hard work in terms of implementing the existing arrangements between the Opera-tions Centre of Russia's EMERCOM (Ministry for Emergency Situations) and its foreign counterparts. More spe-cifically, this means exchanging contact details for keeping in touch on a 24-h basis; exchanging templates for early warnings and requests/offers for assistance; exchanging information during an emergency, where appropriate; conducting com-munications exercises on an agreed basis; and ena-bling operation staff to spend time at the operational center of the other partner's service in order to gain practical experience.
- Exchanging information on lessons learnt from terrorist attacks.
- Inviting experts, on a case-by-case basis, to specific technical workshops and symposia on civil protection issues.
- Inviting observers, on a case-by-case basis, to specific exercises organized by the partner countries.
- Facilitating mutual assistance in search and rescue operations for submarines, ships and air-craft in emergency situations.

Hopefully, a steady implementation of this rather ambitious agenda could substantially change the security environment in the Arctic region in a positive way.

11.5 Conclusions

The nature and roles of military power in the Arctic have been radically changed over the last quarter of a century. In contrast with the Cold War era, when military power was a coercive instrument in a global military confrontation between two superpowers and capitalist and socialist systems, it now has principally new functions.

First and foremost, the purpose of military power is to ascertain coastal states' sovereignty over their EEZs and continental shelves in the region, including disputable areas. Although the probability of an armed conflict between Arctic and non-Arctic players because of maritime disputes is not very high, the military power is still viewed as a tool to prevent such disputes from escalating to a dangerous phase.

Protection of the Arctic countries' economic interests in the North, including mineral and bio-resources, fighting smuggling and poaching, is also seen as another important mission of the armed forces.

Recently, illegal migration has become another security challenge for the Arctic states, which can be effectively met with the help of not only police, but also border and coast guards.

Moreover, present-day military structures should be prepared to prevent potential terrorist attacks against critical industrial and infrastructural objects, including oil and platforms, nuclear plants, and nuclear waste storages.

The military should be also ready to fulfill some dual-use functions, such as SAR operations, monitoring air and maritime spaces, providing navigation safety, and mitigating natural and man-made catastrophes.

Interestingly, Arctic research has become one of the important missions of the military because the Arctic states' air forces and navies have unique technical capabilities for doing this.

Finally, the military carry some symbolic functions. For example, in the case of the Nordic countries, military power can symbolize their Nordic solidarity (NORDEFCO project). For Sweden, its armed forces and a rather developed military-industrial complex are symbols of self-sufficiency and self-reliance in security affairs, the guarantee of its non-aligned status. For Russia, deployment of significant forces in the region and development of the military infrastructure in the High North shows that it retains its great power status and still has world-class military capabilities.

However, these new roles do not preclude military power from fulfilling its traditional functions, such as protection of national territory, power projection, deterrence, containment, etc. For this reason, maintaining strategic nuclear and conventional capabilities as well as their modernization remain important priorities for the US, Russian, and Nordic military policies in the High North.

While some media, politicians, and strategic analysts have portrayed the changes in the military capabilities of the Arctic states as significant military build-ups and even a renewed arms race in the region, the real picture is far from these apocalyptical scenarios. We can speak only about limited modernization and increases or changes in equipment, force levels, and force structure. Some of these changes—such as the creation of new Russian and Norwegian Arctic units strengthening of the Canadian Rangers, commissioning more sophisticated and better armed warships, the establishment of new existing command structures or moving existing ones to the north—have little or nothing to do with power projection into the disputed area; rather they are for patrolling and protecting recognized national territories that are becoming more accessible, including for illegal activities, including overfishing, poaching, smuggling, and uncontrolled migration. Other changes, such as the modernization of US, British, and Russian strategic nuclear forces, may have more to do with maintaining a deterrent potential rather than with developing offensive capabilities.

However, an increase of military capabilities in a region potentially pregnant with maritime and natural resource disputes gives some reasons for concern. In order to prevent potential conflicts and avoid misunderstandings, the Arctic states should be clear about their military policies and doctrines and should include CSBMs in their bilateral or multilateral relations in the Arctic.

The changing nature of military power entails its new uses and roles that should be reflected in military/security doctrines and taken into account by the expert/academic community. In general, these changes are conducive to the regional cooperation between the military, especially in areas such as SAR, emergency situations, air and maritime safety, charting safe maritime routes, cartography, etc.

References

Anishchuk A (2013) Russia needs Arctic presence to guard against U.S. threat: Putin. Chicago Tribune, 3 December. http://www.chicagotribune.com/news/sns-rt-us-al-russia-putin-arctic-20131203,0,5499339.story. Accessed 11 Dec 2016

Bender J (2014) The most isolated US military base could get a lot more important. http://www.businessinsider.com/importance-of-us-thule-air-base-2014–11. Accessed 14 Jan 2017

Commission of the European Communities (2005) Roadmaps to the EU–Russia common spaces. Commission of the European Communities, Brussels. http://europa.eu.int/comm/external_relations/russia. Accessed 11 Dec 2016

Conley H (ed) (2013) The new foreign policy frontier. U.S. interests and actors in the Arctic. CSIS, Washington, DC

Conley H, Kraut J (2010) U.S. strategic interests in the Arctic. CSIS, Washington, DC

Danish Ministry of Foreign Affairs (2011) Kingdom of Denmark strategy for the Arctic 2011–2020. http://um.dk/en/~/media/UM/English-site/Documents/Politics-and-diplomacy/Arktis_Rapport_UK_210x270_Final_Web.ashx. Accessed 11 Dec 2016

Declaration concerning the prevention of unregulated high seas fishing in the Central Arctic Ocean (2015). https://www.regjeringen.no/globalassets/departementene/ud/vedlegg/folkerett/declaration-on-arctic-fisheries-16-july-2015.pdf. Accessed 6 Jan 2017

De Souza M (2015) Canadian navy delays opening of crucial Arctic facility to 2018. Toronto Sun. http://www.torontosun.com/2015/03/02/canadian-navy-delays-opening-of-crucial-arctic-facility-to-2018. Accessed 14 Jan 2017

Dugin A (1991) Misterii Evrazii [The Mysteries of Eurasia]. Moscow: Arktogeia

Fonseca JR (2015) Arctic coast guard forum established, 31 Oct. http://www.marinelink.com/news/established-arctic-guard400213. Accessed 3 Jan 2017

Government of Canada (2010) Statement on Canada's Arctic foreign policy. Exercising sovereignty and promoting northern strategy abroad. http://www.international.gc.ca/polar-polaire/assets/pdfs/CAFP_booklet-PECA_livret-eng.pdf. Accessed 11 Dec 2016

Government of Canada (2016) Shipbuilding projects to equip the royal Canadian navy and the Canadian coast guard—national shipbuilding strategy: February 2012–December 2015 status report. http://www.tpsgc-pwgsc.gc.ca/app-acq/amd-dp/mer-sea/sncn-nss/rapport-report-20151231-3-eng.html#a1. Accessed 14 Jan 2017

Gudev P (2014) Rossiya i SShA v Arktike: Obshie Interesy i Protivorechiya [Russia and the USA in the Arctic: Common Interests and Tensions]. Institute of World Economy and International Relations, Moscow. http://sites.miis.edu/russianinitiative/files/2014/03/ARCTIC.pdf. Accessed 6 Jan 2017 (in Russian)

Jakobsson M, Mayer L, Coakley B et al (2012) The International Bathymetric Chart of the Arctic Ocean (IBCAO) Version 3.0. Geophys Res Lett 39(L12609):1–6. https://www.ngdc.noaa.gov/mgg/bathymetry/arctic/2012GL052219.pdf. Accessed 8 Jan 2017

Khramchikhin A (2011) Voyenno-Politicheskaya Situatsiya v Arktike i Stsenarii Vozmozhnykh Konfliktov [Military-Political Situation in the Arctic and Scenarios of Possible Conflicts]. Arktika i Sever [The Arctic and the North]. 2:1–15. http://narfu.ru/upload/iblock/7f9/grjopufvqlvzveqamx.pdf. Accessed 11 Dec 2016 (in Russian)

Khramchikhin A (2013) Stanet li Arktika Teatrom Voennykh Deistviy po Poslednemu Peredelu Mira [Will the Arctic Become a Military Theater for the Final Re-Division of the World?]. Arktika I Sever [The Arctic and the North]. 10:52–60. http://narfu.ru/upload/iblock/c9d/5.pdf. Accessed 10 Jan 2017 (in Russian)

Klimenko E (2016) Russia's Arctic security policy. Still quiet in the high north? SIPRI (SIPRI Policy Paper No. 45), Stockholm

Kochemasov YV, Morgunov BA, Solomatin VI (2009) *Ekologo-ekonomicheskaya Otsenka Perspectivy Razvitiya Arktiki* [Ecological-economic Assessment of Perspectives of the Arctic's Development]. http://www.perspektivy.info/rus/ekob/arktika_perspektivy_razvitija_2009–04–24.htm. Accessed 11 Dec 2016 (in Russian)

Konyshev V, Sergunin A (2014a) Is Russia a revisionist military power in the Arctic? Defense and Secur Anal 3:1–13

Konyshev V, Sergunin A (2014b) Russian military strategies in the high north. In: Heininen L (ed) Security and sovereignty in the North Atlantic. Palgrave Macmillan, Basingstoke, pp 80–99

Lasserre F, Le Roy J, Garon R (2012) Is there an arms race in the Arctic? J Mil Strateg Stud 14(3–4):43–57

Laruelle M (2014) Russia's Arctic Strategies and the Future of the Far North. M.E. Sharpe, Inc., Armonk, NY

McGwin K (2013) Danish navy to build new Arctic ship. http://arcticjournal.com/politics/313/danish-navy-build-new-arctic-ship. Accessed 3 Jan 2017

McGwin K (2015) American sentenced in narwhal smuggling ring. The Arctic J (12 Jan). http://arcticjournal.com/climate/1248/american-sentenced-narwhal-smuggling-ring. Accessed 3 Jan 2017

Medvedev D (2008) Osnovy Gosudarstvennoi Politiki Rossiiskoi Federatsii v Arktike na Period do 2020 Goda I Dal'neishuiu Perspektivu [Foundations of the State Policy of the Russian Federation in the Arctic up to and Beyond 2020]. http://www.rg.ru/2009/03/30/arktika-osnovy-dok.html. Accessed 11 Dec 2016 (in Russian)

Mikhailov A, Voloshin V (2012) Voennyi Atomny Batiskaf 'Losharik' Ispytali v Arktike [The Nuclear-Powered Military Bathyscaphe 'Losharik' was tried in the Arctic], Izvestiya, 27 Oct. http://izvestia.ru/news/538268. Accessed 6 Jan 2017 (in Russian)

Minister of Public Works and Government Services Canada (2009) Canada's northern strategy: our north, our heritage, our future. http://www.northernstrategy.gc.ca/cns/cns.pdf. Accessed 11 Dec 2016

National Defense and Canadian Forces (2008) Canada first defense strategy. Available at: http://www.forces.gc.ca/site/pri/first-premier/June18_0910_CFDS_english_low-res.pdf. Accessed 11 Dec 2016
Norwegian Ministry of Foreign Affairs (2009) New building blocks in the north, the next step in the government's high north strategy. http://www.regjeringen.no/upload/UD/Vedlegg/Nordområdene/new_building_blocks_in_the_north.pdf. Accessed 11 Dec 2016
President of the United States (2013) National strategy for the Arctic region (May). http://www.whitehouse.gov/sites/default/files/docs/nat_arctic_strategy.pdf. Accessed 11 Dec 2016
Prime Minister's Office (2013) Finland's strategy for the Arctic region. http://narfu.ru/aan/Encyclopedia_Arctic/Encyclopedia_Finland_arctic_strategy_2013.pdf. Accessed 16 Dec 2016
Putin V (2013) Strategiya Razvitiya Arkticheskoi Zony Rossiyskoi Federatsiii Obespecheniya Natsional'noi Bezopasnosti na Period do 2020 Goda [The Strategy for the Development of the Arctic Zone of the Russian Federation and Ensuring National Security for the Period up to 2020]. Approved by President Vladimir Putin on 20 Feb 2013. http://www.правительство.рф/docs/22846. Accessed 11 Dec 2016 (in Russian)
Stoltenberg T (2009) Nordic Cooperation on foreign and security policy. Norwegian Ministry of Foreign Affairs, Oslo
Syring JD (2013) Ballistic missile defense overview. https://cryptome.org/2013/10/MDASpaceMissileBrief2013.pdf. Accessed 11 Jan 2017
Tebin P (2015) The Doomsday ships: new U.S. Ballistic missile submarine program. http://russiancouncil.ru/en/inner/?id_4=6070#top-content. Accessed 10 Jan 2017
The International Bering Sea Forum (2006) Issues. http://www.beringseaforum.org/issues.html. Accessed 11 Dec 2016
The Norwegian Ministry of Defense (2012) The new Norwegian long-term defense plan—summary. http://www.regjeringen.no/nb/dokumentarkiv/stoltenberg-ii/fd/Nyheter-og-pressemeldinger/Nyheter/2012/the-new-norwegian-long-term-defence-plan.html?id=678767. Accessed 12 June 2016
The Norwegian Ministry of Defense (2014) Future acquisitions for the Norwegian armed forces 2014–2022. http://www.regjeringen.no/upload/FD/Temadokumenter/Acquisitions-2014-2022_mars-2014.pdf. Accessed 12 June 2016
U.S. Army (2016a). 1st Stryker brigade combat team (28 Apr). Available at: https://www.army.mil/article/166990/. Accessed 10 Jan 2017
U.S. Army (2016b). Northern warfare training center. Available at: https://www.army.mil/article/170432/. Accessed 10 Jan 2017
U.S. Coast Guard (2013) Arctic strategy. http://www.uscg.mil/seniorleadership/DOCS/CG_Arctic_Strategy.pdf. Accessed 16 Dec 2016
U.S. Department of Defense (2010) Ballistic missile defense review report. https://www.defense.gov/Portals/1/features/defenseReviews/BMDR/BMDR_as_of_26JAN10_0630_for_web.pdf. Accessed 11 Jan 2017
U.S. Navy (2016a) Fleet Ballistic missile submarines—SSBN. America's Navy, 12 Jan. Available at: http://www.navy.mil/navydata/fact_display.asp%3Fcid%3D4100&tid%3D200&ct%3D4. Accessed 10 Jan 2017
U.S. Navy (2016b) Navy submarines arrive in Arctic for ICEX 2016. America's Navy, 24 June. Available at: http://www.navy.mil/submit/display.asp?story_id=93648. Accessed 10 Jan 2017
Vasiliev A (2012) Russia's approaches to international cooperation in the Arctic. Arctic Herald 1:12–27
Wezeman S (2012) Military capabilities in the Arctic. SIPRI, Stockholm
Woolf A (2012) Conventional prompt global strike and long-range Ballistic missiles: background and issues. Congressional Research Service, Washington, DC. http://fas.org/sgp/crs/nuke/R41464.pdf. Accessed 10 Jan 2017
Woolf A (2016) U.S. strategic nuclear forces: background, developments, and issues. Congressional Research Service, Washington, DC. https://fas.org/sgp/crs/nuke/RL33640.pdf. Accessed 10 Jan 2017

Zagorsky A (2013) Arkticheskie ucheniya Severnogoplota [The Arctic exercises of the Northern Fleet]. http://www.imemo.ru/ru/publ/comments/2013/comm_2013_053.pdf. Accessed 11 Dec 2016 (in Russian)
Zilanov VK (2016) The Russian experience of IUU-extraction of marine living resources in the Arctic seas' enclaves. Arctic Herald 4(19):44–49 (in Russian)
Zysk K (2008) Russian military power and the Arctic. Russian Foreign Policy (EU–Russia Center's Review) 8:80–86

Chapter 12
The Rise of China in the Emergence of a New Arctic Order

Xing Li and Bo Peng

12.1 Introduction: The Arctic Region in the Era of the Rise of Emerging Powers

Since 2000, the world has undergone a major transformation in international relations and in the international political economy, brought about largely by the rise of China. This development has created a spill-over effect and has had an impact on global issues such as manufacturing, trade, resources, security, the environment, the balance of power, and the world order. Some of these issues are closely connected with the Arctic region in one way or another, including scientific exploration, climate changes, eco-systems, transportation, and energy.

Today, the world is witnessing great power rivalry combined with a movement of flux and reflux in international relations, in which China or other emerging powers and the US-led existing powers are intertwined in a constant process of shaping and reshaping the world order and regional orders. These developments occur in the nexus of national interest, resource potential, geopolitical and geoeconomic value, and potential conflicts.

In the last few years, questions pertaining to Arctic affairs and governance have increasingly been attracting global attention, both in the media and in academia. Many of the current debates on the Arctic are insufficiently informed by the existing literature on governance and order within the field of Arctic-related international relations. Against the background of new emerging powers and new geopolitical and geoeconomic dynamics, we attempt to explore the factors that are leading to the emergence of an "Arctic order". One of the central debates on the relationship

X. Li (✉) · B. Peng
Aalborg University, Aalborg, Denmark
e-mail: xing@cgs.aau.dk

B. Peng
e-mail: peng@cgs.aau.dk

M. Finger and L. Heininen (eds.), *The GlobalArctic Handbook*,
https://doi.org/10.1007/978-3-319-91995-9_12

between the existing US-led world order and the rise of China as one of the key emerging powers has largely focused on whether China is to be seen as a "status quo" power or as a "revisionist" power in relation to the established "rules of game" under the existing world order. The existing powers in general and the US in particular perceive many of China's foreign policy orientations and ways of behaving as "revisionist" when it comes to the defined rules, norms, values, and systems. It is on the basis of these established rules, norms, values, and systems that China is often judged to be either a challenger or a contributor to the existing world order and, accordingly, as a destabilizing or stabilizing influence.

The objective of this chapter is to map out Beijing's relationship with the Arctic and to interpret China's increasing interest in this region. Methodologically, we place the discussion of Arctic order and governance within the larger context of world order and governance, because the dynamics of Arctic affairs are exerting increasing influence at the global level. During the last three decades, the impact of the Arctic region on regional and global governance has been felt worldwide, especially from political, economic, and environmental perspectives (Ye 2014). Specifically, Arctic-related climate and marine issues are emerging with tremendous impacts on the world environment and on global ecosystems. The potential commercialization of the Arctic region with the opening of the Northern Sea Route will no doubt change the conventional global geoeconomic and geopolitical relationship patterns.

Our main conclusion is that the Arctic is a multipolar "region" whose order and governance is maintained not only by the states within the region and by its regional "regime" alone, but also by multiple global actors intertwined in a constant process of shaping and reshaping the Arctic order in the nexus of national interest, regional orientation, shared economic and political agendas, security concerns and potential conflicts.

12.1.1 The Arctic in Transformation

The Arctic region is a very interesting test case or laboratory in which to study the emergence of "a new world order". In other words, the Arctic is an emerging international stage on which established and rising international actors are maneuvering to increase their stakes bilaterally and multilaterally so that they can shape the development of an "Arctic order". What makes the Arctic region particularly compelling within IR and IPE is the fact that no defined and established order of Arctic governance currently exists; instead, an ongoing global process is currently shaping the rules of the game in the Artic. Against this backdrop, it is not difficult to understand why this region is becoming one of the centers of gravity of world politics, economy, and security. The Arctic is attracting increasing international attention for a number of reasons, including climate change, resource and transportation potentials, political-economic globalization, and indigenous cultures and heritage.

Climate change is widely regarded as the fundamental factor that influences the dynamics of the Arctic region. Almost all of the related literature takes climate change as its point of departure for analysis of Arctic issues. In particular, uncontrollable climate change can be observed in the melting of Arctic ice. On one hand, melting Arctic ice is providing new opportunities for potential economic benefits, including the opening of an alternative sea route (Ebinger and Zambetakis 2009). On the other hand, it will certainly cause environmental security issues, including severe impact on ice-dependent flora and fauna and a distressing effect on the indigenous peoples of the region (Stokke 2011).

The possible commercialization of the Northern Sea Route (NSR) is one of the most conspicuous dynamics in the Arctic Region. According to Margaret Blunden, "with the world's maritime transport system at the forefront of globalization, the emergence of a new sea lane would have global consequences" (Blunden 2012). Motivations for opening the NSR relate to the fact that the current trade routes between Europe and Asia (from the Suez Canal to the Strait of Malacca) are "highly vulnerable both to congestion and to deliberate or accidental disruption" (Blunden 2012). The NSR can cut approximately 7000 nautical miles from the conventional sea route, which would lead to numerous economic benefits, not only because of the reduction in travel time, but also due to the dramatic decrease in the consumption of fuel (Nakagawa 2011). Other potential benefits include the expansion of human activities, such as "operations in untouched fisheries and the development of undersea resources" (Nakagawa 2011).

The natural resources of the Arctic region make this area highly attractive on a global scale. According to the US Geological Survey (USGS), "the Arctic contains up to 30% of the world's undiscovered gas and 13% of the world's undiscovered oil resources" (Campbell 2012). It is worth mentioning that the largest deposits are found in the Russian Arctic offshore territory. Whereas Saudi Arabia's current proven oil reserves amount to 260 billion barrels, "the territory claimed by Moscow could contain as much as 586 billion barrels of oil—although these deposits are unproven" (Borgerson 2008). In addition, the Arctic region also offers bountiful biological resources, such as millions of migratory birds, a variety of marine mammals and some major herds of reindeer. These resources provide plenty of living resources for the indigenous peoples and for commercial exploitation and also make this region highly significant in terms of global biodiversity.

The combination of the above geoeconomic factors is gradually turning the Arctic region into an emerging geopolitical and geoeconomic center in which, due to the region's potential geopolitical and geo-economic benefits, a number of international powers are "staking their claim to the Arctic" (Howard 2009). The struggles among these powers to access the Arctic and to gain as much as possible in the region is evident at both regional and global levels.

At the regional level, as Borgerson put it, "global warming has given birth to a new scramble for territory and resources among the five Arctic powers" (Borgerson 2008). Russia, the US, Canada, Denmark, and Norway are all anxious to establish their claims in this region, and have attempted to do so in myriad ways, including planting national flags and building new Arctic naval patrol vessels. In the meantime, these five Arctic countries also cooperate with each other in their common interest. In 2008, they jointly issued the Ilulissat Declaration, in which they argued that "the existing legal framework provided a solid foundation for responsible management by the five coastal states; there was no need for any new comprehensive legal regime" (Blunden 2012). The declaration indicated that these five Arctic coastal countries wished to impede others from sharing the benefits of the region.

At the global level, European and East Asian countries constantly emphasize the importance of the Arctic region with regard to their own politics and economy. Given the fact that maritime transportation is one of the most critical underpinnings of the dynamic economic development of Asia, the political agendas of the three largest economies in East Asia—China, Japan, and South Korea—have prioritized the stability and security of sea routes. More importantly, these Asian economic giants have gained observer status in the Arctic Council, allowing them to be involved in the processes that are determining the future development of the Arctic region (Milne 2013). With the increasing participation of non-Arctic countries' in Arctic affairs, the Circumpolar North has become increasingly important in world politics. Therefore, it is increasingly difficult to ignore "the growing geostrategic importance of this region and the growing emphasis on innovations in governance and co-management of the Arctic issues" (Heininen 2005).

12.2 What Is Arctic Governance and What Is an Emerging Arctic Order?

Used in both political and academic terms, the concept of "governance" refers, in a general sense, to the task of running a government or of managing affairs of various kinds. The core notion refers to "processes of governing, whether undertaken by a government, market or network, whether over a family, tribe, formal or informal organization or territory and whether through the laws, norms, power or language" (Bevirk 2013: 1). It covers "the processes of interaction and decision-making among the actors involved in a collective problem that lead to the creation, reinforcement, or reproduction of social norms and institutions" (Hufty 2011: 405). In applying the concept of "governance", one is immediately brought into contact with other related domains, such as regimes, rules, norms, values, and laws, etc., upon which governance is supposed to be based, structured, sustained, and regulated and to which it is held accountable.

When narrowly applied by IR scholars and policy-makers, the concept of governance is Westphalian and state-centered. However, as James Rosenau has argued, "global governance refers to more than the formal institutions and organizations through which the management of international affairs is or is not sustained. The United Nations system and national governments are surely central to the conduct of global governance, but they are only part of the picture" (Rosenau 1995: 13).

In the era of globalization and transnational capitalism, global problems are moving beyond the capabilities of individual states (Weiss 2014: 32). This situation requires an overarching system of public and private institutions which are "valid or active in a given issue area of world politics" (Biemann et al. 2009: 15). In global governance, a number of state actors and non-state actors strive to produce a global pattern of rule in the absence of a world government (Bevir 2009). In other words, global governance can be regarded as an international process in which a variety of stakeholders interact to form consensus and to generate guidelines and agreements for the promotion of collective action and international cooperation. Boughton and Bradford's theoretical understanding of an ideal of global governance can usefully be applied in interpreting the opportunities and challenges of Arctic governance:

> The ideal of global governance is a process of cooperative leadership that brings together national governments, multilateral public agencies, and civil society to achieve commonly accepted goals. It provides strategic direction and then marshals collective energies to address global challenges. To be effective, it must be inclusive, dynamic, and able to span national and sectoral boundaries and interests. It should operate through soft rather than hard power. It should be more democratic than authoritarian, more openly political than bureaucratic, and more integrated than specialized. (Boughton and Bradford 2007: 11)

12.2.1 Arctic Governance and Emerging Powers

On one hand, the Arctic states (Canada, Denmark, Finland, Iceland, Norway, Russia, Sweden, and the United States) have a natural presence in the Arctic region and a central position in Arctic governance, which is one of the historical characteristics of the Arctic. On the other hand, the varying international institutional affiliations of the Arctic states, and other major powers including the US, the EU and China, intersect across different contexts, issues, and interests (see Fig. 12.1).

As the highest level of inter-state forum, the Arctic Council plays an indispensable role in Arctic governance. Two thousand and sixteen marked the 20th anniversary of the Arctic Council, which was established on September 19, 1996 by the Ottawa Declaration.[1] So far, the Council has functioned as the formal and official international "governance" organ for the Arctic region. Its agenda is run and maintained by senior representative Arctic officials, and it comprises working groups, task forces, and expert groups. It holds regular ministerial meetings for the permanent members, including not-for-profit corporations (AIA), indigenous peoples organizations (ICC and RAIPON), and non-governmental organizations (SC). It can be argued that the architecture of the Arctic Council has met some of the requirements of the ideal mode of global governance that Boughton and Bradford defined as "a process of cooperative leadership that brings together national governments, multilateral public agencies, and civil society to achieve commonly accepted goals" (Boughton and Bradford 2007: 11).

---

[1]The Ottawa Declaration lists Canada, Denmark, Finland, Iceland, Norway, Russia, Sweden, and the United States as members of the Arctic Council. It also includes six organizations representing Arctic indigenous peoples as permanent participants: the Aleut International Association, the Arctic Athabaskan Council, Gwich'in Council International, the Inuit Circumpolar Council, Russian Association of Indigenous Peoples of the North and the Saami Council.

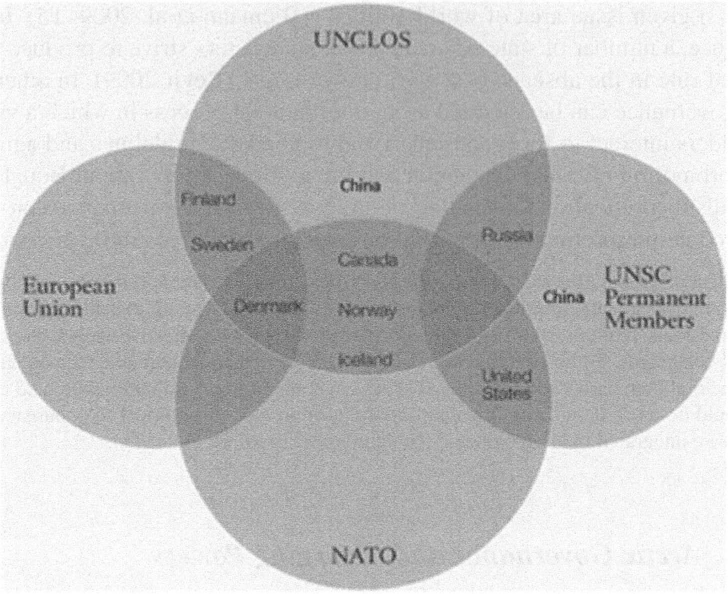

Fig. 12.1 Overlapping and intertwined relationships among the Arctic stakeholders

However, the limitations of the Arctic Council are becoming increasingly salient to the issue of binding effect and the issue of representativeness. On one hand, as mentioned above, the Arctic Council is an inter-state forum rather than an international organization, which largely limits its binding effect. The Council has a limited mandate and its decisions are reached by consensus. Specifically, the Arctic Council is still an entity that yields scientific assessments and technical recommendations through its working groups. There is a shared understanding that the Arctic Council can best serve the realm of governance by providing policy recommendations rather than by becoming a regulatory body. As Timo Koivurova argued, "no serious effort has been taken to go beyond the existing paradigm of producing non-binding technical guidance or fairly abstract policy recommendations" (Koivurova 2009: 148). The Arctic Council was founded as a platform without legal status for the deliberation of regional concerns regarding sustainable development and environmental protection. This situation is bound to influence the effectiveness of the Arctic Council in "providing strategic direction and marshaling collective energies" (Boughton and Bradford 2007: 11) to address regional challenges.

On the other hand, with regard to the issue of representation in Arctic affairs, emerging powers such as China are not fully integrated in Arctic governance, despite the dramatic increase in their role and presence in the Arctic area. The current structure of Arctic governance does not fully reflect the dynamics and development of the global economy. The eight member states in the Arctic Council unquestionably occupy the top of the representative structure, mainly due to their unique geographical positions. By contrast, other major economies (France, Germany, China, Japan and South Korea, etc.) are only granted observer status. In terms of the economic implications of potential resource exploration and shipping route usage, these observers play a limited role in Arctic governance and are excluded from the process of "decision-making at all levels within the Arctic Council" (Graczyk and Koivurova 2014). The observers face the following limitations: (1) they only engage in the Arctic Council at the level of Working Groups; (2) they can propose projects only through an Arctic state or a permanent participant; and (3) their financial contribution to the given project must not exceed financing from Arctic States (Arctic Council 2016). These rules imply that even though they are of great significance to the regional development of the region, the observers have limited opportunities to influence the shape of Arctic governance through the existing Arctic Council. This will inevitably affect the legitimacy of the Arctic Council in Arctic governance in the long term. In other words, if the Arctic Council is to play a more effective role in governing the Arctic region, it must be more "inclusive, dynamic, and able to span national and sectoral boundaries and interests" (Boughton and Bradford 2007: 11).

It is obvious that the Arctic Council must adapt to the new reality of the rapidly increasing political and economic importance of the Arctic Region. Therefore, the role of China as a rising power in Arctic affairs and governance is an important issue for the foreign policy, transnational relations, and economic development of the Arctic region. As Jakobson and Peng noted, "Now, even though the Arctic is not a foreign policy priority, China's growing interest in the region raises concern—even alarm—in the international community about China's intentions" (Jakobson and Peng 2012: 1).

At a time when many regions of the world are facing growing animosity and even conflict over territories and resources, the Arctic can offer an alternative order in which nations cooperate to fashion multilateral, cooperative, and peaceful governance. The "Arctic order" is expected to be based not on military strength or on competition to control resources, but on the multilateral pursuit of common interests.

When it comes to Arctic affairs, Arctic and Asian countries (China, Japan, South Korea, Singapore, and India) are intertwined with each other in terms of scientific, political, transnational, and economic relations. The Arctic is receiving increasing international scientific and political attention as a result of climate change and political–economic globalization. With the "rise of the emerging powers" and "global responses" (Christensen and Li 2016), Asian powers in general and China in particular are growing in economic and political influence and thus see themselves as natural participants and stakeholders in regions around the world, including the Arctic. Consequently, the existing Arctic powers have to relate to these rising Asian powers. Asian interests in the region are driven by climate change concerns because Arctic climate change affects Asian weather patterns and thus agriculture and food security. Likewise, the rising Asian powers depend on the global supply of energy and raw materials, as well as on global shipping, and this dependency results in an interest in Arctic energy, raw materials, and new shipping lanes. Therefore, the Arctic will constitute a new dimension in the relationship between the Arctic states and Asia and may give the small Nordic states in particular privileged scientific, political, and economic access to rising and existing Asian powers.

The Arctic littoral countries must begin to develop a better understanding of the strategic interests of Chinese and other key Asian stakeholders and to identify areas for future collaboration in Arctic affairs. The many areas of overlapping concerns and opportunities include the question of how peace and stability can be ensured in the Arctic region, the role of international maritime law in guiding conduct in Arctic shipping, freedom of navigation, Arctic environmental issues, the development of equitable and sustainable resource extraction, energy resources, fishing, and Arctic disaster management.

12.3 What Impact Has the Rise of China Had on Global Governance?

Since the implementation of its "reform and opening up policy" in the late 1970s, China has gradually experienced a historical transformation from a socialist planned economy to a market economy with Chinese characteristics, and it has been integrated into world politics and the world economy. In the past four decades, China's real GDP growth has averaged about 9% per year, and the country has lifted 400 million people out of poverty. The Chinese achievement is acknowledged by the World Bank as an "unprecedented development miracle in human history" (see Fig. 12.2).

The impact of China's economic rise is being felt worldwide. In 2009, according to the International Energy Agency (IEA), China became the world's largest energy consumer, consuming 2.252 billion tons of oil equivalents. This was about 4% more

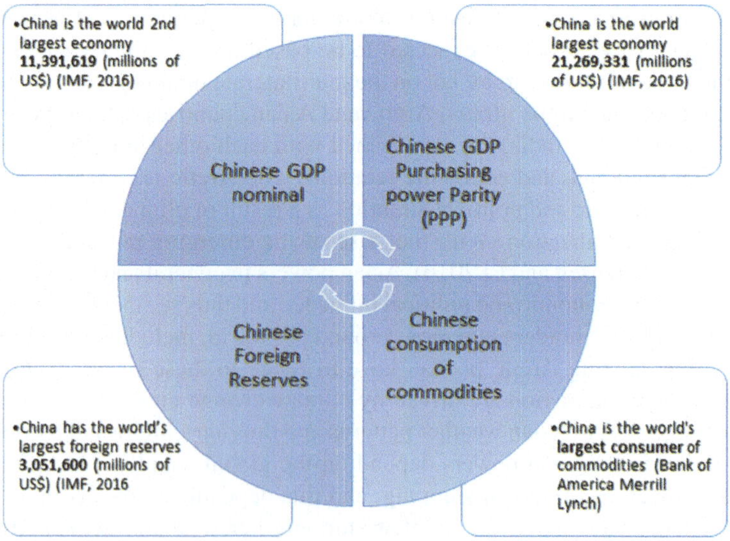

Fig. 12.2 Some of China's current macro-economic data

than the US, whose energy consumption was 2.17 billion tons of oil equivalents (Swartz and Oster 2010). In 2010, China displaced the US as the largest manufacturing nation; its share of world manufacturing was 18.9%, which was slight higher than the US share of 18.1% (Meckstroth 2015). In 2013, China overtook the USA as the world's largest trading nation, as its trade surplus rose 12.8% in 2013 to almost $260 billion (Monaghan 2014). In 2014, according to the IMF, China became the world's largest economy in terms of purchasing power parity (PPP). The total wealth of China was estimated at $17.6 trillion, which was closely followed by the US figure of $17.4 trillion (Bird 2014). In 2015, the Chinese currency renminbi (RMB) was included in the basket of currencies that make up the IMF's Special Drawing Right (SDR). In the revised SDR basket, the Chinese RMB weight is 10.92%, which ranks third after the US dollar (41.73%) and the Euro (30.93%) (IMF Survey 2015). All in all, China has clearly become an indispensable force in the international politics and economy and a key factor in the transformation of global governance.

The economic rise of China and its spill-over effects are being felt worldwide and have been comprehensively documented. The world is closely following Beijing's changing political, strategic, and economic links with the developed and developing world (Lanteigne 2013). Meanwhile, China is showing unprecedented enthusiasm for participation in global governance. In 2015, Chinese President Xi Jinping pointed out that "the global governance system is built and shared by the world, not monopolized by a single country. China certainly has no intention to do so. China is involved in building the current international system, and has always done its part to uphold the international order and system with the UN as its core and the purposes and principles of the UN charter as its foundation" (The Wall Street Journal 2015). In other words, the rise of China is not only manifest in "hard power" areas such as economy, technology, and military, but also in "soft power" areas such as institutions, norms, and values. China's institutional and normative contributions to global governance are gradually being recognized.

In its approach to emerging alternative political/economic institutions, China attaches great importance to multilateralism in global governance. China has always insisted that, given the rise of a number of non-traditional security issues, the UN should play a central role in global security governance. Beijing's position is that "as the most universal, representative and authoritative inter-governmental organization, the UN is the best platform for multilateralism and for collective action to cope with various threats and challenges" (Zhang and Feng 2011). Moreover, China's increasing advocacy of multilateralism is reflected in a number of China-led sub-regional, regional and trans-regional multilateral institutions and mechanisms, such as China-ASEAN Sub-Regional Cooperation, the Shanghai Cooperation Organization (SCO), BRICS, the Conference on Interaction and Confidence Building Measures in Asia (CICA) and the Boao Forum. In addition, China actively participates in several bilateral and multilateral political, economic, and security dialogues. These institutions and mechanisms not only promote the democracy of international relations, but also strengthen the collaboration and coordination between developing countries in efforts to solve the non-traditional security issues which could moderate the negative effects of the US monopoly of power in global and regional institutions.

The G20 is one of the most important economic institutions to have emerged in the wake of the 2008 Global Financial Crisis and has become the most significant forum of global economic governance and cooperation. As the second largest economy and the largest developing country, China is playing a role in bridging the gap between developed and developing countries by promoting the exchange of ideas and the realization of cooperation (Lin and Li 2014). Meanwhile, China has also launched several economic and financial institutions, such as the BRICS' New Development Bank (NDB), the Silk Road Foundation, the Asian Infrastructure Investment Bank (AIIB), and a number of bilateral and multilateral free trade agreements (FTAs). These Beijing-led institutions have demonstrated China's willingness to shoulder more responsibilities to maintain global economic stability, which will not only improve the representation of developing countries in global economic governance, but also enhance the efficiency of the existing multilateral institutions through competition. More importantly, these Chinese initiatives are generally regarded as essential complements rather than challenges to the existing architecture of global financial governance (Dollar 2015).

In its approach political/economic ideas rooted in global governance, Chinese primary political ideation in global governance involves "wide consultations, joint contribution and shared benefits" (共商共建共享). In the Conference on Interaction and Confidence-Building Measures in Asia (CICA) in 2014 and in the Boao Forum for Asia in 2015, Chinese President Xi Jinping elaborated the two political ideas of the "New Asian Security Concept"[2] (新亚洲安全观) and the "Community of Common Destiny"[3] (命运共同体). These proposed political norms have provided new opportunities for the world to understand China's views on global governance and security. From the Chinese perspective, security must be universal, equal, and inclusive (FMPRC 2014). In addition, China proposed a new-modality cooperation framework in global governance "to make sure that all countries respect one another and treat each other as equals, to seek win-win cooperation and common development, to pursue common, comprehensive, cooperative and sustainable security, to ensure inclusiveness and mutual learning among civilizations" (Xinhuanet 2015). Given these Chinese political ideas about global governance, it can be argued that Beijing has no intention of overthrowing the existing international governance system, preferring instead to inject it with some "Chinese characteristics."

[2]See President Xi Jinping's speech at the Fourth Summit of the Conference on Interaction and Confidence Building Measures in Asia. http://www.fmprc.gov.cn/mfa_eng/zxxx_662805/t1159951.shtml. Accessed 5 February 2017.

[3]See President Xi Jinping's speech at the Boao Forum for Asia Annual Conference 2015. http://www.china.org.cn/business/2015-03/29/content_35185720.htm. Accessed 5 February 2017.

12.4 Will China's Influence Shape Arctic Governance and an Emerging Arctic Order?

The previous section shows a general trend whereby China is joining a large number of international organizations and is becoming more engaged and integrated in the international community and in global governance (Johnston 2003: 13). As part of a newly emerging area of global governance, Arctic governance is increasingly becoming a laboratory in which myriad relevant international actors are trying to impose their influence and to shape the development of the Arctic region in accordance with their own interests and ideas. Defining China as a "near-Arctic state,"[4] Chinese foreign policy makers deem China to be a natural participant and stakeholder in the Arctic region. Chinese Rear Admiral Yin Zhuo has claimed that the "Arctic belongs to all the people around the world, as no nation has sovereignty over it... China must play an indispensable role in Arctic exploration as we have one-fifth of the world's population" (Economy 2014).

China's growing engagement and influence in the architecture of global governance suggests that Beijing is becoming an indispensable actor in the process of shaping the structure of Arctic governance on the basis of its economic interests, scientific capacity, and political influence, etc. As Linda Jakobsen pointed out, "China has staked its claim to be included in important Arctic discussions. As the polar cap melts, the impact on China's ecosystem and economy could be significant and, as a result, the country has woken up to this new geopolitical challenge" (Jakobsen 2013: 99).

Even China's observer status in the Arctic Council has created anxieties among existing council members such as Canada and Russia about the possible challenges that might emerge from China's engagement in Arctic affairs. Consequently, the existing Arctic Council members (Canada, Denmark, Finland, Iceland, Norway, Russia, Sweden, and the USA) have to relate to this Asian giant, either willingly or unwillingly. Apart from political and economic calculations, China's interest in the Arctic is also driven by climate change concerns as Arctic climate change affects Asian weather patterns and thereby Asian agriculture and food security. Therefore, the Arctic will add a new dimension to the relationship between the Arctic states and China as well as other Asian stakeholders. This situation may give the small Nordic states privileged scientific, political, and economic connections with the rising and existing Asian powers.

Arctic governance is currently defined along the lines of existing international rules such as the UN Convention on the Law of the Sea or the IMO Polar Code, but also in terms of participation in the region's primary governance institution, the Arctic Council (AC). All in all, the Arctic will be a new dimension in the relationship between Asian powers and traditional Arctic States, with various consequences for the regional order in the Arctic. In particular, the Arctic can offer new paths to

[4]China defines itself as a "near-Arctic state", and also identifies itself as an Arctic "stakeholder". See SIPRI, http://www.sipri.org/media/pressreleases/2012/arcticchinapr.

scientific, economic and political interaction between small Nordic states and Asian stakeholders in the Arctic.

Given Beijing's heavy involvement in important Arctic economic stakes related to shipping, investments, and resource utilization, China has been generally accepted as an increasingly important economic actor as well as a legitimate stakeholder in the Arctic. As one of the largest trading nations in the world, it is heavily dependent on the global supply of energy and raw materials and on global ocean transportation. As discussed above, with most of its import and export dependent on maritime transportation and with rising tensions in the South China Sea and the Somali Sea, China might become the greatest consumer of the formerly closed shipping route in the north (Li and Bertelsen 2013).

The possible opening of the NSR has both economic and security implications for China. In 2013, China overtook the US to become the largest trading nation in the world, which was described by Beijing as a "landmark milestone" for the country (Monaghan 2014). With 90% of global trade carried by sea, it is easy to understand the importance of the Northern Sea Route (NSR), which could save approximately 6400 km on the voyage from Shanghai to Hamburg, compared with the southern passage through the Strait of Malacca and the Suez Canal. It would be even shorter were it possible to break the ice across the North Pole for commercial purpose (*The Economist* September 1, 2012). Moreover, China has been the largest energy consumer since 2009, and China's economy is largely reliant on its foreign trade, with energy imports accounting for a much of that trade. In recent years, the so-called "Malacca Dilemma" has raised concerns among Chinese decision makers about energy security, and this has increased the significance of a possible alternative sea route through the Arctic region (Zhang 2011).

China is also one of the few countries to have defined energy as a "national security" concern and thus as a priority in the foreign policy-making process. In recent years, China has closely followed the evolution of Arctic affairs, and it is playing an increasingly active role in Arctic-related scientific, political, transnational and economic relations. China's growing presence in the Arctic is seen as connected with Beijing's interest in possible untapped Arctic resources as well as in alternative ocean transportation routes. As an emerging stakeholder without a formal place or role in the Arctic's political and legal setup and driven by its shortage of energy resources and by recurrent problems with its maritime transport in the Indian Ocean, China is struggling to find ways to safeguard its right to access Arctic resources and maritime transport routes. The globally shared consensus is that Chinese foreign policy calculation and actions will generate a huge impact on the agenda of global geopolitics and geo-economics in the Arctic region in the years to come (see Fig. 12.3).

China's interest in securing permanent observer status within the Arctic Council and in its growing icebreaker capabilities has spurred reactions from Arctic states that range from warm enthusiasm to extreme caution. Many analysts assume that China is a revisionist territorial actor that is motivated by resource concerns and that could potentially dominate the Arctic Council. Western commentators tend to view China's Arctic ambitions with greater apprehension than they view those of any other state.

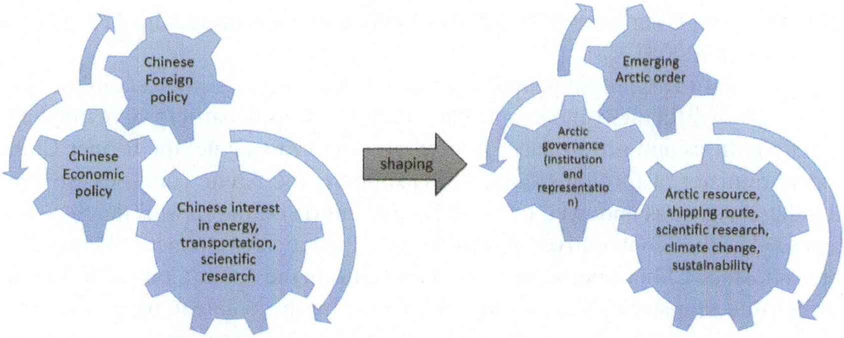

Fig. 12.3 The nexus between China's interest and an emerging Arctic order

It is clear that "scientific diplomacy" alone will not secure China's permanent member state seat in the most influential and important Arctic organization—the Arctic Council. Nevertheless, scientific diplomacy is the most resilient and receptive approach to Arctic policy formation and to debunking the "myth and misperception" of China's Arctic motivation (Su and Lanteigne 2015). Not only does Beijing's Arctic policy increasingly focus on Arctic exploitation in cooperation with Arctic states, but it also sees international law and Arctic governance as a useful platform for China's involvement in Arctic affairs. In brief, China's long-term Arctic policies are designed to avoid provoking misperceptions and negative responses from the Arctic states while also avoiding the exclusion of China from rightful access to its Arctic stakes.

To palliate global anxiety about China's Arctic policy, China has been trying to reduce what is often perceived as geoeconomic calculation and has instead prioritized science diplomacy in the Arctic and Antarctic research; that is, scientific exploration and research collaboration. It is widely acknowledged that China has become a major contributor to Arctic scientific research (Jakobson 2010; Zhu 2009; Spears 2009). In the past two decades, China has dedicated itself to conducting research on the Arctic Ocean's ice cover, environment, and climate and has conducted five Arctic and 28 Antarctic expeditions, in which the world's largest non-nuclear icebreaker *Xuelong* (Snow Dragon) played an important role. China's research position was further strengthened by the installation of the Arctic Yellow River Station in Norway's Svalbard archipelago as an attempt to pursue ecological and environment monitoring. China reached an agreement with the Finnish company Aker Arctic Technology to deploy a second large icebreaker in 2014. In size, the new icebreaker is second only to the MV Xuelong.[5] Furthermore, Beijing's policy objective has been to increase its Arctic representativeness through taking part in the Arctic-related international cooperation in scientific research. Among other moves, it has joined the Arctic Science Committee (1996), the Ny-Ålesund Science Managers' Committee (2005), and

[5]The MV Xuelong was built in 1993 at Kherson Shipyard in Ukraine and was originally converted from an Arctic cargo ship to a polar scientific research and re-supply vessel.

the International Polar Year (IPY) (2007–2008), and it has hosted the Arctic Science Summit Week (2005).

Considering the Chinese geopolitical and geo-economics interests in the Arctic region, it is not difficult to understand China's aspirations to participate in the shaping of Arctic politics and governance. According to Hu Zhengyue, the former Chinese assistant minister of foreign affairs, China supports the Arctic countries' sovereign and judicial rights as endowed by the UN Convention on the Law of the Sea, which is regarded as the "constitutive foundation for the Arctic governance" (Young 2016). Moreover, since China's permanent observer status in the Arctic Council was granted at the Kiruna Ministerial Meeting in 2013, China has been able to participate proactively in the meetings of the Arctic Council and make constructive contributions to Arctic governance; for instance, by proposing projects, providing financial support, and submitting written statements (Arctic Council 2016).

12.5 Conclusion: Issues to Be Resolved

In this chapter we have has argued that the strategic importance of the Arctic region has been growing since the beginning of the 21st century and will be on the agenda of global politics, economics, and security in the years to come, involving both the states in the region and major world powers outside the region. It is expected that competition will eventually develop regarding energy resources, maritime transportation security, and territorial claims on northern sea areas and shelves. In other words, there appears to be a global consensus that the Arctic region represents a huge economic and geostrategic asset that, given the value of the resources in the region and the strategic importance of the Arctic, has the potential to become a source of political conflict and military competition.

Despite consensus about China's interest and heavy stake in the Arctic region, many important issues remain unresolved concerning China's role in the Arctic. First, China's commercial interest in the Northern Sea Route is inevitably intertwined with its international relations with other major powers, primarily the USA, Russia, and Canada. Second, China's commercial interests and business activities in exploring Arctic resources are also limited by market prices and climate change, and they are both implicitly and explicitly connected with the situation in the East and South China Sea. Third, China's concrete "Arctic policy" remains unclear. Although China has declared itself an important Arctic stakeholder and a "near-Arctic" state, Beijing has still not published an official Arctic white paper (policy statement). The major question facing countries in the Arctic region is whether China's interest in the frozen region is "more driven by opportunism than by a long-term desire to challenge the littoral States' sovereignty" (Lasserre et al. 2015).

Finally, the authors regard the Arctic as an issue area; it is a testing ground in which China is experimenting with various strategies and tactics to exert influence on the reform of existing Arctic governance and to imprint Chinese characteristics on an emerging Arctic order.

References

Arctic Council (2016) Observers. http://www.arctic-council.org/index.php/en/about-us/arctic-council/observers. Accessed 28 July 2016

Bevir M (2009) Key concepts in governance. SAGE Publications Ltd, London

Bevir M (2013) Governance: a very short introduction. Oxford University Press, Oxford, UK

Biemann F, Pattberg P, Asselt H, Zelli F (2009) The fragmentation of global governance architectures: a framework for analysis. Glob Environ Politics 9(4):14–40

Bird M (2014) China just overtook the US as the world's largest economy. http://www.businessinsider.com/china-overtakes-us-as-worlds-largest-economy-2014-10?r=US&IR=T&IR=T. Accessed 27 July 2016

Blunden M (2012) Geopolitics and the Northern Sea route. Int Aff 88(1):115–129

Borgerson SG (2008) Arctic meltdown: the economic and security implications of global warming. Foreign Aff 87(2):63–77

Boughton J, Bradford C (2007) Global governance: new players, new rules. Finance Dev 44(4). http://www.imf.org/external/pubs/ft/fandd/2007/12/boughton.htm. Accessed 28 July 2016

Campbell C (2012) China and the Arctic: objectives and obstacles. US-China Economic and Security Review Commission. http://origin.www.uscc.gov/sites/default/files/Research/China-and-the-Arctic_Apr2012.pdf. Accessed 1 Aug 2016

Christensen SF, Li X (2016) Emerging powers, emerging markets, emerging societies: global responses. Palgrave Macmillan, London

Dollar D (2015) China's rise as a regional and global power: the AIIB and the 'One Belt One Road'. https://www.brookings.edu/research/chinas-rise-as-a-regional-and-global-power-the-aiib-and-the-one-belt-one-road/. Accessed 4 Aug 2016

Ebinger CK, Zambetakis E (2009) The geopolitics of the Arctic melt. Int Aff 85(6):1215–1232

Economy E (2014) The four drivers of Beijing's Emerging Arctic play and what the world needs to do. Forbes. http://www.forbes.com/sites/elizabetheconomy/2014/04/04/the-four-drivers-of-beijings-emerging-arctic-play-and-what-the-world-needs-to-do/#1e66a0270d8d. Accessed 5 Feb 2017

FMPRC (2014) New Asian security concept for new progress in security cooperation. Ministry of Foreign Affairs of the People's Republic of China. http://www.fmprc.gov.cn/mfa_eng/zxxx_662805/t1159951.shtml. Accessed 4 Aug 2016

Graczyk P, Koivurova T (2014) A new era in the Arctic Council's external relations? Broader consequences of the Nuuk observer rules for Arctic governance. Polar Rec 50(254):225–236

Heininen L (2005) Impacts of globalization, and the circumpolar North in world politics. Polar Geogr 29(2):91–102

Howard R (2009) The Arctic gold rush: the new race for tomorrow natural resources. Continuum Publishing Corporation, London

Hufty M (2011) Investigating policy processes: the governance analytical framework (GAF). In: Wiesmann U (ed) Research for sustainable development: foundations, experiences, and perspectives. Geographica Bernensia, Bern, pp 403–424

IMF (2015) Chinese Renminbi to be included in IMF's special drawing right basket. https://www.imf.org/external/pubs/ft/survey/so/2015/new120115a.htm. Accessed 27 July 2016

Jakobson L (2010) China prepares for an ice-free Arctic. SPIRI Insights Peace Secur 2010(2). https://www.google.dk/url?sa=t&rct=j&q=&esrc=s&source=web&cd=1&ved=0ahUKEwiU872Bpr3WAhVG2RoKHYDxBM8QFggoMAA&url=https://www.sipri.org/sites/default/files/files/insight/SIPRIInsight1002.pdf&usg=AFQjCNGBO9KhQrGRNO6WQEBfZ3yj13iX_g. Accessed 24 Sept 2017

Jakobson L (2013) China wants to be heard on Arctic issues. Glob Asia 8(4):99–101

Jakobson L, Peng J (2012) China's Arctic aspirations. SIPRI Policy Paper, 2012/34

Johnston AI (2003) Is China a status quo power? Int Secur 27(4):5–56

Koivurova T (2009) Limits and possibilities of the Arctic Council in a rapidly changing scene of Arctic governance. Polar Rec 46(237):146–156

Lanteigne M (2013) Chinese foreign policy: an introduction. Routledge, London

Lasserre F, Alexeeva OV, Huang L (2015) China's strategy in the Arctic: threatening or opportunistic? Polar Record. http://corpus.ulaval.ca/jspui/bitstream/20.500.11794/876/1/China%20Arctic%20opportunistic%20Polar%20Record%20OA%20FL%20LH%202015.pdf. Accessed 5 Aug 2016

Li X, Bertelsen RG (2013) The drivers of Chinese Arctic interests: political stability and energy and transportation security. In: Heininen L (ed) Arctic yearbook 2013. Northern Research Forum, Iceland, pp 53–68

Lin H, Li X (2014) G20 and G2: Sino-US relations as an institutional cooperation game? In: Li X (ed) The BRICS and beyond: the political economy of the emergence of a new world order. Ashgate, Farnham, England, pp 111–128

Meckstroth D (2015) China solidifies its position as the world's largest manufacturer. https://www.mapi.net/blog/2015/09/china-solidifies-its-position-world%E2%80%99s-largest-manufacturer. Accessed 27 July 2016

Milne R (2013) China wins observer status in Arctic Council. https://www.ft.com/content/b665723c-bd3e-11e2-890a-00144feab7de. Accessed 9 Aug 2016

Monaghan A (2014) China surpasses US as world's largest trading nation. https://www.theguardian.com/business/2014/jan/10/china-surpasses-us-world-largest-trading-nation. Accessed 27 July 2016

Nakagawa M (2011) Northeast passage (Northern Sea Route). http://www1.american.edu/ted/ice/northeast-passage.htm. Accessed 27 July 2016

Rosenau J (1995) Governance in the twenty-first century. Glob Gov 1(1):13–43

Spears KJ (2009) China and the Arctic: the awakening snow dragon. China Brief 9(6)

Stokke OS (2011) Environmental security in the Arctic: the case for multilevel governance. Int J 66(4):835–848

Su P, Lanteigne M (2015) China's developing Arctic policies: myths and misconceptions. J China Int Relat 3(1):1–25

Swartz S, Oster S (2010) China tops US in energy use. http://www.wsj.com/articles/SB10001424052748703720504575376712353150310. Accessed 27 July 2016

The Economist (2012) Snow dragons: as the Arctic melts, Asia shudders at the risks but slavers at the opportunities. http://www.economist.com/node/21561891. Accessed 24 Sept 2017

The Wall Street Journal (2015) Full transcript: interview with Chinese President Xi Jinping. http://www.wsj.com/articles/full-transcript-interview-with-chinese-president-xi-jinping-1442894700. Accessed 28 July 2016

Weiss T (2014) Global governance: Why? What? Whither?. Polity Press, Cambridge

Xinhuanet (2015) Full text of Chinese President's speech at Boao Forum for Asia. http://news.xinhuanet.com/english/2015-03/29/c_134106145.htm. Accessed 4 Aug 2016

Ye J (2014) China's role in Arctic affairs in the context of global governance. Strateg Anal 38(6):913–916

Young O (2016) Governing the Arctic Ocean. Mar Policy. http://dx.doi.org/10.1016/j.marpol.2016.04.038. Accessed 10 Aug 2016

Zhang Z (2011) China's energy security, the Malacca Dilemma and responses. Energy Policy 39(12):7612–7615

Zhang G, Feng Y (2011) China's UN diplomacy: 1971–2011. Strateg Anal 35(6):973–981
Zhu Y (2009) We are already one of the strongest countries to conduct polar scientific expedition [我们已经是极地考察最强国之一]. http://discover.news.163.com/09/1025/08/5MF6TM59000125LI.html. Accessed 29 July 2016

Chapter 13
Special Features of Arctic Geopolitics—A Potential Asset for World Politics

Lassi Heininen

13.1 Introduction

The current state of international politics is often interpreted as turbulent and unstable. The world's geopolitical situation has been influenced by constant warfare, such as regional conflicts and wars in the Middle East and Asia, growing military tension and threat of nuclear weapons, and even nuclear war (on the Korean Peninsula), as well as global terrorist attacks, counterattacks by authorities, and fearful responses by the general population. This provides excuses for an arms race, a growth in military expenses, constant fights against international terror, and the persistent presence of NATO member states to enlarge Western governing structures. On the one hand, there is growing geopolitical tension and lack of trust between Russia, and the US and the European Union (EU). This is a result of Russia annexing the Crimea and the war in the Eastern Ukraine. Furthermore, Russia is seen as untrustworthy and domineering in its foreign policy. The associated growing tension, as well as Western sanctions and Russian countersanctions, could lead to further action if threats escalate into emerging conflicts between Russia and the West.

On the other hand, there is also a lack of trust between the US and its NATO allies, causing uncertainties. This can be seen in President Trump's support for protectionism and stopping free-trade negotiations, as well as distancing the US from multi-lateral cooperation, including NATO's principle of mutual assistance, based on Article 5 of the NATO Charter. Therefore, there is no guarantee of firm commitments

L. Heininen (✉)
University of Lapland, Rovaniemi, Finland
e-mail: lassi.heininen@ulapland.fi

L. Heininen
Tongji University, Shanghai, China

M. Finger and L. Heininen (eds.), *The GlobalArctic Handbook*,
https://doi.org/10.1007/978-3-319-91995-9_13

from the US for solidarity and the "responsibility to protect" doctrine. Instead, there are potential risks for a trade war situation.

Recent uncommon instabilities in world politics (e.g., *TIME* Jan. 23, 2017, 16–17, 60) might mean that traditional ad hoc alliances and strategic partnerships, such as the EU-Russian and Russian-Turkish partnerships, the US-Russian-Turkish axis, and even the NATO partnerships, will no longer function. Furthermore, the "US hegemony in security, trade and the promotion of values provided stability for the global economy" might be over (Bremmer 2017). One interpretation of these shifts in global politics posits that, due to growing populism in several democratic states with elections (as manifested by the Brexit vote in the UK and the 2016 US presidential election), power will be decentralized "away from central state actors toward local officials, at the expense of international cooperation. This anger undermines the authority of supranational organizations—the EU, NATO, [and] the UN" (Bremmer 2016/2017).

These findings are important parts of the bigger geopolitical picture, which reflects world politics, the global economy and world trade, as well as global relationships between nations and people. However, this picture is not complete. On the one hand, there are good arguments for fewer wars, less poverty, and more stability and prosperity than 20 years ago. On the other hand, there are also common interests between the major powers of China, the EU, and the US and Russia in particular. Such common interests include the nuclear deal with Iran (although it has been criticized), joint efforts to end the Syrian war, and stopping the growing military threat from North Korea. All these major powers want to find a solution via negotiations and support the UN Security Council's resolution for North Korea to stop its constant nuclear and missile tests. There are also common concerns about environmental degradation, particularly the global impacts of climate change. Finally, there are common interests between the US and Russia, and all Arctic states (including Canada and the Nordic countries), regarding the Arctic region and Arctic affairs.

Interestingly, East-West tensions due to the Ukrainian crisis or the Syrian war have not affected the high geopolitical stability and international cooperation in the Arctic region. There have been suggestions that Western sanctions have hampered, but not stopped, Arctic cooperation in regional institutions. By the end of 2017, East-West tensions did not have any harmful effect on international Arctic cooperation. Geopolitical Arctic stability, based on the civilianization of intergovernmental cooperation on environmental protection and science, national region-building, and circumpolar cooperation by indigenous peoples' organizations and sub-national governments (e.g., Heininen 2004) remains resilient.

In this chapter, I argue that the globalized Arctic is an exceptional political space in world politics and international relations, based on intensive international, functional cooperation and high geopolitical stability (see Byers 2017; Käpylä and Mikkola, in this volume). Furthermore, the post-Cold War Arctic can influence current turbulent world politics (e.g., Heininen 2016; NGP Yearbook 2012). This stability does not result from either the classical approach of Great-Game geopolitics or the Hobbesian zero-sum approach. It results from applying a critical and constructivist approach to geopolitics (e.g., Lamy et al. 2013). It combines Gorbachev's (1987) realist concept of the eight Arctic states as a "zone of peace," Arctic globalization, and critical approaches of (state) sovereignty and traditional powers by local, regional and global (non-state) actors, emphasize immaterial values and that the environment matters. This hypothesis of Arctic geopolitics as a potential asset for world politics is inspired by Chancellor Angela Merkel's (2011) speech, in which she said that solidarity matters the most and stated that nation's political legitimacy comes from that it reformulates globalization, as well as has global responsibility.

The chapter will draw a holistic picture of Arctic geopolitics and international relations/IR in the (post) post-Cold War period. Its aim is to examine and analyze how the high geopolitical stability of the Arctic was achieved and been maintained, and why it seems to be resilient. The chapter will first briefly discuss the recent transformation of Arctic geopolitics from classical to critical. Second, it will identify, examine, and analyze common interests between the Arctic states, with an ultimate aim to transform from confrontation to cooperation. Third, it will discuss relevant features of Arctic geopolitics and security, which can be seen as prerequisites for high stability. Finally, it will conclude by asking if there are special features that could be new themes of global Arctic geopolitics, based on the common interests and prerequisites.

13.2 Transformation from Classical to Critical Geopolitics

This section briefly discusses Arctic geopolitics, particularly its transformation from classical to critical geopolitics, and a need for more comprehensive coverage of its past and present. When assessing post-Cold War Arctic geopolitics, there is ambivalence about how to define *geopolitics*. For example, issues regarding indigenous people, the challenges of climate change, or non-Arctic states becoming observers on the Arctic Council, are usually considered matters of governance, not geopolitics. Unlike, a Russian long-range strategic bomber flying through Arctic international air space, or a military exercise, are interpreted to mean that "Geopolitics is back!". Actually, these all examples deal with geopolitics broadly understood.

Behind is the fact that popular geopolitics is more than written text and the way how media reports about the Arctic. For example, the Russian expedition in August 2007 to the bottom of the North Pole mattered in terms of how big an audience read about the situation and defined geopolitics.[1]

There seems to be confusion about Arctic geopolitics in terms of how to interpret and define it. There is also a need to clarify that current Arctic geopolitics is not only about mainstream interpretations of (classical) geopolitics, but also the reasons for, and consequences from, the recent significant transformation of the Arctic. The two major discourses of post-Cold War Arctic geopolitics—as either a peace zone

[1] So far, this misunderstanding has had less harmful consequences, mostly causing misinformation and confusion, but it has tested the established international Arctic cooperation and might put its stability at stake (e.g., Heininen 2016).

or a resource race with emerging conflicts—closely deal with traditional geopolitics and international relations, and emphasize the role of Arctic states. However, there are some new approaches to looking at Arctic geopolitics. First, environment matters, and Arctic geopolitics and security are affected by substantial (environmental) challenges, such as long-range pollution, rapid climate change (e.g., Hoogensen Gjörv et al. 2013; Heininen 2013), and the temptation of mass-scale resource utilization based on globalization demands. Second, new dynamics have globalized Arctic geopolitics (e.g., Heininen and Southcott 2010; Keil and Knecht 2017), with multidimensional, worldwide implications. Third, Arctic geopolitics is influenced by an increasingly dense network of transnational actors, including indigenous peoples' (organizations) that emphasize their cultural and political identities, sub-national governments in charge of regional development, non-governmental organizations (NGOs) with concerns and ambitions to shape the discourse, and academic communities producing new knowledge and shaping our understanding of the region.

Finally, Arctic geopolitics and international relations of the post-Cold War era have a certain dualism of military nuclear weapons systems alongside high geopolitical stability based on international, institutional cooperation. There are neither armed conflicts nor real disputes, but instead a functional cooperation in several fields. There is also an established dialogue between the Arctic states and other Arctic actors, as well as between Arctic actors and those outside the region. This provides room for discussing governance and sovereignty on different faces, images and perceptions of the post-Cold War Arctic (e.g., Murray and Nuttall 2014; Steinberg et al. 2015; Powell and Dodds 2014), as well as concentrating on institutional points of view, international treaties and institutions such as the Arctic Council (e.g., Raspotnik 2016; Yearbook of Polar Law 2013).

These notions indicate the need for a holistic approach and a substantial description of the global Arctic, particularly concerning resource geopolitics, to support the critical point of view of geopolitics. In contrast to classical approaches to geopolitics as a determined dogma more than an analysis, a holistic geopolitical description of the current, globalized Arctic includes comprehensive coverage of factors and identities of both the Arctic's Cold War legacy and the contemporary, broad, and new approaches to geopolitics (e.g., Heininen 2016; Raspotnik 2016). Based on the main themes or trends of geopolitics and international relations (Heininen 2004), new, special features of the post-Cold War Arctic geopolitics are based more on factors from a critical approach to geopolitics than from a classical one (physical space and natural resources, and power and the state). On the one hand, these new factors are based on immaterial values such as identities, images, knowledge, confidence, and the environment. On the other hand, there are several relevant actors, such as people and societies, which have their own interests, in addition to those of the state. It also matters if these interests are common, competitive, or conflicting.

According to a critical approach to geopolitics, politicizing the Arctic as a physical space means including its identity, knowledge, indigenous people, and growing concern with the state of the environment and the climate as geopolitical factors. Knowledge of the climate and environment is closely related to indigenous people and civil societies as a relevant form of power. Although the media is sometimes full of mis-/disinformation, conspiracy theories, manipulation and falsification, sophisticated environmental power is based on knowledge and communications. In Arctic research and policy-shaping, traditional or indigenous environmental knowledge (TEK) is well recognized by Western science (e.g., Toyama Conference Statement 2015; Agreement on Enhancing International Arctic Scientific Cooperation 2017).

This multi-dimensional geopolitical approach was seen in the Viking Age (approximately 800–1050 AD), characterized by the mobility and expansion of the Scandinavian people (Ahola and Frog 2014, 21). However, geopolitics is rarely applied to historical eras prior to the existence of states. In addition to the classical geopolitical factors of physical space and place(s), and powerful armed forces the Vikings introduced new factors that could be included in critical geopolitics. The Vikings saw themselves as great communicators with distinct identities, knowledge and technological skills (ship-building and navigation), who were highly mobile. In particular, this mobility was a special factor and precondition for geopolitics, and became the main means by which it was possible for the Vikings to spread their influence over large areas of the northern Atlantic and Europe, politicizing these regions. This mobility and connectivity later allowed for building and maintain communication lines for supplies, which established the domination of Vikings societies in these regions (Heininen et al. 2014).

Current Arctic geopolitics is often misunderstood, particularly by the media and policy-makers, to mean classical approaches to geopolitics. In truth, post-Cold War Arctic geopolitics is multi-functional. Although the legacy of Cold War geopolitics is still present in terms of strong military structures, new actors and factors of critical geopolitics play a bigger role in Arctic geopolitics. However, there is a growing need for a paradigm shift in approaching geopolitics, how to interpret politics and power, and go beyond the classical approach.

13.3 Common Interests for High Arctic Stability

Despite different perceptions, discourses and approaches, the post-Cold War Arctic has high geopolitical stability. This is based on several factors of classical and critical approaches, as well as institutional, international cooperation started by the Arctic states, and supported by non-Arctic state actors (indigenous peoples' organizations, NGOs, sub-national governments, and civil societies). As a result, the post-Cold War Arctic is more politicized, but less militarized, so there are no armed conflicts or serious disputes over national borders. Instead, there is functional cooperation on several fields and dialogue between the Arctic states and other local actors, and between Arctic actors and those from outside the region.

The high stability and cooperation in the Arctic region have a solid foundation and are affirmed by the eight Arctic states. The first preamble of the joint ministerial meeting declarations of the Arctic Council reaffirms "the commitment to maintain peace, stability and constructive cooperation in the Arctic" (e.g., Iqaluit Declaration 2015; Fairbanks Declaration 2017). The five coastal states of the Arctic Ocean declared to "strengthen this cooperation [over the Arctic Ocean], which is based on mutual trust and transparency" after the confusion of how to interpret the Russian expedition to the bottom of the North Pole in the summer of 2007 (e.g., Heininen 2016, 25–26; Ilulissat Declaration 2008). This stability also seems resilient, since it has been maintained despite recent, turbulent international politics (e.g., Clifford 2017).

On the one hand, the high stability of the Arctic region is due to common interests of the Arctic states to decrease military tension and increase political stability by transforming from the Cold War period to functional environmental cooperation. On the other hand, certain features of Arctic geopolitics were prerequisites for a transformation. The Arctic states had common interests to functionally cooperate across borders and maintain good (formal and informal) bilateral and multilateral dialogues between the states and other stakeholders, and strengthen the Arctic Council as the leading policy-shaping body for the Arctic and Arctic cooperation (e.g., *Arctic Yearbook* 2016).

Although there is a need for further and deeper research on the matter, there were and still are several common interests between the Arctic states: Decreasing military tensions and increasing political stability; starting transboundary, functional (expert and scientific) cooperation for environmental protection, with an aim to expand to other fields; starting modern region-building with states as major actors; supporting circumpolar cooperation between indigenous peoples/nations, sub-national governments, and other non-state actors; enhancing the Arctic as a workshop for international scientific research, based on the history of exploration; and enhancing and developing sustainable, long-term business relations and economic development.

In this section, I will examine and analyze these common interests between the Arctic states for confidence-building in the post-Cold War Arctic region.

13.3.1 Decrease Military Tension and Increase Political Stability

The first and most fundamental common interest between the Arctic states at the end of the Cold War period was, and still is, to decrease military tension and increase political stability in the Arctic between the former rivals of the Cold War period (e.g., Heininen 2004). This is the ultimate aim for Arctic cooperation, and requires confidence-building measures. Transboundary cooperation and trade across national borders may be the best means to achieve this objective because they require and promote interdependence. Correspondingly, the best way to strengthen interdependence is by building mutual confidence and trust between parties. Confidence is neither automatic nor immediate; it must be earned and slowly implemented.

Parties that become rivals, such as the Soviet Union versus the US and other NATO member states during the Cold War Arctic period, need a special method to build confidence. According to Mitrany's theory of functionalism (1975), the most efficient method is functional cooperation in areas of low politics, such as sports (the US ping-pong diplomacy toward the Peoples' Republic of China; the 2018 Winter Olympic Games diplomacy between North and South Korea), culture (the Soviet Red Army Chorus concerts in the US), and environmental protection (transboundary cooperation for Arctic environmental protection).

To summarize, the first, most fundamental common interest between the Arctic states to start building confidence—decreasing military tension and increasing political stability—has been successful so far.

13.3.2 *Transboundary Cooperation on Environmental Protection*

Transboundary and functional (expert and scientific) cooperation for environmental protection and assessment in the Arctic was the origin of Arctic cooperation in the 1990s, as well as the main pillar of the Arctic Council. The idea behind the theory of functionalism is that successful environmental protection cooperation could be expanded to other fields.

Functional cooperation on Arctic environmental protection is based on the Murmansk Speech by then-Soviet President Mikhail Gorbachev (1987). The speech inspired Finland to follow the speech's initiative regarding Arctic environmental protection. The first heyday was in June 1991, when the Arctic Environmental Protection Strategy (AEPS) was signed by the Arctic states' environment ministers in their first meeting (Rovaniemi Declaration 1991). The second heyday was in September 1996, when the Arctic Council was established (Ottawa Declaration 1996), when institutionalized, intergovernmental Arctic cooperation began.[2]

On the one hand, long-range air and water pollution became a trigger for environmental awakening in the entire northern region, which created growing pressure on the Arctic states. Due to the migration of pollutants, such as DDT, heavy metals, organic pollutants and radioactivity (particularly from the Chernobyl nuclear plant accident) (e.g., AMAP 2002), indigenous peoples (particularly the Saami and Inuit), environmental organizations, scientists and scholars, and civil societies became concerned about the state of the environment in the entire northern region. They pushed the governments of the Arctic states to do something in response. By adopting new, innovative practices (AEPS) and establishing new structures (Arctic Council), the Arctic states also took back control of the region, which was in accord with the interests of all the Arctic states, despite the dividing line of the Cold War (see Heininen 2013). On the other hand, the first international treaties for environmental protection,

[2]Interestingly, the concept of the eight Arctic states was formulated by the Rovaniemi process and confirmed by the AEPS in 1991, when the Arctic was geopolitically defined by the Arctic Circle, not the Arctic Ocean and its rim lands. Here, the Ilulissat Ministerial, defined by the five coastal states of the Arctic Ocean, can be interpreted as a throwback to the original, stricter definition of the Arctic to mean the Arctic Ocean.

such as the London Dumping Convention in 1972 and the Polar Bear Agreement in 1973, were enforced. Furthermore, the environment, including its protection, became a new field of governing for the Nordic countries when, for example, the ministry of the environment was established.

The so-called Rovaniemi process led to the AEPS and the successful establishment of the AC. First, it broadened into international cooperation in several fields and became permanent. Second, since environmental awareness in the entire northern region grew, the environment and the climate have mattered more and become an important factor in Arctic geopolitics and security. In particular, there are larger environmental challenges and the so-called wicked problems in the region, such as a rapidly warming climate. Therefore, the Arctic region is now considered an environmental linchpin when it comes to long-range pollution and climate change (e.g., AHDR 2004). It is also seen as a barometer of climate change, since according to annual climate reports the region is warming twice as fast as the rest of the planet (e.g., Global Climate Report 2017). As a result, the Arctic ecosystem is experiencing melting sea ice, rising sea levels, and rapid and extreme weather changes. Finally, institutionalized international cooperation on environmental protection can be the first conscious step toward political stability, which was reached within a rather short timeframe. Therefore, the theory of functionalism worked well in this context.

To summarize, the Arctic states were encouraged by indigenous peoples, environmental organizations, and civil societies to start transboundary cooperation for environmental protection. The functional expert and scientific cooperation for assessing the environment has expanded into other fields. By having this as the main objective of more institutionalized cooperation, the Arctic states could build mutual trust between former rivals, which strengthened geopolitical stability.

13.3.3 Region-Building with States as Major Actors

Modern region-building with states as major actors is the third common interest, which is related to the first two common interests. Examples of this include establishing the Arctic Council and the Barents Euro-Arctic Region. The Arctic states build up the region as another method to decrease military tension and increase political stability, as well as to institutionalize intergovernmental cooperation in the post-Cold War Arctic. Environmental awareness, transboundary cooperation for environmental protection, Nordic cooperation between five of the Arctic states, and supporting circumpolar cooperation by indigenous peoples and sub-national governments made it possible for Arctic states to apply modern region-building methods in the region.

On the one hand, the Arctic states wanted to build up the region, due to new uncertain challenges and threats, such as globalization, pollution, mass-scale utilization, climate change, and growing interests of newcomers to the Arctic region. On the other hand, the Arctic states were worried about losing control of their northernmost regions, including the entire circumpolar northern region, and wanted to keep firm state sovereignty firm (e.g., Heininen et al. 1995). This could be interpreted as a hegemonic approach, particularly by the five coastal states regarding the Arctic Ocean and its resources. Indeed, states are still the major actors and power users in the region, Arctic politics, and international Arctic cooperation. Major institutions, such as the Arctic Council, are intergovernmental forums or hybrid organizations, in which unified states have power. Furthermore, Arctic geopolitics, particularly resource geopolitics, is still about state sovereignty, not the least due to the fact that most of the operating oil and gas companies are State-Owned Enterprises, SOEs (of the five coastal states) (also Finger and Krivorotov in this volume).

To summarize, region-building is very much a common interest of the Arctic states, as their strong support of the Arctic Council clearly indicates (e.g., Fairbanks Declaration 2017). It has proven successful as one of the main trends of post-Cold War Arctic geopolitics and international relations.

13.3.4 Circumpolar Cooperation by Non-state Actors

Supporting and encouraging circumpolar cooperation between Arctic indigenous peoples, sub-national governments, and other non-state local and regional actors is closely related to region-building. Correspondingly, these non-state actors supported environmental protection and gave a legacy to region-building.

Since the 1980s, there has been increasing trans-boundary circumpolar Arctic cooperation by indigenous peoples' organizations, sub-national governments, and environmental NGOs. They together with a few scientists and scholars were the forerunners for environmental protection, mostly regarding nuclear safety. This circumpolar cooperation is supported by para-diplomacy such as in Greenland (Ackren 2014). Similarly, the Baltic Sea regional cooperation among cities, universities, and civilian organizations inspired Norway to initiate the European Arctic cooperation and provide a model for the Barents Euro-Arctic Region in the early 1990s. There are certain benefits and lessons to be learned from the Arctic, Baltic Sea, and North Atlantic cooperative regions, such as: "Solutions and policies cannot be only national or regional, either there is a joint regional strategy or not" (Heininen 2017a).

Inter-regional cooperation is currently threatened by the growing influence of major powers from Europe and Asia through their observer-country positions on the Arctic Council. This influence does not weaken the unified state system in general or the Arctic states' sovereignty, but it might weaken the position and influence of Arctic indigenous peoples (and their organizations) and sub-national governments within the region.

To summarize, circumpolar cooperation is one of the common interests of the Arctic states. It promoted human development and democracy in the post-Cold War Arctic, and was an important factor behind the paradigm shift from confrontation to cooperation (Heininen et al. 1995; Östreng 1999, 16–17). Correspondingly, this cooperation of non-state actors benefits from high stability, since big geopolitical changes in the region, such as Greenland's self-governing status, will most probably be peacefully executed with full agreement of both a national government and a home rule sub-national government.

13.3.5 The Arctic as a Workshop for International Scientific Research

There is a firm agreement to maintain and enhance the Arctic as a workshop for international scientific research. This agreement also allows the scientific community to have more influence within the trans-national Arctic policy-making environment. Science and research has long been one of the main fields (of low politics) of transboundary cooperation. This attitude that knowledge matters is prevalent in the Arctic concerning difficult problems, particularly climate change. It is no wonder that knowledge creation is one of the main functions of the Arctic Council and its working groups (see *Arctic Yearbook* 2016). International scientific cooperation plays an important role in Arctic geopolitics and international relations, as the Agreement on Enhancing International Arctic Scientific Cooperation (2017) shows. Correspondingly, high geopolitical stability is a precondition for sustaining Arctic research (Toyama Conference Statement 2015).

Arctic research is becoming more interdisciplinary, due to multi-functional, complicated, and complex problems, such as pollution and rapid climate change. This trend is supported by the ICARP processes every ten years and the global International Polar Years (e.g., IPY 2006–2007). Arctic research is further supported by the interplay between science, politics and business, in particular the interaction and interdependency pattern of politics and science (see, Östreng 2010, 39–89), as well as between Western science and traditional/indigenous knowledge. This also means that the scientific community has reached a higher position of influence "within the Arctic Council's sub-governmental policymaking environment than at the domestic level" (Forbis and Hayhoe 2018). In addition, there are non-state academic actors, such as IASC and the University of the Arctic, and annual academic gatherings, such as Calotte Academy and Arctic Science Summit Weeks. These stakeholders comprise a so-called *epistemic community*, which successfully negotiated the Stockholm Convention on Persistent Organic Pollutants, signed in 2001 (e.g., Downie and Fenge 2003; Byers 2017, 391–392).

To summarize, the newly accepted scientific agreement clearly states that scientific cooperation is a common interest of the Arctic states. Therefore, research and stability in the Arctic are tightly combined. The Arctic states might even consider the Arctic for de facto peaceful purposes, or the same kind of status as the Antarctic has under the Antarctic Treaty System (Antarctic Treaty 1959).

13.3.6 Economic Activities and Sustainable Business

The last common interest is enhancing and developing sustainable, long-term business relations and economic activities in the Arctic, according to the main mission of the Arctic Economic Council. Furthermore, economic activities are prioritized by the Arctic states in their national policies and strategies (Heininen 2011).

Cooperation and stability are usually preconditions for sustainable, long-term business relations and trade. Furthermore, economic cooperation and trade across national borders increase interdependence, which correspondingly decreases the probability of war. After WWII, free trade was viewed as a peace project, and supported and enhanced by intergovernmental (Western) institutions such as the General Agreement on Tariffs and Trade (GATT) and the World Trade Organization (WTO). Recent development in the Arctic between the Arctic states, supported by Arctic Council observer countries, has increased economic activities, particularly regarding energy and mineral resources, and transportation. Examples include Norwegian offshore oil drilling and Russian onshore gas drilling and liquefied natural gas exporting. Therefore, there is an increasing number of business actors from the Arctic and non-Arctic countries, such as state-owned enterprises (Statoil, Gazprom, and CNCP), sovereign wealth funds (Government Pension Fund of Norway and National Welfare Fund of Russia), and platforms for international dialogue with different stakeholders (World Economic Forum, Arctic Circle Assembly, Arctic Frontiers, and St. Petersburg Economic Forum).

Where there are economic activities, there will naturally be competition. If there is competition without stability and clear guidelines, there is a danger of uncontrolled competition for resources. Since uncontrolled interest is not in accord with the interests of the Arctic states, particularly the five coastal states (Ilulissat Declaration 2008), sustainable business and trade, and long-term business relations are preferred. This is important, since there is a need for new, sustainable economic activities based on renewable resources, such as clean tech, communication, and tourism. In spite of the strategic importance of energy security, opening new oil and gas fields in the Barents Sea and around Alaska, and opening northern sea routes, the reality does not support the continual hype of Arctic hydrocarbons. At the very least, the peak might be over, due to harsh conditions, high cost, and high environmental risks. There are also PR risks in supporting traditional hydrocarbon economies, due to increasing global attention and scrutiny. There is also the potential impact on global climate change, habitat degradation, and community health and welfare, as well as apprehensions about offshore drilling.

As a summary, economic and business activities in the Arctic comprise one of the largest interests of the Arctic states. Business and trade support transboundary cooperation and increase interdependence, which was one of the reasons for the Arctic states to establish the Arctic Economic Council in 2014. Stability is a precondition for sustainable and long-term business relations, **and correspondingly sustainable business relations increase** the stability of the Arctic region.

To conclude this section, these are among the current common interests between the Arctic states. Some were in place when the Arctic states started transboundary cooperation that consequently transformed the region from a military theater of the Cold War into a zone of peace. This does not mean that the Arctic region was defined as a security community based on Karl Deutch's traditional Cold War concept. However, **the region is** becoming a regional security complex, if the changes in security premises of the Arctic states transferred into a paradigm shift of traditional national security (Heininen 2016, 27–30).

In the next section, I examine and discuss features of Arctic geopolitics, which made it possible to define and materialize these common interests. I interpret them as prerequisites for the transformation from confrontation to cooperation, and further to building stability.

13.4 Prerequisites for High Arctic Stability

The common interests and high stability of the post-Cold War Arctic would not be present without a joint understanding of the value and importance of high stability, built on mutual confidence (which is correspondingly based on cooperation) that is beneficial for all parties (e.g., Heininen 2016; Byers 2017; *Arctic Yearbook* 2016). The high stability is manmade, since the Arctic states consciously keep the Arctic region and its affairs out of crises to maintain stability (Heininen 2017b). This understanding, or Arctic consensus (Zagorski 2017), is present, even to the extent that there is rational thinking within the Arctic states of how much each would lose, if any of them rocked the boat and damaged the cooperation. The concept behind this is *spillover*, which is the main idea of the process of functional cooperation as states "become more embedded in an integration process, the benefits of cooperation and the costs of withdrawing from cooperative ventures increase" (Lamy et al. 2013, 91–92). The short history of the post-Cold War Arctic supports this assertion of the theory of functionalism.

There were some features of Arctic geopolitics and security that can be interpreted as prerequisites for the current state of high stability. They made functional cooperation (for environmental protection) possible, promoted increased transboundary cooperation, and made the transformation possible. These preconditions then allowed the Arctic states to carefully consider the risk and costs if they would loose this stability. Therefore, they created opportunities for the Arctic states to recognize common interests and understand the benefits of them, and thus to strengthen the cooperation.

Among these prerequisites for international cooperation and high geopolitical stability: Global nuclear deterrence of the two superpowers as the original nature of the military; no (armed) conflicts or serious disputes over state sovereignty; a high degree of international legal certainty based on firm state sovereignty (such as UNCLOS); devolution and other soft ways of governance; and separation of issue areas and flexible agenda setting.

13.4.1 Original Nature of the Military

The first prerequisite is the original nature of the Cold War-era military, and the nuclear deterrence of the two superpowers—the Soviet Union and the USA—which is now part of its legacy. The heavy military structures in the Arctic region mostly consist of Russian and American nuclear weapon systems, such as a strategic nuclear submarine (SSBN) with the capability to reach 150–200 targets by nuclear warheads, and the command, control, communication and intelligence systems (e.g., Heininen 2010, 238–245; Wezeman 2012). They were constructed and deployed during the Cold War, due to the Arctic's geographical location at the top of the globe **with short distances between the continents**, as well as that of the Arctic Ocean with sea ice, to make it possible to reach the maximum numbers of targets within the Northern Hemisphere (e.g., Miller 1986; Heininen 2010). Although these nuclear weapons (particularly SSBNs) could be used against targets within the region, they are still, first and foremost, for global deterrence. This is the ultimate aim of nuclear weapons, as well as for global balance between Russia and the US. This means that **in the global context** the US and Russia, have competitive military and security-political interests. For example, the new US National Defense Strategy identifies "the re-emergence of long-term, strategic competition" with Russia and China (e.g., *Japan Times* 2018, 4–5).

In conclusion, transboundary intergovernmental cooperation, excluding military-security (e.g., the Ottawa Declaration 1996), region-building, and high geopolitical stability were possible in the Arctic states despite heavy military structures deployed in the region due to the original global nature of those structures. This also explains the main reason why there are no nuclear disarmament or specific arms-control agreements for the Arctic, although then-Soviet president Gorbachev and then-US President Ronald Reagan agreed in principle at the Reykjavik Summit in 1986 to demolish all nuclear weapons in the world.

13.4.2 No (Armed) Conflicts or Disputes of Sovereignty

The second prerequisite was that there were no (armed) conflicts or serious disputes of state sovereignty or disagreements of national borders between the Arctic states after WWII.

After Russia and Norway signed a delimitation treaty on the Barents Sea in 2009, there are only a few disputes between the Arctic states in the entire region. They are mostly between NATO member states, such as the long-term dispute over the Northwest Passage between Canada and the US, which does not seem easy to resolve, and the dispute over Hans Island between Canada and the Kingdom of Denmark (Byers 2013, 10–16, 56–91, 128–170). The small number of border disputes in the maritime Arctic is despite false mainstream interpretations that there are serious disputes and that some may escalate into more serious conflicts. The reality is that the state sovereignty of the Arctic states is firm and respected by all states, including the observer countries.

However, there are environmental conflicts between different stakeholders and disagreements between Arctic indigenous peoples and governments on how to use land and water resources. There are also adopted and implemented agreements on the use of land and waters on the one hand (particularly in Alaska and the Canadian Arctic), and self-governing areas by indigenous people, such as Nunavut in Canada and the new status of the Greenland Home Rule, on the other hand. A new trend seems to be that Arctic indigenous peoples' organizations prefer to redefine the traditional (state) sovereignty to mean resource sovereignty, based on special principles of resource development with "the rights and responsibilities of all indigenous peoples" (Inuit Declaration 2009).

13.4.3 High Degree of Legal Certainty

The next prerequisite is closely related to the previous one. There is a high degree of international legal certainty in the Arctic region. It is based on the two-found development of a firm state sovereignty and national legalization (of the Arctic states), and international legally binding agreements, particularly the UN Convention of the Law of the Seas (UNCLOS).

This high legal certainty is strengthened in the following ways. First, UNCLOS has clear rules on maritime sovereignty and rights to utilize resources, although there is room for interpretations of how to define an ice-free-Arctic Ocean. Also, all parties must respect the convention and follow its rules (for more details, see Byers 2013). Second, the five Arctic Ocean coastal states declared that, although the Arctic Ocean faces significant changes, the Law of the Sea consists of a needed legal framework, as well as giving the coastal states stewardship for protection. Therefore, there is "no need to develop a new comprehensive international legal regime to govern the Arctic Ocean." (Ilulissat Declaration 2008) Third, there are three legally binding agreements between the Arctic states concerning the region,[3] as well as the International Code for Ships Operating in Polar Waters (the Polar Code) by the International Maritime Organization (IMO) in 2017. Although these agreements were negotiated under the auspices of the Arctic Council, they are treaties between the Arctic states, which is a clear evidence of firm state sovereignty and common interests between the eight states. Fourth, the Ilulissat Declaration was followed by the Fisheries Agreement on the Central Arctic Ocean, signed in December 2017. It prohibits illegal fisheries in the basin of the Arctic Ocean. It is a new kind of global agreement concerning the Arctic, since it was negotiated between the five coastal states, Iceland, and the EU

---

[3]Agreement on Cooperation on Aeronautical and Maritime Search and Rescue in the Arctic (accepted in 2011), Agreement on Cooperation on Maritime Oil Pollution Preparedness and Response (accepted in 2013), and Agreement on Enhancing International Arctic Scientific Cooperation (accepted in 2017).

(representing Finland and Sweden), as well as political entities from outside of the region (China, Japan, and South Korea).

To conclude the discussion of these two prerequisites, there are no (armed) conflicts or serious disputes over state sovereignty borders, or any real benefit for use of military force between the Arctic states. Instead, there is a high degree of international legal certainty, which constitutes solid ground for strengthening cross-border connections and stability.

13.4.4 Devolution and Soft Ways of Governance

Devolution and other soft ways to govern, particularly the Nordic model of governance, is the fourth prerequisite for international Arctic cooperation and the region's high stability. This is due to the exceptionalities and special nature of Arctic governance (e.g., *Arctic Yearbook* 2015, 13–20). The value of this is recognized and promoted by the "development of innovative political and legal arrangements that meet the needs of the residents of the circumpolar North without rupturing the larger political systems in which the region is embedded," as formulated by AHDR, emphasizing success stories (2004, 236–237; *Arctic Yearbook* 2017). The idea behind this comes from the above-mentioned concern for the environment by indigenous peoples and civil societies, and supported by Nordic governance, or devolution as a part of democracy and welfare-state models. Environmental politics has also become a new field of foreign policy for Nordic countries, such as the bilateral cooperation on environmental protection between Finland and the Soviet Union, and between Norway and the Soviet Union already in the Cold War era.

Cooperation and integration of the Nordic countries, including three self-governing areas—the Faroe islands, Greenland and Åland, created special models for making and shaping policy, and Nordic-style governance (such as democracy, welfare state, civil society, and peace) (e.g., Kangas and Palme 2005; Kurunmäki and Strang 2010; Archer and Joenniemi 2003). The process of self-determination and self-governing among Arctic indigenous people is closely related to this, such as the above-mentioned land-claim agreements and the self-governing home-rule governments of Greenland and the Faroe Islands (e.g., Bertelsen 2014). As a significant geopolitical change, the latter agreement was peacefully executed, with full agreement of a state and a sub-national government.

To conclude, Nordic cooperation and models as soft ways of governance are identity markers in world politics. Furthermore, they can be interpreted as one of the prerequisites for common interests among the Arctic states, and even as the core for international Arctic cooperation and region-building (see Heininen 2015).

13.4.5 Flexible Agenda Setting

The final prerequisite for common interests is that the Arctic states have separate issue areas and remain flexible in keeping them so. This is mostly done by leaving highly political issues, particular those dealing with military-security and fisheries, out of the joint agenda of the Arctic Council. A footnote to the Ottawa Declaration (1996) states: "The Arctic Council should not deal with matters related to military security."

By consciously separating issue areas, "they are less affected by tensions or breakdowns in other areas," and agendas can be set based on common interests and shared problems (Byers 2017, 388, 391). Furthermore, despite the mutual agreement that the Arctic Council (consisting of the "Arctic Eight" and the Permanent Participants [A8+]) is the main high-level forum for intergovernmental cooperation, there is a practice of creating coalitions among the Arctic states, and the Arctic states and the observer countries, for handling certain issues. The best example of this is the "Arctic Five" (A5), i.e. the five Arctic coastal states—Canada, Kingdom of Denmark, Norway, the Russian Federation and the U.S.A.—that formed to agree on governing the Arctic Ocean (Ilulissat Declaration 2008). Another coalition of 10 countries co-signed the Fisheries Agreement in 2017.

The wisdom behind this is to emphasize the joint will between the Arctic states to maintain the status quo and decrease military tension and increase political stability. This was clearly expressed by the first preamble of the declaration of the ministerial meeting of the Arctic Council in May 2017, at which the foreign ministers of all eight Arctic states were present for the first time: "[R]eaffirming the commitment to maintain peace, stability, and constructive cooperation in the Arctic" (Fairbanks Declaration 2017). These words are taken as speech act, which, along with texts, (re)shape the reality and construct geopolitics (e.g., Moisio 2003). Unlike (new) realism argues, based on Alexander Wendt's claim that "anarchy is what states make of it" (Lamy et al. 2013, 124), the Arctic states believe they can (re)construct their reality of post-Cold War Arctic geopolitics. This is a constructivist approach, since the "forces of power go beyond material; they also can be ideational of discursive" (Lamy et al. 2013, 125). This acceptance of ideas and "power based on knowledge" is more than "power based on force" and control over state sovereignty.

The Arctic has "remained largely insulated from wider geopolitical issues" following the conflicting events of 2014 (Clifford 2017). This means there are chances to deepen cooperation in existing fields and broaden the scope of cooperation into new issue areas, as well as increase human capital for confidence-building (Heininen 2016).

To conclude this section, these prerequisites, **together with** the common interests of the Arctic states, and the understanding of their benefits, consist of a solid ground for high geopolitical stability of the post-Cold War Arctic. This is a proper foundation on which to build in the future, since it has already passed the first tests and shown itself to be resilient. In the next section, I conclude these findings and briefly define which special features could be recognized as potential new themes of Arctic geopolitics and international relations, based on these common interests and prerequisites.

13.5 Conclusion

The Arctic region plays a key role in the global ecosystem and bio-geophysical processes that are heavily affected by pollution, climate change, and other global changes. The Arctic region is also closely integrated with the global economy and energy security dynamics, so it relates to world politics. Long-range pollution was the first trigger for a wake-up call to the Arctic states, due to a growing concern about the environment by non-state actors, such as indigenous peoples, NGOs, and scientists and scholars to start functionally cooperating for environmental protection. This was followed by climate warming as another trigger, which caused wide concern about rapid environmental changes and their local and global socio-economic and political consequences. There are difficult challenges and problems, and the warming Arctic is a potential tipping point for the entire planet. This means that the (post) post-Cold War Arctic is globalized, and that the environment matters a great deal in Arctic geopolitics.

Since then, there have been significant changes in Arctic geopolitics, from the confrontation of the Cold War period to the current state of transboundary cooperation. The region is not over-run by military conflicts or disarmament. Instead, transboundary cooperation is peaceful, with high geopolitical stability. However, heavy Russian and American, military, nuclear-weapon systems are still deployed in the region. International cooperation between the Arctic states and Arctic Council observer countries continues, and has handled growing tension between Russia and the West.

The common interests of the Arctic states to decrease military tension and increase political stability via functional cooperation on environmental protection are the basis for the current successful geopolitical situation in the Arctic. Furthermore, long-term business relations, as well as international scientific cooperation, will be needed to maintain cooperation and high stability for sustainable economic development in the region. However, certain features of Arctic geopolitics, such as the original nature of the military, firm state sovereignty, a high degree of legal certainty, and flexible agenda setting, were prerequisites for the ability to decrease military tension and increase political stability.

Following from this, the first conclusion of this article is that the high geopolitical stability of the Arctic is conscious, manmade, and resilient. Another conclusion is that the state of post-Cold War Arctic geopolitics has certain special features, based on the common interests of the Arctic states and the prerequisites for transboundary cooperation. Among them are: Arctic geopolitics and security are combined and closely related to the environment; the Arctic is exceptional in its current geopolitical situation in the current world political climate filled with uncommon instabilities;

and the Arctic has become part of global, multi-dimensional (political, economic, technological, environmental, and societal) change, which has significant worldwide implications.

These special features could have potential as new avenues of geopolitical study of the global Arctic. For example, to (re)place the Arctic within the context of global, multi-dimensional change and explore worldwide implications would possibly define the global Arctic in a new geopolitical context. Alternatively, the highly stable Arctic region could be become a potential asset to reformulate a state of world politics with uncommon instability. Finally, a global, stable Arctic region could have the same kind of de facto status as the Antarctic as a global commons for peaceful and scientific purposes under its governorship by the Antarctic Treaty System.

References

Ackren M (2014) Greenlandic paradiplomatic relations. Security and sovereignty in the North Atlantic. Palgrave Macmillan, Palgrave Pivot, Basingstoke, pp 42–61

Agreement on Enhancing International Arctic Scientific Cooperation (2017) Signed May 11, 2017 by the Arctic states

AHDR (2004) Arctic Human Development Report, AHDR 2004. Stefansson Arctic Institute, Akureyri

Ahola J, Frog (2014) Approaching the Viking age in Finland. An introduction. In: Frog JA, Tolley C (eds) Fibula, fabula, fact: the viking age in Finland (pp. 21–84). Studia Fennica Historica, Vantaa. Finnish Literature Society

Antarctic Treaty (1959) Treaty signed in Washington on 1 December 1959 by twelve countries. Available at: http://www.ats.aq/e/ats_treaty.htm

Archer C, Joenniemi P (2003) The Nordic peace. Ashgate, Aldershot

Arctic Monitoring and Assessment Program (2002) Arctic pollution: persistent organic pollutants, heavy metals, radioactivity, human health, changing pathways (Oslo)

Arctic Yearbook 2015—Arctic governance and governing. In: Heininen L, Exner-Pirot H, Plouffe J (eds) Thematic network on geopolitics and security, and northern research forum. Available at: http://www.arcticyearbook.com

Arctic Yearbook 2016—The Arctic council—20 years of cooperation & policy-shaping. In: Heininen L, Exner-Pirot H, Plouffe J (eds) Thematic network on geopolitics and security, and Northern research forum. Available at: http://www.arcticyearbook.com

Arctic Yearbook 2017—Change and innovation in the arctic: policy, society & environment. In: Heininen L, Exner-Pirot H, Plouffe J (eds) Thematic network on geopolitics and security, and northern research forum. Available at: http://www.arcticyearbook.com

Bertelsen R (2014) Devolution and withdrawal: Denmark and the North Atlantic. In: Heininen L (ed) Security and sovereignty in the North Atlantic. Palgrave, London, pp 9–27

Bremmer I (2016/2017) The era of American global leadership is over. Here's what comes next. TIME Magazine, 26 Dec 2016/2 Jan 2017, 15–17

Bremmer I (2017) The top global risks for 2017, a year of geopolitical recession. TIME Magazine, 16 Jan 2017

Byers M (2013) International law and the Arctic. Cambridge Studies in International and Comparative Law, Cambridge, UK

Byers M (2017) Crises and international cooperation: an Arctic case study. Int Relat 31(4):375–402

Clifford R (2017) How has cooperation in the Arctic survived Western-Russian geopolitical tension? http://pagebuilder.arctic.arcpublishing.com/pb/voices/analysis/2017/01/04/how-has-cooperation-in-the-arctic-survived-western-russian-geopolitical-tension/
Downie D, Fenge T (2003) Northern lights against POPs: combating toxic threats in the Arctic. McGill-Queens University Press, Montreal
Fairbanks Declaration (2017) The Fairbanks declaration on the occasion of the tenth ministerial meeting of the Arctic Council
Forbis R Jr, Hayhoe K (2018) Does Arctic governance hold the key to achieving climate policy targets? Environ Res Lett 13:020201. Available at: https://doi.org/10.1088/1748-9326/aaa359
Gorbachev M (1987) The speech in Murmansk at the ceremonial meeting of the occasion of the presentation of the Order of Lenin and the Gold Star Medal to the city of Murmansk. 1 Oct 1987. Novosti Press Agency, Moscow
Heininen L (2004) Circumpolar international relations and geopolitics. In: AHDR (Arctic Human Development Report), pp 207–225. Stefansson Arctic Institute, Akureyri
Heininen L (2010) Globalization and security in the circumpolar North. In: Heininen L, Southcott C (eds) Globalization and the Circumpolar North. University of Alaska Press, Fairbanks, pp 221–264
Heininen L (2011) Arctic strategies and policies—inventory and comparative study. The Northern Research Forum & University of Lapland, Akureyri
Heininen L (2013) Politicization of the environment: Environmental politics and security in the circumpolar north. In: Zellen BS (ed) The fast-changing arctic: rethinking Arctic security for a warmer world. University of Calgary Press, Calgary, pp 35–55
Heininen L (2015) The Nordic countries and the Arctic—Why do the Nordic countries not take joint action? Keynote presentation. Security and Governance in the Globalized Arctic: Nordic and International Perspectives. Matchpoints Seminar at Aarhus University, 12–13 Nov 2015
Heininen L (2016) Security of the global Arctic in transformation—Changes in problem definition of security. In: Heininen L (ed) Future security of the Global Arctic. State policy, economic security and climate, pp 12–34. Palgrave Macmillan, Palgrave Pivot, Basingstoke
Heininen L (2017a) The Arctic, Baltic and North-Atlantic "cooperative regions" in "Wider Northern Europe": similarities and differences. J Baltic Stud 48(4):435–450
Heininen L (2017b) From theory to action—How to transform the (geo)political stability of the Arctic into a peaceful change. In: International conference of representatives of the Arctic state-members of the Arctic council, observer states and foreign scientific community Northern Sea route towards strategic and equal partnership in the Arctic, pp 15–21. Anadyr-Pavek, 30 Aug–1 Sept 2016. Security Council of the Russian Federation, Moscow
Heininen L, Southcott C (eds) (2010) The Circumpolar North and globalization. University of Alaska Press, Fairbanks
Heininen L, Jalonen O-P, Käkönen J (1995) Expanding the Northern dimension. Tampere Peace Research Institute. Research Report No. 61. University of Tampere, Tampere
Heininen L, Ahola J, Frog (2014) Geopolitics of the viking age?—actors, factors and space. In Ahola J, Frog, Tolley C (eds) Fibula, fabula, fact: the viking age in Finland, pp 296–320. Studia Fennica Historica. Finnish Literature Society, Vantaa
Hoogensen Gjörv G, Bazely DR, Goloviznina M, Tanentzap AJ (eds) (2013) Environmental and human security in the Arctic. Routledge and Earthscan, Tonbridge, Kent
Ilulissat Declaration (2008) Arctic Ocean conference, 28 May 2008. Available at: http://arcticcouncil.org/filearchive/Ilulissat-declaration.pdf
Inuit Declaration (2009) A circumpolar inuit declaration on sovereignty in the Arctic. Adopted by the Inuit Circumpolar Council
Iqaluit Declaration (2015) Adopted at the ministerial meeting of the Arctic council in May 2015 in Iqaluit, Nunavut, Canada. Available at: http://www.arctic-council.org
Japan Times (2018) U.S. defense strategy calls China, Russia to threats. 21 Jan 2018, pp 4–5

Kangas O, Palme J (2005) Social policy and economic development in the Nordic countries: an introduction. In: Social policy and economic development in the Nordic Countries. Palgrave Macmillan, London, pp 1–16
Keil K, Knecht S (eds) (2017) Governing Arctic change. Global perspectives. Palgrave Macmillan, London
Kurunmäki J, Strang J (2010) Rhetorics of Nordic democracy. Finnish Literature Society, Helsinki
Lamy SL, Baylis J, Smith S, Owens P (2013) Introduction to global politics, 2nd ed. Oxford University Press, New York, Oxford
Merkel A (2011) Speech by Federal Chancellor Angela Merkel at the world economic forum annual meeting 2011 in Davos, 28 Jan 2011. https://www.bundesregierung.de/ContentArchiv/EN/Archiv17/Reden/2011/2011-02-18-bkin-davos.html [28.3.2018, 11:42]
Miller S (1986) The maritime strategy and geopolitics in the high North
Mitrany D (1975) The functional theory of politics. London School of Economics & Political Science
Moisio S (2003) Geopolitiikka kamppailuna. In: Harle V, Moisio S (eds) Muuttuva geopolitikka. Gaudeamus, Tampere, pp 93–109
Murray RW, Nuttall AD (eds) (2014) International relations and the Arctic. Understanding policy and governance. Cambria Press, Amherst, NY
NGP Yearbook (2012) Sustainable development in the Arctic region though peace and stability. Nordia Geogr Publ 40:4. Geographical Society of Northern Finland, Tornio
Östreng W (1999) National security and international environmental cooperation in the Arctic—the case of the Northern Sea Route. Kluwer Academic Publishers, Dordrecht
Östreng W (2010) Science without boundaries. Interdisciplinarity in research, society, and politics. University Press of America, Plymouth, UK
Ottawa Declaration (1996) Declaration on the establishment of the Arctic Council, 19th day of September 1996 in Ottawa, Canada. Available at: http://www.arctic-council.org/establ.asp
Powell RC, Dodds K (eds) (2014) Polar geopolitics: knowledges, resources and legal regimes. Edward Elgar, Cheltenham, UK and Northampton, MA
Raspotnik A (2016) The European Union and its Northern frontier: European geopolitics and its Arctic context. Ph.D. Universität zu Köln, Deutschland
Rovaniemi Declaration (1991) Signed by the eight Arctic Nations, 14 June 1991, Rovaniemi, Finland. (mimeo)
Steinberg PE, Tasch J, Gerhardt H (eds) (2015) Contesting the Arctic, politics and imaginaries in the circumpolar North. I.B. Tauris Publishers, London
TIME Magazine. 23 Jan 2017, pp 16–17, 60
Toyama Conference Statement (2015) Integrating Arctic research: a roadmap for the future. ASSW 2015 in Toyama, 30 Apr 2015. (mimeo)
Wezeman ST (2012) Military capabilities in the Arctic. SIPRI background paper
Yearbook of Polar Law (2013) Volume 5. In: Alfredsson G, Koivurova T, Stepien A (eds). Martinus Nijhoff Publishers, Leiden, Boston
Zagorski A (2017) Arctic consensus. Arctic Herald, Inf Anal J 1(20):42–47

Chapter 14
Diplomacy and Paradiplomacy in the North Atlantic and the Arctic—A Comparative Approach

Maria Ackrén

14.1 Introduction

Paradiplomacy is the concept that sub-regional governments use to handle international relations. In recent decades, sub-regional jurisdictions have been given more self-rule according to the right of self-determination and various voices from different actors have become increasingly crucial in the international debate (Bartmann 2006; Lecours 2008). The present chapter sheds light on the recent developments in the North Atlantic and the Arctic regarding paradiplomatic relations, with a focus on the sub-regional jurisdictions of the Faroe Islands, Greenland, Nunavut, and Svalbard. Although these regions differ in terms of how they are regulated by national and international jurisdictions, all of these cases apart from Svalbard are also heading towards even more self-government in relation to their respective metropolitan state. Svalbard can be seen as a reverse case, where a kind of 'reverse paradiplomacy', or rather normal diplomacy, is at hand. I will elaborate on this subject in a separate section about this region.

Paradiplomacy has become a reaction of currents of change within international relations, where globalization and the rise of transnational regimes are in the forefront. Globalization has expanded economic space and interlinked states and other actors in a complex web of interactions (Bartmann 2006). In such a setting, sub-national jurisdictions try to meet the challenges of flux in all directions. Paradiplomacy can best be understood as a field of international interaction apart from the conventional channels of international diplomacy, where various players are involved with different objectives and acting at various levels of governance (Bartmann 2006). Although regional governments have approached international relations in different ways, three overall layers can be distinguished: the first layer

M. Ackrén (✉)
University of Greenland, Nuuk, Greenland
e-mail: maac@uni.gl

M. Finger and L. Heininen (eds.), *The GlobalArctic Handbook*,
https://doi.org/10.1007/978-3-319-91995-9_14

corresponds to economic issues, the second involves cooperation within various fields (such as culture, education, and technical matters), and the third layer focuses on political considerations as an identity marker within international relations (Lecours 2008). Within the economic field, the regional government aims to develop an international presence in order to attract foreign investments into the region, and to promote target markets for export. In the Arctic context, this refers specifically to the areas of fishing, hunting, and extractive industries. The second layer involves a lot of different areas, where culture and identity play a major role in order to cooperate within these issues. This layer also relates to matters of environmental concerns and therefore does not focus on economic gain. The third layer corresponds to the legal competencies for the region to enter into international agreements or international organizations as a member of its own (Ackrén 2014b). Usually only state actors are members of international agreements or international organizations, but some regions with an extended self-rule might conclude membership of their own.

If the foreign action of a sub-state is modest, in that it only deals with cooperation, a fairly informal information-sharing process between the state level and the regional level may be sufficient. However, if the foreign action involves more issues at hand and if the sub-national jurisdiction can enter into its own agreements and membership, then the relationship between region and state might include genuine consultation and coordination (Lecours 2008). The relative distribution of power between the states and the structure of the international political system gives the sub-regional entities an option to acquire and enhance powers to defend their citizens by external balancing. By forming alliances with other states, the state might reduce the resources needed for its own security because the state can rely on its allies' resources. However, it is not without risks. Sub-regional entities as part of their security policy, must also find allies, but they are not able to offer substantial human forces or expensive military equipment. Their strength in international relations is that they possess something (such as natural resources) that other states or regions might find interesting (Jákupsstovu and Berg 2012). The sub-regional entities, or what Rezvani (2014) calls partially independent territories, can have foreign powers, but never the power of the external use of the military. The process of sharing and dividing sovereignty between the region and the state can deliver substantial advantages that may not be realized through full independence by the region (Rezvani 2014). When a sovereign state is associated with a partial independent territory, the core state or metropolitan state no longer has a monopoly on legitimate control within its international legal space. Final authority and control over some issues is shared and divided within the partially independent union rather than being exclusively controlled by the core state (Rezvani 2014).

Paradiplomacy is not a new concept, having originally been used to refer to the overwhelming economic external activities of American federal states and Canadian provinces. From the early 1960s onwards, the paradiplomatic era began as a consequence of economic growth, dynamism, and activities of sovereign players on the international stage (Dickson 2014; Paquin and Lachapelle 2005). During the 1980s, the European regions started to attract foreign investors through their own initiatives and in the 1990s formal steps towards diplomatic relationships were taken; we are now experiencing sub-regional governments forming their own foreign policies (Criekemans 2008). Three variables or explaining factors stand in the forefront of the expansion of paradiplomacy: the nation-state crisis and globalization; nationalism and nation-building; and the internalization processes (Paquin and Lachapelle 2005). Regions often emulate state-like qualities and, in some cases, use international relations to build up the polity domestically. Therefore, sub-regional jurisdictions can be seen to 'mimic' the practices and discourses of state diplomacy to legitimize their international presence (Dickson 2014).

The underlying objective for sub-regional governments seeking their own identity is recognition. This need for recognition and legitimization could explain why the development of paradiplomacy by sub-state nationalist movements remains a priority. It might also explain why these same nationalist movements often develop a more forceful paradiplomatic position (Paquin and Lachapelle 2005).

14.2 Paradiplomacy in the Arctic

The Arctic has become a topical region in recent decades due to climate change and the situation of being a last resort for natural resources. In the 20th century, we have seen the Circumpolar North become characterized as either a place of decolonization to more metropolitan centers or as a complex mosaic of homelands to peoples with their unique and pre-modern identities (Heininen 2014). During the present century, the main themes to emerge have been the post-Cold War era increasing circumpolar cooperation by regional non-state actors, region-building by Arctic states, and a new kind of relationship between the region and the outside world (Heininen 2014).

Approximately four million people live in the Arctic and close to half a million of them constitute indigenous peoples from a great variety of cultural and linguistic groups. Native people in the Arctic share a common history of attempted forced assimilation into their various metropolitan states and a lack of recognition of their rights on the land they have lived on in the past (Haftendorn 2012). Currently, all indigenous peoples in the Circumpolar Arctic region are aiming for the economic and social development of their communities at the same time as preserving traditional ways of living (Haftendorn 2012). The Saami and the Inuit have actively contributed to the evolving global discourse that is empowering individuals and indigenous peoples to give a voice in international relations, through such fora as the Arctic Council and the regional associations of the people themselves (the Inuit Circumpolar Council, the Saami Council, etc.) (Hough 2013).

Through the incorporation of new non-state actors, the traditional political system upon which international relations has played out is being reorganized and redefined. The boundaries of who governs and how governance proceeds are finding new ways to attain institutional sovereignty (Shadian 2010). For example, Inuit political aspirations for self-determination have encompassed a stewardship approach or rights over historical claimed territories and the resources that accompany a particular region. This also relates to issues of the environment as a means for achieving sustainable development (Shadian 2010).

14.3 The Choice of Cases

As mentioned, this chapter focuses on four regions of the North: the Faroe Islands, Greenland, Nunavut, and Svalbard. The Faroe Islands and Greenland are self-governing areas underneath the Danish realm; Nunavut represents a territory within the Canadian federation; and Svalbard belongs under Norwegian sovereignty, but is regulated by an international treaty of over 40 signatory states. The choice of cases can be seen as two similar cases (the Faroe Islands and Greenland) at one end, and two different cases (Nunavut and Svalbard) at the other. This follows the logic of fuzzy-set analysis, where combinations of similar and different cases are possible (see Ragin 2000), or, as Gerring put it, diverse cases (see Gerring 2007). Choosing cases always raises questions about the researcher's motivations and interests. The motives behind might be both theoretical and empirical grounded, but, as many authors within methodology suggest, they should always be representative for the matter in question (Platt 1992; Gerring 2007; Ragin 1992). All cases are situated either in the North Atlantic or the Arctic and they all lie at the subnational level of the state of which they constitute a part. Another similarity is that they also act within international relations, but with different competences—and this is exactly the issue of investigation.

The Faroe Islands have recently addressed an Arctic strategy of their own from 2013 and both the Faroe Islands and Greenland have the ability to enter into bilateral international relations. Nunavut received the right to self-government in 1999, since which time it has tried to develop its self-government with Greenland as a model. Svalbard is a peculiar case; it falls under Norwegian sovereignty, but is at the same time regulated by an international treaty that allows all the signatories to utilize the islands for science and exploration and exploitation. Below, each region is explored in a more in-depth study.

14.3.1 The Faroe Islands

Up until the 21st century, the Faroe Islands were protected by the Danish state and its ability to dispose military and societal security. During the Cold War, the Faroe Islands were leaning on both NATO and Denmark for security reasons. Today, the Faroe Islands' policy is to join international organizations and sign bilateral agreements with countries of importance (Jákupsstovu and Berg 2012).

The question of national island identity has encouraged an international projection of islands in the establishment of paradiplomatic offices. The Faroe Islands are an example of this, where a mix of identity and functional interests make paradiplomatic relations attractive (Bartmann 2006). The issue of home rule has dominated Faroese politics throughout the postwar period and, in the late 1990s, even independence was a focal issue (Ackrén 2006); in fact, it remains a question that is repeatedly raised in Faroese politics. The Faroe Islands maintain representative offices abroad in London, Brussels, Copenhagen, Reykjavik, and Moscow. The 2005 law on the Islands' external relations gives the Faroese Government the right to enter into negotiations and conclude treaties with other states and international organizations without previous consent from Denmark regarding all areas that are underneath Faroese authorities (Bartmann 2006; Adler-Nissen 2014). This means that the Faroe Islands can negotiate and conclude bilateral agreements with other states and organizations of vital importance for the Faroese context.

In 1973, when Denmark entered the European Community (EC), the Faroe Islands remained outside the EC in order to protect its political autonomy in relation to fisheries. The Faroe Islands is functioning as a 'third country' in relation to the EU. The EU plays an important role both as an export market and as a reference for a political future outside the Danish realm (Adler-Nissen 2014). The Faroe Islands have bilateral agreements with the EU regarding fisheries and free trade.

In 1998, the Faroese Government established its own diplomatic representation in Brussels and, in 2000, the Faroese Parliament set up the Faroese Treaty Commission to draft a constitution for a sovereign Faroe Island state in a 'free association' with Denmark (Adler-Nissen 2014).

In April 2002, the Faroe Islands took over responsibility for search and rescue (SAR) in the Faroese sea territory and established a MRCC (Maritime Rescue Coordination Centre) in Tórshavn, the Faroe Islands. This center cooperates with similar centers in Aberdeen, Scotland and Reykjavik, Iceland. In case of accidents on the sea, the Danish military is also ready to help from its command center in Nuuk, Greenland. A new civic law was passed in the Faroese Parliament (*Løgtingið*) in May 2012 (Jákupsstovu and Berg 2012) and relates to all Faroese emergency and civic security services in case of a security-related crisis or potential crisis. Each minister will be responsible for preparing for emergencies related to his/her portfolio and is in charge of coordinating and advising the locals. This means that the Faroe Islands have been developing their own societal and security policy, a field that used to be shared with the Danish state. The development took its toll in the late 1990s and there has been a clear shift in the institutional context of safety and security policies for the Faroes, as the local administration is in the process of taking full responsibility for a policy field it used to share with Danish state authorities and partly also with NATO forces (Jákupsstovu and Berg 2012).

The treaties made in 2003 and 2006 in the Danish Parliament led to the Faroese having responsibility for human and societal security in the islands since 2007. The police force is an exception, since it is still under Danish control. The issues related to the other areas of responsibility are divided between three ministries: the Ministry of Fisheries, the Ministry of Interior, and the Ministry of Foreign Affairs (Jákupsstovu and Berg 2012).

The introduction of Home Rule led to the Faroese taking responsibility for infrastructure, communication, energy, environment, fire services, the police (until 1958, when it became Danish again), and the health system. Subsequently, the following areas were taken over: coast guard in 1976, media in 2006, contingency planning in 2007, and weather forecasting in 2009. Air traffic is under Danish rule, but Iceland is the executor of this field and radiological security is a joint matter between Denmark and the Faroe Islands (Jákupsstovu and Berg 2012).

The Faroe Islands have taken charge of security on the sea and are planning to take over similar responsibilities in the air. The next step will probably be to transfer the police forces to the Faroes again, giving them the formal responsibility for security on land (Jákupsstovu and Berg 2012).

The Faroe Islands often attract criticism for the way they conduct whaling, but this is actually not subject to international control as it targets small species of whales (mainly pilot whales and some dolphin species) that the International Whaling Commission (IWC) does not account for. Therefore, there are no legal mechanisms currently available to prevent this form of hunting (Adler-Nissen 2014).

Additionally, the Faroe Islands are active within Nordic cooperation, as well as the other autonomous regions in the Nordic context (Greenland and the Åland Islands). The Faroe Islands take part in both the Nordic Council and the Nordic Council of Ministers; the main difference between the two is that the Nordic Council does not make any binding decisions and acts largely as an advisory body to the Nordic Council of Ministers. All Nordic autonomous territories have two seats each in the Nordic Council. These seats are filled with members elected from the representatives of the regional parliaments (Stephan 2014).

14.3.2 Greenland

Greenland has become a focal point due to the vast natural resources that the region possesses. This means that Greenland has become increasingly aware of its ability to act within international relations as an independent country (see, e.g., Bertelsen 2014; Ackrén 2014b).

Like the Faroe Islands, Greenland belongs to the Danish Kingdom and when Denmark speaks on behalf of the whole country, Greenland and the Faroe Islands are often included, but not always. However, the complex web of competencies between the state and the autonomous regions constituting the Danish realm is usually under dispute. Some competencies are directly within Danish control (for example, foreign and security policy), some are joint matters (competencies shared by the state and the autonomous region), and others are totally in the hands of the autonomous region (see, e.g., Ackrén and Jakobsen 2015). This sometimes leads to confusion on an international level among other states, especially in cases where foreign investors are at hand. Asian states, for example, are not always sure which Government they should be in contact with regarding extractive industries. Here both the Greenlandic and Danish Governments should clarify which rules are applicable. According to the

Greenlandic Self-Government Act of 2009 and the Mineral Act from 2010, Greenland has the sole responsibility over its own natural resources.

Greenland also has the ability to enter into bilateral treaties with other states, especially in those areas that are of vital importance for Greenland. Greenland can also enter into international organizations, where matters of Greenlandic concern are in focus (Lov om Grønlands Selvstyre 2009; Ackrén 2014b).

Regarding EU relations, in 1973 Greenland was forced to join together with Denmark in the European Community (EC). Greenland was then an integral part of Denmark and had not yet received Home Rule, which meant that all major decisions were made in Copenhagen. Due to the negative turnout in the Greenlandic referendum in 1972, while Denmark voted in favor of membership, Greenland was dissatisfied with the situation. After introduction of Home Rule in 1979, Greenland saw its chance to negotiate for withdrawal from the EC. A new referendum was held in 1982 and three years later Greenland left the EC and became an Overseas Country and Territory (OCT) underneath the EC and later the EU (Gad 2013). In 2006, a joint declaration between EU and Greenland was signed regarding fisheries, and other agreements were also struck in relation to minerals, transportation, and climate research in a partnership agreement (Gad 2013).

Greenland now has its own diplomatic representation offices in Copenhagen, Brussels, and Washington DC (*Naalakkersuisut*). The 1951 agreement between Denmark/Greenland and the USA regarding Thule Air Base in the North of Greenland is still running and it has been upgraded after 2004 with the Igaliku agreement, which also includes environmental protection (Ackrén and Jakobsen 2015). Greenland does not have a coast guard, so the Danish armed forces perform the tasks of civil protection, search and rescue missions, controlling territorial waters, fisheries inspections and surveillance, and maritime clean-up (Ackrén and Jakobsen 2015).

As a self-governing region, Greenland takes part in various international organizations as its own nation. One of the most important is the Inuit Circumpolar Council (ICC), which is an umbrella organization for Inuit affairs across the Arctic including Alaska, Canada, and Russia. Another important organization is the Arctic Council, where Greenland and the Faroe Islands participate with the Danish delegation. Greenland has also been a member of the Nordic Council and the Nordic Council of Ministers since 1984. In 1985, a West-Nordic initiative was launched as cooperation between the Faroe Islands, Greenland and Iceland. This cooperation was later extended to what has become known as the NORA (North Atlantic Cooperation) including North and coastal West Norway (Ackrén 2014a).

There is an ongoing territorial dispute between Canada and Denmark/Greenland regarding Hans Island, a tiny rock located between Greenland and Ellesmere Island. This island has been claimed by both Canada and Denmark for the past 30 years. Even the 1973 Danish-Canadian delineation of the sea between Greenland and Ellesmere Island neglected to take Hans Island into account. However, the dispute of Hans Island is quite obscure and a solution was suggested by Professor Michael Buyers for both parties to split the tiny rock into two parts or to govern it as a condominium (Hough 2013).

A new coalition was established 27 October 2016 in Greenland between the parties of *Siumut* (Forward-Social Democrats), *Inuit Ataqatigiit* (Inuit Community-leftwing) and *Partii Naleraq* (Indicator-leftwing), since the former coalition broke down between *Siumut*, *Atassut* and *Demokraatit*. The new coalition is now more left-oriented and will continue as government until the next election. The new coalition agreement for 2016–2018 states that international relations will be strengthened and that Greenland will begin negotiations with Denmark in order to give Greenland a better voice within these matters. Regarding *Pituffik* (Thule Air Base), Greenland wants a new evaluation regarding the contracts for the base. The cooperation with the EU within trade, business, natural resources, and the energy sector should be strengthened. Another aim is to have diplomatic missions in the nearest neighboring countries and strategic relations regarding trade. An ongoing goal is to take over immigration policy. Competence for Greenlandic airspace should also be taken over by 2018 and Greenland should become more active in the protection of Greenlandic waters. This could lead to stricter requirements for sailing in these waters (Koalitionsaftale 2016–2018: *Lighed, Tryghed og Udvikling*). A proposal for a Greenlandic constitution will also be fulfilled during the current election period.

14.3.3 Nunavut

In Canada, indigenous peoples have been engaged for decades in a steady process of almost entirely peaceful decolonization in their path toward seeking recognition of their original sovereignty and adjustment of their collective relationships to state institutions. Inuit diplomacy in the North of Canada has been focused on the process of internal decolonization through land claim (modern treaty) negotiations, and the establishment of self-governing entities throughout Inuit territories (Abele and Rodon 2007).

The North West Territory (NWT) previously included Nunavut, but conflicts and disputes in the territory emerged during the 1970s and 1980s between various indigenous groups and other Canadians. In 1982, a referendum was held, which favored dismantling NWT. This meant that a development toward the creation of Nunavut was coming to reality. The NWT sponsored the continuing process of the Nunavut Constitutional Forum and Western Constitutional Forum to debate and to decide on the new constitutional arrangements for both halves of the NWT. The Nunavut land claims negotiations led to the formal creation of the region in April 1999 (Jull 2001). Nunavut has a public government and the territory also has ownership of the surface and subsurface lands, meaning that Nunavut has the right to exploit the natural resources (Abele and Rodon 2007). Nunavut is the largest but least populated territory in Canada, with approximately 33,000 residents, 84 percent of whom are Inuit (Hough 2013).

Regarding paradiplomacy, Canada's most prominent international relations are with the United States; these are of vital importance for Canada's provinces and territories to access the US market. In 2002, less than 0.5% of Nunavut's exports went to the USA. However, this trade is important for the Canadian Inuit living in Nunavut, since it is a market for their musk ox and caribou meat (Mouafo 2004). Nunavut also holds a lot of diamonds, acting as the fifth largest diamond producer in the world (Shadian 2010). Canadian federalism provides subnational governments with great flexibility to engage in international relations. Therefore, Canadian provinces and territories are among the most active in paradiplomacy at various levels of government. The most commonly used mechanism is memoranda of understanding (MoU) between Canada and the USA. In Canada, ministers and other senior officials from the provinces and territories are authorized to enter into international agreements on behalf of their governments. The premier of Nunavut also holds the portfolio of intergovernmental and international relations, as well as indigenous affairs (Mouafo 2004).

The ultimate goals of the paradiplomatic relations with the USA are promoting trade, developing sound investment policies, establishing companies, attracting tourists, and generating jobs. In the North of Canada, Alaska is the partner; here, energy issues, environment and climate change, mining, conservation issues, and indigenous issues are at the forefront of discussions. Most of the paradiplomatic work also goes through other institutions and organizations, such as the Arctic Institute of North America, the Arctic Council, Inuit Circumpolar Council (ICC) and the Northern Forum (Mouafo 2004). Another important NGO is the Inuit Tapiriit Kanatami (ITK), which represents all the Inuit in Canada. The organization has an advocacy role and takes part in international forums (Abele and Rodon 2007).

Foreign policy for the Circumpolar North is largely underneath the federal level, since this is seen as a strategic issue of the Arctic in regard to national defense, national sovereignty, environmental protection, and mineral and energy resources. In 2000, the Canadian Federal Government appointed a Canadian Ambassador for Circumpolar Affairs with the Department of Foreign Affairs in Ottawa, as part of the Northern Dimension Foreign Policy (NDFP) vision in Canada, for the Circumpolar North (Mouafo 2004). However, there has always been cooperation between the Canadian military and the Inuit peoples in the North regarding security policy. Most of the settlements are located in the neighborhood of the former Distant Early Warning (DEW) Line, which is now called the North Warning Stations. The Canadian Rangers have been operating in Inuit lands for more than 60 years and serve as an essential bridge between Inuit communities and outside military forces that train and operate in the Arctic. Inuit serving the Rangers are encouraged to use their indigenous skills and knowledge for search and rescue and emergency responses and helping with local events. The Canadian military is also an important business partner for Inuit communities. Department of Defense (DND) projects in Nunavut have provided opportunities for businesses (Evalik and Lyall 2010).

In Nunavut, Inuit participation is also used in the clean-up of former Distant Early Warning (DEW) Line sites and one of the corporations, Nunavut Tunngavik Incorporated, has been a partner in this matter. This has led to increased employment and training opportunities in the region (Evalik and Lyall 2010). There are also other examples of successful projects with the military (for more details, see Evalik and Lyall 2010).

14.3.4 Svalbard

Svalbard was first discovered by the Dutch explorer Willem Barents in 1596. In 1613, the Muscovy Company, an English whaling operation, claimed exclusive rights to the islands. These claims were contested by the then-united Kingdom of Denmark–Norway (Grydehøj et al. 2012). In 1870, the Swedish–Norwegian Council of Ministers considered annexing Svalbard and setting up a colony on the islands; however, the Russians opposed this idea and it was never realized. Shortly after Norway gained independence from Sweden in 1905, the former sought clarification on the Svalbard issue. Since the archipelago had been a *terra nullius*, several countries could claim an interest in Svalbard (Grydehøj et al. 2012).

Following the First World War, the Svalbard Treaty was signed in Paris in 1920. This gave Norway sovereignty over the archipelago, but all other signatories to the Treaty acquired the right to fish, hunt, trap, set up mines and commercial activities, and acquire mineral rights in Svalbard (Anderson 2009). There were 14 original signatory states, with the Soviet Union and Germany eventually signing the treaty in 1924 and 1925, respectively. Today there are over 40 signatory states included in the Treaty (Grydehøj et al. 2012).

Norway has the right to regulate Svalbard's activities but not to profit from them. Any taxes paid in Svalbard must be spent there, which makes the tax level lower than in Norway proper. No one is allowed to use Svalbard for military purposes, since the region is demilitarized and there is a special treaty obligation for Norway to preserve the natural environment. Up to 60% of Svalbard is covered by nature reserves (Anderson 2009).

Norway's interpretation of the Treaty is that it applies only to the islands and their immediate waters: the wider 200-nautical mile exclusive economic zone (laid out in the UN Convention of the Law of the Sea) belongs to Norway because Norway sees it as an extension of the Norwegian continental shelf. Thus, the fisheries surrounding the islands and rights to the sea bottom belong to Norway. However, Russia does not recognize this interpretation, and other signatories do not exactly agree either, since all signatories enjoy equal rights in these waters (Anderson 2009; see also Numminen 2011).

There have been several disputes regarding the continental shelf, fisheries, mining areas, and even some security matters during the decades in relation to Svalbard (for details, see Pedersen 2008). As an example related to security issues, Russia interprets the provision around demilitarization as a de jure clause. However, seen from a Norwegian perspective, Article 9 in the Svalbard Treaty regulating this matter does not entail absolute prohibition against Norwegian military activity on Svalbard. This means that, according to the Norwegian perspective, Norway can visit the archipelago with naval or coastal guard vessels, have military aircraft land or even have Norwegian military personnel temporarily in the region (Åtland and Pedersen 2008).

In the 1990s, Svalbard became the arena for an increasing number of space-related activities. The Norwegian and international space research community, as well as the aerospace and telecommunications industries, saw a potential of Svalbard as a platform for high-latitude atmospheric research and commercial satellite projects (Åtland and Pedersen 2008). Several research stations are now placed in Svalbard to measure atmospheric changes and climate changes with various international staff.

Svalbard policymaking rarely involves anyone other than high-level governmental officials pursuing their national interests. In 2005, the Norwegian Government initiated *nordomorådedialogene* (dialogues of the North territories, own translation) as bilateral consultations with the United States, the United Kingdom, Germany, France, Canada, and the EU. Instead of intentions to solve some of the issues, Norway received collective objections on its standpoint regarding the Svalbard issue (Pedersen 2008).

Norway's means of exercising sovereignty over Svalbard has been through the office of the Governor (*Sysselmannen*), who reports to the Ministry of Justice in Norway. The governor is responsible for areas such as, police, environmental protection, notarial duties, travel and tourism, coordination, public information, transport, and contact with the foreign settlements in Svalbard (Grydehøj et al. 2012; The Governor of Svalbard).

The main forum for inter-state cooperation in the Arctic is, as mentioned earlier, the Arctic Council. The Council has eight members, all of which are Arctic states. Added to this, the Council accommodates permanent participants in the form of indigenous associations in the Arctic and non-Arctic states as observers (12 states since 2013) (Numminen 2011; The Arctic Council). All these states are also signatories of the Svalbard Treaty. This means that Svalbard is usually handled through diplomatic forums of national interests from the states, but cooperation and peace are often the trademarks of the relationship. The disputes are handled through diplomatic relations and have not yet escalated into any military threats. Recently, however, the Arctic has seen increased military activity from Russia and Canada and the USA. Norway is also moving its military northwards, so some tensions may arise in the Arctic waters in the future.

14.4 Comparative Lessons of Paradiplomacy in the Arctic

The four regions under investigation are all of international interest, in one way or another. The Faroe Islands, Greenland, and Nunavut have all developed into self-governing territories with major competencies, both internally and internationally. All regions participate in international relations through diplomatic missions abroad, bilateral agreements with other states and regions, as members in international organizations, and they even have clear Arctic agendas. The Svalbard case can be seen as a 'reverse case' of paradiplomacy or just as a case of diplomacy, since this case is part of high-level diplomacy between sovereign states. The purpose of the local administration, under which the governor of the island is located, is only to showcase Norwegian sovereignty, but in reality the island is governed by all signatory states. However, Norway and Russia are the two countries that have commercial interests on

Table 14.1 Overview of the regions

Region	Diplomatic missions abroad	Membership in international organizations	Bilateral agreements	Sovereign power
Faroe Islands	London, Brussels, Copenhagen, Reykjavik, Moscow	AC, FAO (Associate Member), IMO (Associate Member), UNESCO (Associate Member), NC, NIB, NAMMCO, UPU	**Fisheries**: Iceland-Faroe Islands, Russia-Faroe Islands, EU-Faroe Islands, Faroe Islands-Greenland **Free Trade Agreements**: Faroe Islands-EU, Faroe Islands-Switzerland, Faroe Islands-Norway, Faroe Islands-Iceland	Denmark
Greenland	Brussels, Copenhagen, Washington DC	AC, ICC, NC, NIB, IMF, UPU	**Fisheries**: EU-Greenland, Iceland-Greenland, Norway-Greenland, Greenland-Faroe Islands, Greenland-Russia **Partnership Agreement**: EU-Greenland **Joint Agreements**: Denmark/Greenland-USA	Denmark
Nunavut		ICC	**MoU**: Nunavut-Greenland, Nunavut-Manitoba	Canada
Svalbard	–	–	–	Norway, but international treaty

Source Rezvani (2014); CIA World Factbook: https://www.cia.gov/library/publications/the-world-factbook/; The Government of the Faroe Islands: http://www.government.fo/foreign-relations/the-faroe-islands-in-the-international-community/; Department of Executive and Intergovernmental Affairs: http://gn-curriculum.ub8.outcrop.com/eia/information/intergovernmental-affairs-0

the island through mining, although other signatory states are carrying out research and scientific measurements through their research stations.

Table 14.1 shows an overview over the international relations regarding these regions.

As the table shows, the Faroe Islands seem to be the most involved in international relations. The archipelago has five representative offices abroad (London, Brussels, Copenhagen, Reykjavik, and Moscow), all of which are significantly important in relation to fishing policy and free trade. The Faroe Islands also take part in several international organizations, either as a full member or as an associate member. Several bilateral agreements have been struck between the Faroe Islands and neighboring entities, such as Iceland, Norway, Greenland, Russia, and the EU. Free trade agreements have also been entered into with the EU, Switzerland, Iceland, and Norway. Greenland is also engaged in multilateral participation in international organizations and has three representative offices abroad to date (Brussels, Copenhagen, and Washington DC). However, the new coalition agreement states that several other representative offices might be underway in the closest neighboring countries; Iceland and Canada are likely relevant options in this regard. Greenland has also several fishery agreements with other countries and regions and a partnership agreement with the EU. Within the security field, Greenland is also a signatory of the joint agreement between Denmark and the USA. Several other agreements are related to this matter. Nunavut does not have the same powers as the Faroe Islands and Greenland, but within the Canadian flexible system there are several possibilities for Nunavut to act in the international arena. However, this competence does not seem to be a priority for Nunavut at the moment. Svalbard can be seen as a subject for Norway, which still acts as the sovereign over the territory; at the same time, Svalbard is seen as an object for international collaboration on a high-level diplomatic arena.

References

Abele F, Rodon T (2007) Inuit diplomacy in the global ERA: the strengths of multilateral internationalism. Can Foreign Policy J 13:45–63

Ackrén M (2006) The Faroe Islands' option for independence. Island Stud J 1:221–238

Ackrén M (2014a) Paradiplomacy in Greenland. In: Grydehøj A et al (eds) Paradiplomacy. Centre Maurits Coppieters, Brussels, pp 64–76

Ackrén M (2014b) Greenlandic Paradiplomatic relations. In: Heininen L (ed) Security and sovereignty in the North Atlantic. Palgrave Macmillan, Basingstoke, pp 42–61

Ackrén M, Jakobsen U (2015) Greenland as a self-governing sub-national territory in international relations: past, current and future perspectives. Polar Record 51:404–412

Adler-Nissen R (2014) The Faroe Islands: independence dreams, globalist separatism and the Europeanization of postcolonial home rule. Cooperation and Conflict 49:55–79

Anderson A (2009) The strange case of Svalbard. In: Anderson A (ed) After the ice: life, death and politics in the new Arctic. Virgin Books, USA, pp 122–131

Åtland K, Pedersen T (2008) The Svalbard Archipelago in Russian security policy: overcoming the legacy of fear—or reproducing it? Eur Secur 17:227–251

Bartmann B (2006) In or out: sub-national island jurisdictions and the antechamber of paradiplomacy. Round Table: Commonwealth Jo Int Aff 95:541–559

Bertelsen R (2014) Devolution and withdrawal: Denmark and the North Atlantic, 1800–2100. In: Heininen L (ed) Security and sovereignty in the North Atlantic. Palgrave Macmillan, Basingstoke, pp 9–27

Criekemans D (2008) Are the boundaries between paradiplomacy and diplomacy watering down? Preliminary findings and hypotheses from a comparative study of some regions with legislative power and small states. In: Paper presented at the 2nd global international studies conference: what keeps us apart, what keeps us together? International order, justice, value. Ljubljana, Faculty of Social Sciences, University of Ljubljana, Slovenia

Dickson F (2014) The Internationalisation of regions: paradiplomacy or multi-level governance? Geogr Compass 8:689–700

Evalik C, Charlie L (2010) Canadian sovereignty, the military and infrastructure development in the Inuit homeland. Submission to the senate standing committee on national security and defence, December 2010. Kitikmeot Inuit Association, Nunavut

Gad UP (2013) Greenland projecting sovereignty—Denmark protecting sovereignty away. In: Adler-Nissen R, Gad UP (eds) European integration and postcolonial sovereignty games, pp 217–234. Routledge, London and New York

Gerring J (2007) Case study research—principles and practices. Cambridge University Press, Cambridge

Grydehøj A, Grydehøj A, Ackrén M (2012) The Globalization of the Arctic: negotiating sovereignty and building communities in Svalbard, Norway. Island Stud J 7:99–118

Haftendorn H (2012) The case for Arctic governance: the Arctic puzzle. Institute of International Affairs—Centre for Arctic Policy Studies, University of Iceland, Reykjavik

Heininen L (2014) Northern geopolitics: actors, interests and processes in the circumpolar Arctic. In: Powell RC, Dodds K (eds) Polar geopolitics? Knowledge, resources and legal regimes. Edward Elgar Publishing, Cheltenham, UK and Northampton, USA, pp 241–258

Hough P (2013) International politics of the Arctic: coming in from the cold. Routledge, London and New York

Jákupsstovu í B, Regin B (2012) The Faroe Islands' security policy in a process of devolution. Stjórnmál & Stjórnsýsla 2:413–430

Jull P (2001) Negotiating nationhood, renegotiating nationhood: Canada's Nunavut and Nunavut's Canada. In: Paper presented at re-thinking indigenous self-determination: an international conference on the theory and practice of indigenous self-determination, 25–28 Sept 2001, School of Political Science & International Studies, University of Queensland, Brisbane, Australia

Lecours A (2008) Political issues of paradiplomacy: lessons from the developed world. Discussion Papers in Diplomacy, Netherlands Institute of International Relations "Clingendael"

Lov om Grønlands Selvstyre (2009) Lov nr. 473 af 12. juni 2009. Udgivet af Bureau for Inatsisartut

Mouafo D (2004) Regional dynamics in Canada–United States Relations. In: Paper presented at the 2004 conference of the Canadian Political Science Association, University of Manitoba, Winnipeg, 3 June 2004

Numminen L (2011) A history and functioning of the Spitsbergen treaty. In: Wallis D, Arnold S (eds) The Spitsbergen treaty—Multilateral governance in the Arctic, pp 7–20. Arctic papers, vol 1. Applied International Law Network

Paquin S, Lachapelle G (2005) Why do sub-states and regions practice international relations? In: Lachapelle G, Paquin S (eds) Mastering globalization: new sub-states' governance and strategies. Routledge, London and New York, pp 77–89

Pedersen T (2008) The dynamics of Svalbard diplomacy. Diplomacy Statecraft 19:236–262

Platt J (1992) Cases of cases…of cases. In: Ragin C, Becker HS (eds) What is a case? Exploring the foundations of social inquiry. Cambridge University Press, Cambridge, pp 21–52

Ragin CC (1992) Introduction: cases of 'what is a case?'. In: Ragin CC, Becker HS (eds) What is a case? Exploring the foundations of social inquiry, pp 1–17. Cambridge University Press, Cambridge

Ragin CC (2000) Fuzzy-set social science. The University of Chicago Press, Chicago and London

Rezvani DA (2014) Surpassing the Sovereign state: the wealth, self-rule, and security advantages of partially independent territories. Oxford University Press, Oxford

Shadian J (2010) From states to polities: reconceptualizing sovereignty through Inuit governance. Eur J Int Relat 16:485–510
Stephan S (2014) Making autonomies matter: sub-state actor accommodation in the Nordic council and the Nordic council of ministers. European Diversity and Autonomy Papers, EDAP 03/2014, EURAC Research. http://www.eurac.edu/edap

Web Pages

CIA World Factbook. https://www.cia.gov/library/publications/the-world-factbook/
Department of Executive and Intergovernmental Affairs. http://gn-curriculum.ub8.outcrop.com/eia/information/intergovernmental-affairs-0
Koalitionsaftale 2016–2018: Lighed, Tryghed og Udvikling. http://naalakkersuisut.gl/~/media/Nanoq/Files/Attached%20Files/Naalakkersuisut/DK/Koalitionsaftaler/Koalitionsaftale_S_IA_PN_2016_2018.pdf
Naalakkersuisut. http://naalakkersuisut.gl/da/Naalakkersuisut
The Arctic Council. http://www.arctic-council.org/
The Government of the Faroe Islands. http://www.government.fo/foreign-relations/the-faroe-islands-in-the-international-community/
The Governor of Svalbard. http://www.sysselmannen.no/en/Toppmeny/The-Governor-of-Svalbard/The-Governors-Tasks/

Chapter 15
The Arctic as a Laboratory of Global Governance: The Case of Knowledge-Based Cooperation and Science Diplomacy

Rasmus Gjedssø Bertelsen

15.1 Introduction: The Arctic Giving Back to the International System

The take-away message from this chapter is that while the Arctic has historically been heavily influenced by the international political, economic, and security system, the area may now offer important lessons for managing very complex and dangerous systemic processes of power transition. Popular and even academic writing on the Arctic has described international and geopolitical attention to the Arctic as something new (Rosen 2016). This new attention is supposed to have been driven by climate change, which has made natural resources and shipping routes more accessible. There have also been alarmist writings about a sudden new risk of conflict in the Arctic over these natural resources (Bittner 2016). Such "presentism" (Halliday 2001) concerning the Arctic has clouded the research and policy lessons of the Arctic.

The Arctic has been an integrated part of a Western-led international political, economic, and security system for centuries, and questions of peace and war in that system have deeply affected the Arctic and its people. The North Atlantic Arctic was a part of medieval Europe. Both the North American Arctic and the Pacific Arctic were deeply influenced by the expansion of Europe (including the Russia of Peter the Great) in the 1600s and 1700s. Many parts of the Arctic were greatly affected by the World Wars and the Cold War, politically, economically, and socially. Today, the Arctic is deeply affected by the great systemic forces of our time: the ongoing struggles following the Cold War, power transition with the rise of Asia (the subject of chapters in this book by Kopra and Li and Peng), and globalization with the compression of time and space in interactions between societies (Harvey 1989) around the world (Heininen and Southcott 2010).

R. G. Bertelsen (✉)
UiT-the Arctic University of Norway, Tromsø, Norway
e-mail: rasmus.bertelsen@uit.no

M. Finger and L. Heininen (eds.), *The GlobalArctic Handbook*,
https://doi.org/10.1007/978-3-319-91995-9_15

However, an important difference between the present and previous centuries is that the Arctic may hold important lessons for the world for handling both the ongoing systemic struggles over the post-Cold War place of Russia and the power transition with the rise (or return) of China and other emerging economies, especially in Asia (see Kopra, and Li and Peng in this volume). This possible learning from the Arctic is a significant change from when the Arctic was the dependent variable of international systemic developments.

The end of the Cold War brought political freedoms and new socio-economic opportunities to many, but also initiated struggles over Russia's place in the international system (and Russia's domestic political and economic system). This has had significant consequences and even violent conflict, for instance, in Georgia in 2008, in Ukraine since 2014 (Mearsheimer 2014), and most recently in Syria. Power transition is one of the most complex and dangerous recurrent processes in international politics. For centuries, power transition caused by the decline of the status quo hegemon and/or catch-up by a challenger rising power has led to extremely violent and costly large-scale human suffering. The challenger may well fail and a third power may end up benefitting from the struggles of hegemon and challenger. Examples of such power transition conflicts are the Napoleonic Wars, the two World Wars, and the Cold War. Perhaps the only peaceful power transition has been that from British hegemony to American during the first half and mid-1900s (Organski 1958).

My argument in this chapter is that the Arctic is being deeply affected both by post-Cold War struggles over Russia's international position and by power transition with the return of China (see Li and Peng's chapter). However, there are processes and institutions in the Arctic that hold lessons of broader international value for managing both these current international systemic challenges, which on a broad international level hold the risk of large-scale conflict and human suffering. The lessons of the Arctic concern the role of transnational knowledge relations embedded in scientific (including all fields of research, culture, society, nature, technology, health, etc.) and other knowledge-based collaboration. Such activity is usually termed science diplomacy (The Royal Society and AAAS 2010), which is also the term used here. I look at a larger sense of knowledge-based collaboration between academia, business, civil society, and government inspired by the triple-helix concept (Etzkowitz 2008). However, there are usually more than three helixes intertwined because of the key role of indigenous peoples and civil society in the Arctic and outside.

Transnational knowledge cooperation (such as public science diplomacy or triple-helix knowledge cooperation) is a dimension of the question of the social relevance of science. The debate regarding the social relevance of science focuses on questions as science addressing societal problems that can—as in the European Union research context—be labeled as grand challenges, such as climate change, socially inclusive societies, and aging societies. Other dimensions of the social relevance of science are public accountability, democratization of science, or business applications of science. Fundamental dimensions of the social relevance of science are the role it plays in decision-making and quality control of science (Scott 2006). These dimensions are all relevant to Arctic science. When looking at the social relevance of science, there can be a partial focus on local activities and local effects, which overlooks the importance of local–global connections through locally embedded knowledge-based activities. Academic, public, private, and civil society knowledge-based activities in the Arctic are illustrative examples of the transnational or local–global dimensions of the social relevance of science. In this chapter I will discuss examples such as the interlinked platforms of the Northern Research Forum Open Assemblies, the University of the Arctic Thematic Network on Geopolitics and Security, and the GlobalArctic.org project, which illustrate the local–global or transnational dimensions of social effects of science in the Arctic.

The Arctic is a geographically vast region with a very small population and a scientifically important environment, which has only become much more scientifically important due to climate change (see also chapter by Kopra). The North and South Poles are the most challenging areas to conduct research, apart from space and perhaps the deep oceans. Therefore, the Arctic (together with the Antarctic, which is outside the scope of this chapter) is a theater of intensive and costly scientific activities. The Arctic is also the scene of much transnational scientific and knowledge-based collaboration between public, private, and civil society actors. This chapter looks at the lessons of this knowledge-based collaboration for larger research and policy questions of global governance (Bertelsen and Rasch 2016).

Thus, the Arctic becomes a laboratory of global governance, which is explored in this chapter. I will start by recalling how the Arctic has, for centuries, been an integrated part of the international political, economic, and security system (Heininen and Southcott 2010). This historical outline sets the scene of an Arctic that is very much a part of this system, has been greatly affected by this system, and may now offer lessons for this system. This outline reminds us of the dangers of succumbing to "presentist" accounts of a forgotten or isolated Arctic discovered by the world as a result of climate change, or the misguided alarmism warning of new risk of local conflicts in the Arctic over natural resources made accessible by climate change.

I will continue this historical introductory overview by outlining how the Arctic is affected by the dominant current international systemic developments, the ongoing struggles over post-Soviet Russia's place in the international system (which is intertwined with struggles of Russia's domestic political and economic system), and power transition, especially with the rise of China and globalization. China's interest(s) in the Arctic have been the topic of much academic, policy and media attention. However, this attention has often seen China's interest(s) in the Arctic as an isolated Arctic question and not a reflection of power transition (as explained in the chapters by Kopra and Li and Peng). During that period, relations between Russia and the West in the Arctic were harmonious, reflecting the general international system. Again, many observers were not cognizant of how good Russia-West relations in the Arctic reflected general good Russia-West relations in the international system. The post-Cold War Russian-Western relationship came close to derailing over Georgia in 2008, and definitely did derail in 2014 over Ukraine's place between Russia and the European Union and NATO (Mearsheimer 2014). This Russian–Western conflict has been accentuated by conflict over and in Syria. Today, rather than being concerned with China's interest(s) in the Arctic, many people are fearful of Russia's actions and plans in the Arctic (Bittner 2016). Unfortunately, the analysis of Russia in the

Arctic in the broader international systemic context is sometimes as faulty as that of China a few years ago.

In light of these two systemically induced potential conflicts in the Arctic, I turn to the place of knowledge-based cooperation in the two settings. In order to address these two situations, I first outline the web of multilateral knowledge-based cooperation in the Arctic (Bertelsen and Rasch 2016). How has knowledge-based cooperation been affected by the Russia–West conflict over Ukraine, and suspicions raised by the rise of China? Concerning the Ukraine crisis, this brief discussion will show that civilian academic and civil society knowledge-based cooperation is quite resistant, while military and commercial knowledge-based cooperation is sensitive to political and economic sanctions. The longer-term development of power transition and the Arctic shows that China has been met with suspicion—also in the Arctic—which is common for rising powers, but that knowledge-based cooperation between academia and civil society is useful for building trust and co-creating knowledge. These China-related lessons encourage expansion of knowledge-based cooperation to commercial activities and investments instead of the traditional focus on investing in extractive industries and natural resources, which holds important lessons for China engaging and sourcing energy and raw materials globally. How transnational knowledge-based cooperation between Western, Russian, Chinese, and other academic, public, private, and civil society actors can contribute to more resilient and informed relations is an important example of the transnational and international dimension of the social relevance of science.

Therefore, the lessons from the Arctic laboratory of transnational knowledge-based relations in international systemic processes are that knowledge-based relations can serve useful roles to manage these processes, which hold a great potential for conflict. The Arctic cases of Russian–Western knowledge-based cooperation and Sino–Arctic knowledge-based cooperation illustrate the possibilities and limitations of such cooperation for managing systemic processes with high conflict potential. There are lessons of value to the academic literature on the role of transnational knowledge-based relations, science diplomacy and epistemic communities, and there are applicable policy lessons for how to design transnational knowledge-based cooperation so it can be used as a strategic tool in concert with other strategic political, economic, military, and other instruments.

15.2 The Arctic in the International System—Past, Present, and Future

To fully understand the Arctic, it is important to not fall victim to "presentism" (which is not uncommon in International relations scholarship without sufficient historical consciousness (Halliday 2001)) of a previously isolated and forgotten Arctic that was "discovered" by the world. The Arctic has been a part of the Western-led international political, economic, and security system for centuries, reflecting this system and deeply affecting Arctic communities (Heininen and Southcott 2010). The Arctic has been part of larger indigenous systems of migration, trade and exchange, which is beyond the scope of this paper and expertise of the author.

A brief overview of the Arctic illustrates this systemic embeddedness. The North Atlantic waters off Northern Norway and Iceland are particularly rich fishing grounds. Fisheries here have been part of the European food system since medieval times, when North Norwegian and Icelandic stock fish fed Catholic Europe, especially during Lent. These historical food system ties are reflected today in a well-known dish called bacalao. One of the oldest private companies of Western capitalism is the Hudson Bay Company, founded in 1670 to bring North American furs to European markets. The "Bay" continues to have a strong presence in Canadian society today, albeit with very different businesses. Northern Swedish iron-ore mining currently supplies 90% of iron-ore mined in the EU. Since the late 1800s, this mining has played a key role in Swedish industrial and export economy, and Sweden has built a socio-technical "mega-system" of mining, transport, energy, communities, and defense, covering Northern Sweden and even reaching into Northern Norway. The strategic role of Northern Swedish iron-ore for German industry led Britain and France to try to block shipments from Narvik along the Norwegian coast in 1940, to which Germany responded with the occupation of Denmark (as a stepping stone) and Norway. This is a clear illustration of the historical integration of the Arctic and its resources into great power strategies and conflicts. The USSR greatly industrialized its Arctic to extract oil, gas, minerals and other resources. Alaska is one of the most important oil-producing states in the USA (Heininen and Southcott 2010).

The alarmism regarding potential local Arctic conflicts over natural resources made accessible by climate change is innocent but misguided. The Arctic has often been the scene of intense conflict, but that conflict has never been locally caused and has always reflected war and peace between the great (Western) powers. The exceptions are the Cod Wars between Britain and Iceland, which also had international dimensions in the development of the law of the sea (Jónsson 1982). In 2005, the Icelandic daily, *Morgunblaðið*, reported that the Icelandic historian Anna Agnarsdóttir doing archival work in Paris had found a note from the French admiralty to the French king in the 1760s or 1770s. France lost Canada (Nouvelle France) in the Seven Years War of 1756–1763 (which was perhaps the first global war), and the French navy was very concerned about having lost its position in the North Atlantic. This introduces the key geostrategic role of the North Atlantic and led the French Admiralty to propose to the French king to offer Louisiana in exchange for Iceland to the king of Denmark–Norway. If the offer had been made and accepted, it would in all likelihood have led to immediate Anglo–French war over Iceland, as Britain could never have accepted a French navy based in Iceland.

The North Atlantic is the bridge between Europe and North America, which has defined its role in global conflict. During both World Wars, which were wars of attempted power transition, Germany tried to cut off Britain from North America through submarine warfare. Britain was equally adamant about maintaining control over this sea-space, which led to immediate British occupation of the Faroe Islands on April 11, 1940 after the German occupation of Denmark on April 9, 1940 and of Iceland on May 10, 1940 with the impending Anglo–French–Polish-Norwegian defeat in Norway. These occupations brought some of the most profound and consequential socio-economic and political changes to both the Faroe Islands and Iceland. The USA relieved Britain of the protection of Iceland on July 7, 1941, six months before the attack on Pearl Harbor (Corgan 2002). The independently acting Danish ambassador in Washington DC, Henrik Kauffmann, signed a defense agreement with the USA concerning Greenland on April 9, 1941, which brought US bases to Greenland (Lidegaard 1996). The Barents Sea, Northern Norway, Lapland, and the Kola Peninsula were the scene of large-scale fighting in Germany's attempt to cut off the USSR from Western supplies by the Murmansk convoys (Huldt 1985).

The Cold War brought enormous militarization of the Arctic for geostrategic and technological reasons. The introduction of long-range flying, international ballistic missiles, and nuclear weapons gave new central importance to the Arctic—albeit for no local reasons. The trans-Polar route is the shortest flight-path between North America and Eurasia, which makes it key in an environment of nuclear weapons, intercontinental ballistic missiles, and strategic bombers. The development of submarine-launched ballistic missiles hiding under the sea ice gave an assured second strike capability. These geostrategic and technological factors led to an enormous deployment of strategic nuclear weapons systems, early warning systems, and conventional forces to protect them. Nowhere was this development more visible than on the Kola Peninsula, which was the gateway of the USSR to the oceans and remains so for Russia (Huldt 1985). This enormous militarization also led to widespread nuclear and other pollution, which has threatened human and environmental security (Gjørv et al. 2014).

As the Cold War defined the Arctic so profoundly, the end of the Cold War created the Arctic we know today. This development was instigated by Mikhail Gorbachev's speech in Murmansk in 1987 that called for the Arctic to be a zone of peace, environmental protection, and scientific collaboration. This call, along with the end of the Cold War and dissolution of the USSR were the main enabling factors for a Circumpolar Arctic, with East-West collaboration in most areas of society. Finland immediately grabbed the opportunity, as an exposed small state, to use non-military means to pursue security policy and instigated the Rovaniemi Process in 1989, which led to the Arctic Environmental Protection Strategy in 1991 as the first intergovernmental Circumpolar Arctic collaboration. Another small state was Norway, which followed suit closely in 1993 with the Barents Euro Arctic Council cooperation. The middle power of Canada played the Finnish initiative forward as the Arctic Council with the Ottawa declaration in 1996. These initiatives combined to create the post-Cold War Circumpolar Arctic with political cooperation in the Arctic Council and scientific cooperation in the International Arctic Science Committee and the International Arctic Social Sciences Association (Bertelsen and Rasch 2016; English 2013; International Arctic Science Committee (IASC) 2015). However, while East and West were busy fighting the Cold War, other parts of the world were reawakened.

15.3 New Challenges in the Arctic: Post-cold War and Power Transition from West to East

Today, many people speak of the "rise of China" when referring to the spectacular economic growth experienced by China since Deng Xiaoping's Open Door policy from 1978 made China the factory of the world, creating both international and domestic dislocations around the world. However, it is perhaps more appropriate to speak of the "return of China", when international economic historical data shows that at the start of the eighteenth and nineteenth centuries, Asia represented 50–60% of global GDP, a number that dropped to less than 20% in 1950. The Asian Development Bank predicts that Asia will have regained its high historical share by 2050 (Asian Development Bank 2011). Li and Peng present this power transition driven by China's spectacular growth, and Kopran outlines the global environmental and climate aspects of this growth elsewhere in this volume. This international economic history of course illustrates that we have experienced a historical window of Western dominance, which in all likelihood is coming to a close. This development is the power transition from Western states to Eastern states, which is one of the dominant intercontinental research and policy questions today (Nye 2011) and key to the argument of the Arctic as the laboratory of global governance here. This Western dominance has deeply affected the Arctic, as outlined above, and its ending is already starting to affect the Arctic as it affects every corner of the world. A China representing 20-plus percent of the world economy in 1700 or 1800 naturally did not affect Western societies as China does today, which is clear from the domestic political disruptions illustrated by Brexit or the election of Donald Trump as US President. The other development today is globalization with the compression of time and space in the interactions of societies around the world, brought on by developments in transport and communication technology, free trade and financial deregulation (Harvey 1989).

The dissolution of the USSR was an enormous socio-economic and political collapse that imposed great human costs on the Russian and post-Soviet populations, while also giving political freedoms and economic opportunities. During the 1990s, Russia's international role was greatly depressed by its socio-economic crisis. Russia (USSR), as an enormous state with fluid borders, has a particular strategic tradition concerning space and time. Usually, Russia has historically been surrounded by economically and technologically more advanced neighbors to the East and West, which have posed great threats to Russia. From the West, Russia was invaded by Napoleon and later Hitler. It is therefore embedded in Russian strategic thinking to think of space as being synonymous with time. Space can be sacrificed for time to exhaust enemies and counter-attack to regain space, as was the case with both Napoleon and Hitler. Therefore, Russia (USSR) is deeply concerned with strategic depth (Cheng 2011).

During Russia's material weakness in the 1990s and early 2000s, and its lack of soft power (an unattractive socio-economic or political model), Central and Eastern European countries joined the EU and NATO, which used membership to stabilize and develop what could otherwise have been a neighborhood of instability and poverty. Thereby Russia gradually lost more and more strategic depth, to which Russia protested. When NATO membership seemed to become a possibility for Georgia in 2008, the result was a short and decisive Russian-Georgian war based on Georgian miscalculation and Russian determination to demonstrate its sphere of influence. In 2013–2014, the EU and Ukraine moved towards an association agreement that set in course the sequence of events with initial refusal by the Ukrainian President Yanukovych, the Euro-Maidan protest deposing President Yanukovych, sparking Russian covert intervention and annexation of Crimea, and intervention and support of a civil war in Eastern Ukraine. The USA and the EU responded with a range of political, commercial, technological, and financial sanctions (which have affected the Arctic, as will be discussed shortly); Russia responded with its own sanctions against European food and other exports (which also affected the North Atlantic Arctic). The annexation of Crimea and the civil war in Eastern Ukraine has brought Russia–West relations to their lowest post-Cold-War point. On the surface, this deep Russia–West crisis is about Ukraine, but the fundamental reasons are the ongoing struggles over post-Soviet Russia's position in the post-Cold War international system, which we will see also affects the Arctic, as systemic developments always have (Mearsheimer 2014).

The Ukraine crisis and strong Western concern about threats of overt or covert Russian aggression against the West in other theaters, most notably in the Baltic countries, together with heightened Russian air and naval activity in various theaters has also caused alarm regarding the Arctic. There has been widespread alarmist reporting of Russian military buildup in the Arctic and warnings of the danger of Russian aggression in the Arctic (Bittner 2016). It is beyond the scope of this chapter to analyze this issue, and I will shortly turn to the aspect of Arctic knowledge-based cooperation between Russia and the West. However, it is instructive to note that Russia has clear strategic concerns in Eastern Europe and the Caucasus, although unacceptable to the West. In comparison, Russia has no similar strategic concerns in the Arctic. Roughly half of the Arctic is Russian. There are enormous oil, gas, and mineral resources on undisputed and well-defined Russian land and EEZ (United States Geological Survey 2008). There are no land or sea delimitation disputes in the Arctic between Russia, Norway, Finland, or the USA, and these countries have usually enjoyed close cooperation in their border regions, while acknowledging the traditional military national security competition in these regions for geostrategic reasons. Russia may fly strategic bombers along the coast of Norway or near Iceland, but it is an open question whether that is strategic signaling between super and great powers or has anything to do with the localities. It can be argued that it was actually the West that brought the Ukraine crisis into the Arctic through financial and technological sanctions of Russian Arctic offshore oil and gas projects, which is a targeted and sophisticated way of hurting specific political-economic interests in Russian society (Rosen 2016).

15.4 The Role of Science Diplomacy in the Post-cold War and Power Transition

The Arctic (and the Antarctic) is home to extensive scientific activities. The Poles are the most difficult and expensive places to conduct science apart from space and perhaps the deep oceans. Therefore, there is much international scientific cooperation in the Arctic, which became Circumpolar after the end of the Cold War when Soviet (and shortly thereafter Russian) scientists and Western scientists could join forces. Circumpolar cooperation—scientific, people-to-people, health, environmental, and so on—made possible by the end of bipolarity is a great example of the international and transnational dimensions of the social relevance of science. There is a well-developed structure of multilateral scientific collaboration in the Arctic, which forms the background for the lessons of knowledge-based collaboration for global governance. The center of this multilateral knowledge structure is the Arctic Council, whose working groups and task forces bring experts together. There is the International Arctic Science Committee, which particularly gathers the natural science community. In parallel, there is the International Arctic Social Sciences Association, which gathers the humanities and social sciences. There is the University of the Arctic network, which involves close to 200 universities inside and outside the Arctic collaborating in more than 40 thematic networks and institutes and exchange faculty and students (Bertelsen and Rasch 2016; International Arctic Science Committee (IASC) 2015).

One example of a long-running international, transnational and transdisciplinary platform, which shows the local–global aspects of the social relevance of Arctic science, is the interlinked platforms of the Northern Research Forum (University of Akureyri Research Centre), the University of the Arctic Thematic Network on Geopolitics and Security (Arctic Politics 2017b), the GlobalArctic.org project (GlobalArctic MIR 2015), and the traveling Ph.D. course of the Calotte Academy (Arctic Politics 2017a). The Northern Research Forum, or NRF, is a platform for dialogue between local and international research and stakeholders from all sectors of society on crucial issues facing Northern communities. It was launched in a speech by then-President of Iceland Ólafur Ragnar Grímsson at the University of Lapland in 1998 and started operations with a secretariat in Akureyri in 1999. The NRF has run open assemblies, which emphasize openness, to relevant stakeholders in Akureyri in 2000, Novgorod in 2002, Yellowknife in 2004, Oulu/Luleå in 2006, Anchorage in 2008, Hveragerdi in 2011, Akureyri in 2013, and Reykjavik in 2015.

The UArctic Thematic Network on Geopolitics and Security is a joint network of the UArctic and the NRF. The network brings together senior and junior scholars from a wide range of universities around the Arctic and wider. Out of the NRF and the thematic network has grown the GlobalArctic.org project, which is a focused research network that looks at the Arctic in a globalized world. The final element of these connected platforms is the Calotte Academy, a week-long Ph.D. course held early each summer that travels around Lapland, Kola Peninsula, Norrbotten, Västerbotten,

Finnmark, and Troms. The academy's pedagogical philosophy involves bringing international Ph.D. students and local stakeholders together in the local communities.

It is clear that these interlinked platforms of the NRF, thematic network, GlobalArctic.org, and the Calotte Academy contribute significantly to the local–global and transnational social relevance of science. These activities connect various Northern communities, stakeholders inside and outside Northern communities, and Northern and international research talent with Northern stakeholders and communities. As such, these activities contribute strongly to the social relevance of science aims of addressing important societal problems, public accountability, democracy, (perhaps to less extent) business application, and definitely decision-making at different levels (Scott 2006).

15.5 Keeping the Circumpolar Arctic Together During the Ukraine Crisis

Knowledge-based cooperation between Russia and the West has been affected by the Ukraine crisis in some sectors, which shows the political and economic vulnerability of such collaboration. I will briefly discuss observations of civilian academia, civil society, and defense-related and industrial knowledge-based cooperation.

Civilian academic cooperation has shown the most resilience in light of political crisis and conflict. Academics on both the Russian and Western sides have continued to show deep commitment to and interest in cooperation and overcoming difficulties brought on from the political level. Research and educational cooperation continue to be the avenues for researchers and educator from both sides to meet and collaborate on Arctic research and teaching. It is clear from conferences, workshops, applications, etc., that these conversations have been largely unaffected by the Ukraine crisis.

The political level can choose to either encourage or sabotage such civilian academic cooperation. Below, I will provide two examples—with a potentially surprising finding to some Western readers—of Canadian sabotage and Russian cooperation.

The University of Northern British Columbia hosted the 8th International Congress of Arctic Social Sciences in May 2014 with hundreds of international scholars. A large delegation of Russian academics was also supposed to attend; however, many of them could not attend because Canada held up their visa applications as part of sanctions against Russia following the annexation of Crimea (Bennett 2014). Such action appears counterproductive from a scientific/diplomatic point of view and perhaps rather reflects the Canadian domestic political significance and use of the Ukraine crisis. Canada also boycotted some Arctic Council working group meetings held in Russia, which was a strong signal in view of Canada's chairmanship of the Arctic Council 2013–2015.

The Russian Federation Security Council has for some years organized an annual international high-level meeting of representatives of Arctic Council member states and observer states. The meeting is usually held for a couple of days somewhere in the Russian Arctic and it used to include official representatives of these countries. The host of the meetings is the secretary of the security council, Nikolay Patrushev, who is also the former director of the security service FSB. For the August 2014 meeting in Naryan-Mar, it must have been clear to the organizers that no Western official participants would participate in light of boycott of Russia. Therefore, the organizers and the Northern Arctic Federal University (NARFU) in Arkhangelsk invited academic representatives, where I was the Danish academic representative. I was invited again for the September 2015 meeting in Arkhangelsk; on this occasion, Western diplomats were returning, led by the Danish ambassador to Moscow. For the 2016 meeting in the Bering Region, Danish academia was represented by Associate Professor Uffe Jakobsen from the University of Copenhagen. I interpret the invitation to Western and Asian academics at the 2014 meeting as a signal from Russia of its determination to continue the dialogue about the Arctic, where Russia sees no conflict with the West, despite conflict in Ukraine. There was a strong Russia media presence at the 2014 meeting, where the foreign academics could remove an impression of isolation of Russia. I interpret the continued invitation of academics to the 2015 and 2016 meetings when diplomats had returned as Russian recognition of the value of including academics in these meetings for resilience of the dialogue to political upheavals.

I also mention defense-related cooperation here as knowledge-based cooperation, although it may not usually be seen as science diplomacy. Defense forces of highly developed countries as the Arctic states and their operations are complex and knowledge-based. Examples include the biannual US-Russian-Norwegian Northern Eagle exercises. Therefore, I propose here that joint exercises should be seen as a case of international knowledge-based cooperation. Defense forces around the Arctic have cooperated increasingly in recent years, including a Northern Chief of Defense forum. For instance, the Norwegian navy had the annual *POMOR* joint exercises in the Barents Sea with the Russian Northern Fleet. The Western states cut off this defense-related cooperation as a sanction for the annexation of Crimea; consequently, for instance, the Norwegian navy no longer conducts joint exercises with the Russian Northern Fleet, but coastguard cooperation continues. Such a boycott may be politically and diplomatically necessary and wise, but it threatens personal and organizational connections between defense forces in the Arctic, which contributes to confidence building and crisis management, and will take time to reestablish afterwards. There is a cost for this kind of sanction against knowledge-based cooperation.

Another area of knowledge-based cooperation that has received little public attention is industrial knowledge-based cooperation. Arctic economies are usually natural resource-based. This fact is very much the case concerning the Russian Arctic economy, which is heavily dependent on especially oil and gas production. Such production is very technology-intensive. Major Western oil companies, such as French Total, American Exxon Mobile, or Norwegian Statoil had been deeply involved in developing oil and gas resources in the Russian Arctic. These activities were hit doubly by Western political, financial, and technological sanctions and by the collapse of oil prices. For instance, Exxon Mobile had invested billions of US dollars in projects in the Kara Sea, which were abandoned. Such projects involve massive use of science and technology between the partners in the project (carefully managed by intellectual property rights), and their abandonment due to a political crisis shows their vulnerability to political risk (Rosen 2016).

This vulnerability of commercial knowledge-based cooperation can partly be offset by the collaboration of academia involved in oil and gas production. On November 19–20, 2014, I participated in the Norway–Russia conference of the Norwegian Research Council entitled "Norge og Russlands samarbeid – Hvor går vi, hvor står vi?" (Norway and Russia's Cooperation—where are we going, where are we now?). The subject was oil resources, energy cooperation, and societal developments in Russia, and there was a strong desire among Norwegian and Russian academics to continue this research and educational cooperation. The educational cooperation in especially industry-targeted double master's degrees, such as in energy management between Nord University and MGIMO University and in Offshore Field Development Technology between University of Stavanger and Gubkin Russian State University of Oil and Gas, demonstrates the temporal perspective, that current economic relations may be taken hostage for political reasons while the interpersonal academic basis of future cooperation is ensured.

15.6 Building Mutual Trust and Confidence with China Under Power Transition

Power transitions often generate strong distrust between the status quo and rising powers, both in the general public and at high-level decision-making levels. An historical example of this is the cartoon mockery of Germany in British newspapers and magazines before World War I. At the strategic level, German investments in strategic infrastructure around the world, such as the Berlin-Baghdad Bahn (railroad) project, caused great concern among British and French strategists about the pre-World War I Middle East (McMeekin 2010). Today, the rise of China has caused distrust and suspicion both at public and high levels, also in the Arctic. Two examples suffice to illustrate how China, as the rising challenger power to the international status quo (which the Arctic states are all part of, also Russia vis-à-vis China), causes suspicion and distrust in the Arctic. The first example is the HUANG Nubo saga in Iceland. HUANG Nubo is a Chinese billionaire real estate investor, who in 2011 proposed to buy 300 km^2 of land to build an eco-resort in Northeast Iceland. HUANG Nubo has a personal affinity for Iceland based on his friendship with some of the first Icelandic students to study at Peking University in the late 1970s, when foreign students reentered China after the Cultural Revolution. This investment proposal caused enormous controversy in Iceland, where wide circles of society were deeply suspicious of his motives. Many refused to believe that this investment would make commercial sense and saw hidden strategic motives behind of securing access to land and resources in the Arctic. HUANG Nubo eventually withdrew his interest in response to both official and public distrust (Higgins 2013). I only have knowledge of the case from media coverage, but what strikes me as a former resident of Iceland and a frequent visitor to China is the mutual incomprehension between the two sides. For a Chinese billionaire with an affinity for Iceland, such a proposal might seem a worthwhile risk, considering the enormous market potential of outbound Chinese tourists (more than 100 million per year), which means that a tiny fraction would make even an eco-resort in a desolate corner of Iceland worthwhile commercially. Even for well-educated Icelandic observers from the tiny economy and market of Iceland, the Chinese numbers may be hard to grasp.

The Kingdom of Denmark represents a similar case of distrust of China concerning investment in the Arctic, with the example of the Isua iron-ore mining prospect in Greenland. Around 2012–2013, when commodity prices were high, there was great optimism in Greenlandic politics that mining could enable a quicker transition to independence from Denmark. The large iron-ore deposit at Isua was the most prominent case. The rights to this prospect were owned by a (then-British) mining company called London Mining Group, which developed and sold international mining projects. The most likely investors for the Isua project would have been Chinese. The project reached the level where the Danish and Greenlandic parliaments passed legislation to allow thousands of—presumably—Chinese mining workers access to Greenland to construct this mine. The debate around this proposal took place in an atmosphere of deep public and some political Danish distrust of Greenland and of China, which poisoned Danish-Greenlandic relations and led the Chinese Ministry of Foreign Affairs to publicly state that there was no Chinese mining activity in Greenland (Breum and Chemnitz 2013). The project never materialized because of collapse in commodity prices. London Mining Group went bankrupt—and was actually acquired by a Hong Kong Chinese company—but that only caught specialist attention (Su and Lanteigne 2015).

Sino-Arctic scientific cooperation stands in contrast to these accounts of distrust and suspicion between Arctic states and communities and China. There is growing Sino-Arctic research cooperation, where the Sino-Nordic Arctic research cooperation is perhaps the best organized in the China-Nordic Arctic Research Center (CNARC), hosted by the Polar Research Institute of China (PRIC) in Shanghai, which is one example of the potential of science diplomacy between status quo and a rising power in the Arctic. CNARC was founded in 2013 and originated in Sino-Icelandic decisions to focus scientific cooperation on Arctic research, which the two sides wisely expanded from the outset to the other Nordic countries. CNARC is a virtual center that brings together Chinese and Nordic institutions including, besides PRIC: Shanghai Jiao Tong University, Tongji University, Shanghai Institutes for International Studies, Ocean University of China and Dalian Maritime University (newcomer), RANNÍS-The Icelandic Center for Research (Research Council), Fridtjof Nansen Institute, Norwegian Polar Institute, Nordic Institute of Asian Studies (University of Copenhagen), Swedish Polar Secretariat, Arctic Centre (University of Lapland), UiT-The Arctic University of Norway (newcomer), and Aarhus University (newcomer). Every year, the center organizes the China Nordic Arctic Cooperation Symposium, which also includes a business roundtable held in China one year and in the Nordic countries the next. CNARC also receives Nordic guest researchers in China, while Chinese guest researchers visit the Nordic member institutions. Based on my own participant observation at the 2014, 2015, 2016 and 2018 CNACS, and as a guest researcher in Shanghai March–April 2016, it seems clear that these activities contribute to co-creation of knowledge and therefore mutual understanding and trust (Bertelsen et al. 2016).

Another example of the benefits of Sino-Arctic science collaboration, rather than commercial investment in land or natural resources, is the China-Iceland Aurora Observatory being built on the farm Kárhóll in Northeast Iceland, the same corner of Iceland as the aborted tourism project. This project is moving forward through collaboration between PRIC, RANNÍS, and local Icelandic partners and seems to have been generally well received by the local and national Icelandic public, based on my own field visits there in June 2014, October 2015, and October 2016.

The very poor reception of the mere interest or possibility of Chinese investment in land and natural resources in Iceland and Greenland (which is reminiscent of the controversies surrounding Chinese extractive industries and agriculture in regions as Africa or Latin America), in comparison with the better reception of science collaboration, suggests the potential for commercial innovation and knowledge-based collaboration beyond typical academic research. As mentioned above, Arctic economies are generally based on natural resources, whether renewable biological or finite mineral resources, which creates a range of environmental, social, and cultural challenges to sustainability. Therefore, there is serious research and policy interest in transforming Arctic economies into more knowledge-based economies (which is also a general desire of natural resource-based economies around the world).

There are a range of possible areas for commercial innovation and knowledge-based cooperation between Nordic Arctic and Chinese partners. Iceland and China collaborate both in the public and private domain on geothermal energy, which can bring green space heating to many Chinese homes, where fossil fuel-based space heating contributes greatly to air pollution in the winter in China. Finnair and public and private Finnish actors have particular strength in developing Sino-Asian-Nordic-Arctic tourism. The mining mega-system knowledge of Swedish state mining company LKAB may be more valuable for Chinese domestic and global sourcing of natural resources than Swedish iron-ore in itself. For instance, Tromsø is a center of cold and blue biotech, which should be of interest to relevant Chinese partners. In the Kingdom of Denmark, the Faroe Islands is a special setting for developing reliable, renewable energy-based micro-grids; this could be of interest to China, which struggles to maintain its enormous national grid. Such innovation and knowledge-based cooperation is likely to be much better received than investments in land or natural resources, the mere idea or possibility of which has raised deep suspicions reflecting the inherent distrust in power transition.

15.7 Conclusion: The International Social Relevance of Arctic Transnational Knowledge Relations

For centuries, the Arctic has been on the receiving end of the effects of international political, economic and systemic developments. These developments have strongly impacted the small Nordic Arctic nations and Arctic communities with little concern for their wishes. Some outside developments have brought economic opportunities, technology, or possibilities for immediate and subsequent self-government or sovereignty, such as World Wars I and II did for Iceland, the Faroe Islands, and Greenland. Other impacts have been deeply detrimental as the nuclear and other pollution from Cold War installations threatening human and environmental security.

The Arctic continues to be an integrated part of global developments and is deeply influenced by the two systemic processes currently ongoing; that is, the continuing struggles and adaptions after the end of the Cold War, especially concerning post-Soviet Russia's international position and domestic political and economic system; and power transition with the rise of especially China and globalization with the compression of time and space in interactions between societies around the world. The ongoing post-Cold War geopolitical competition between the West and Russia flared up in the Ukraine crisis, which also affects the Arctic, not least because of Western political, financial, and technological sanctioning of Russian Arctic offshore energy developments (such horizontal escalation makes sense from a Western point of view). The Arctic is positively and negatively affected by power transition and globalization. Power transition has a high conflict potential reflected in distrust, which is also the case in the Arctic. At the same time, the spectacular economic growth of China and other emerging markets led to increases in commodity prices, which the Arctic also benefitted from.

However, both the challenges of the Ukraine conflict and the distrust of China in power transition points to important Arctic lessons of the potential of knowledge-based collaboration for global governance. The Arctic is a particularly valuable laboratory on this question because it is, in important respects, at the center of the international political and environmental systems and, as such, is a crucial case of global governance, as explained elsewhere in the volume by Li and Peng and Kopra. The eight Arctic states are all old status quo powers of the international system (Russia was weakened after the dissolution of the USSR and is striving to regain its historical status). For centuries, Western great power geopolitics has been played out by those states and the old European great powers, all of which are observers to the Arctic Council. When rising powers, led by China, knock on the door of the Arctic, they are not knocking on the door of a peripheral region, where the status quo powers may or may not defend their influence; they are knocking on the backdoor of the status quo powers.

The resilience of Russian–West Arctic research cooperation at the researcher level during the Ukraine crisis shows the value of such civilian academic research and educational cooperation to maintain these connections during such a conflict. The Russian and Canadian actions I have described show how states can support academic cooperation (Russia) or sabotage it (Canada). Looking at military joint exercises, such as those between the Norwegian navy and the Russian Northern Fleet, which were disbanded from Norwegian side as part of NATO sanctions of defense cooperation with Russia, shows the vulnerability of defense cooperation. The withdrawal of Exxon Mobile from offshore oil and gas projects in the Kara Sea is an example of the vulnerability of commercial, costly science and technology cooperation to political, financial, and technological sanctions.

Sino-Nordic Arctic research collaboration is in clear contrast to aborted or possible Chinese investments in land in Iceland or natural resources in Greenland. Distrust is a structural feature of power transition, and such distrust is also reflected concerning China and the Arctic. The mutual understanding and trust through knowledge co-creation, which appears to be the case of Sino-Nordic Arctic science cooperation, points to the value of research and educational cooperation and knowledge co-creation as valuable tools under power transition. The benefits of such knowledge-based cooperation (and the distrust surrounding possible investments in natural resources and extractive industries) suggest that innovation and knowledge-based cooperation in the commercial sector should be promoted in the Arctic—and other regions.

Researcher-initiated interlinked platforms as the Northern Research Forum, the UArctic Thematic Network on Geopolitics and Security, the GlobalArctic.org project, and the Calotte Academy show how virtual platforms with real-life activities can bring Arctic and outside researchers, talent, and stakeholders from all sectors of society together. These activities show the local–global and transnational dimensions of the social relevance of science, which contributes to manage international systemic processes.

References

Arctic Politics (2017a) Calotte Academy [Homepage of arctic politics]. http://www.arcticpolitics.com/index.php/calotte-academy. Access 30 Mar 2017

Arctic Politics (2017b) Thematic network on geopolitics and security [Homepage of arctic politics]. http://www.arcticpolitics.com/index.php/home. Accessed 30 Mar 2017

Asian Development Bank (2011) Asia 2050: realizing the Asian century. Asian Development Bank, Singapore

Bennett M (2014) 05/28-last update. Reflections on the eight International Congress of Arctic Social Sciences [Homepage of cryopolitics]. https://cryopolitics.com/2014/05/. Accessed 21 Nov 2016

Bertelsen RG, Rasch M (2016) Rigsfællesskabet i multilateralt arktisk forskningssamarbejde. Udenrigs 48–53

Bertelsen RG, Li X, Gregersen MH (2016) Chinese Arctic science diplomacy: an instrument for achieving the Chinese Dream? In Iglesias Sanchez S, Conde Perez E (eds) Global challenges in the Arctic region: sovereignty, environment and geopolitical balance. Ashgate, Abingdon, pp 442–460

Bittner J (2016) 2016/04/09-last update. To whom does the north pole belong? [Homepage of zeit online]. Available: http://www.zeit.de/politik/ausland/2016-04/arctic-russia-america-conflict-zone. 21 Sept 2016

Breum M, Chemnitz J (2013) No, Greenland does not belong to China. The New York Times, New York. 20 Feb 2013
Cheng Y (2011) The power of diplomatic traditions: understanding the logic of Russia's foreign policy in the post-soviet era. Eurasian Rev 4, 25–40 Nov 2011
Corgan MT (2002) Iceland and its alliances: security for a small state. E. Mellen Press, Lewiston
English J (2013) Ice and water: politics, peoples, and the Arctic Council. Allen Lane, Toronto
Etzkowitz H (2008) The triple helix: university-industry-government innovation in action. Routledge, New York
Gjørv GH, Bazely DR, Goloviznina M, Tanentzap AJ (eds) (2014) Environmental and human security in the Arctic. Earthscan, London
GlobalArctic MIR (2015) GlobalArctic [Homepage of GlobalArctic MIR]. Available: http://www.globalarctic.org. 30 Mar 2017
Halliday F (2001) The romance of non-state actors. In: Josselin D, Wallace W (eds) Non-state actors in world politics. Palgrave, Basingstoke, New York, pp 21–37
Harvey D (1989) The condition of postmodernity: an enquiry into the origins of cultural change. Blackwell, Oxford
Heininen L, Southcott C (eds) (2010) Globalization and the circumpolar North. University of Alaska Press, Fairbanks
Higgins A (2013) Teeing off at the edge of the Arctic? A Chinese plan baffles Iceland. The New York Times. 22 Mar 2013
Huldt B (1985) The strategic North. Wash Q 8(3):99–109
International Arctic Science Committee (IASC) (2015) 25 years of international Arctic research cooperation [Homepage of International Arctic Science Committee]. Available: http://iasc25.iasc.info. 21 Nov 2016
Jónsson H (1982) Friends in conflict: the Anglo-Icelandic cod wars and the law of the sea. C. Hurst, London
Lidegaard B (1996) I Kongens navn: Henrik Kauffmann i dansk diplomati 1919–1958. Københavns Universitet
McMeekin S (2010) The Berlin-Baghdad express: the Ottoman Empire and Germany's Bid for world power. Belknap Press, Cambridge
Mearsheimer JJ (2014) Why the Ukraine crisis is the West's fault: the liberal delusions that provoked Putin. Foreign Aff 93(5):77–89
Nye JS Jr (2011) The future of power. PublicAffairs, New York
Organski AFK (1958) World politics. Knopf, New York
Rosen ME (2016, 2016/11/15-last update). The Arctic is the First Stop in the United States Reset with Russia [Homepage of The National Interest]. Available: http://nationalinterest.org/blog/the-buzz/the-arctic-the-first-stop-the-united-states-reset-russia-18409. 21 Sept 2016
Scott A (2006) Peer review and the relevance of science. SPRU-Science and Technology Research, University of Sussex
Su P, Lanteigne M (2015) China's developing Arctic policies: myths and misconceptions. J China Int Relat 3(1):1–25
The Royal Society and AAAS (2010) New frontiers in science diplomacy: navigating the changing balance of power. The Royal Society; AAAS, London
United States Geological Survey (2008) Circum-Arctic resource appraisal: estimates of undiscovered oil and gas North of the Arctic circle. United States Geological Survey, Menlo Park
University of Akureyri Research Centre. No date for last update. Northern Research Forum [Homepage of University of Akureyri Research Centre]. Available: https://www.rha.is/nrf. 30 Mar 2017

Chapter 16
Subnational Tier of Arctic Governance

Alexander Sergunin

16.1 Introduction

In the post-Cold War era, subnational actors in the Arctic-regions, members of federations (in Canada, Russia and the United States), autonomies, administrative units, cities, and municipalities—are actively changing their roles in both policy-making and regional governance. They do not hesitate to build horizontal, non-hierarchical ties with foreign counterparts that have the same legal and political status. The external activities of subnational actors (often called paradiplomacy), which are a concomitant of the process of regionalization/localization, have become typical for many countries of the world, including the Arctic nations. Many Arctic regions and municipalities consider the active development of international contacts not only as an efficient instrument for problem-solving but also for building sustainable development strategies. These actors believe that even their marginal location can be an additional resource or competitive advantage in their development strategies; with international cooperation they can transform themselves from *terra incognita* into attractive places for investment, serving adjacent sea routes, tourism, cultural events, and so on.

Thus, my main purpose is to examine how the Arctic subnational actors affect regional policy-making and governance systems. To attack this research problem, I asked the following questions: Why do subnational actors actively develop their paradiplomacies? What are the most popular capacity-building strategies and functional areas of networking/cooperation? What kind of an institutional framework

This research was done in the framework of the following projects: Arctic PIRE: Promoting Urban Sustainability in the Arctic (funded by the US National Science Foundation) and CANARCT, project no. 257638 (funded by the Research Council of Norway, 04-2016–03-2019).

A. Sergunin (✉)
St. Petersburg State University, Saint Petersburg, Russia
e-mail: sergunin60@mail.ru

M. Finger and L. Heininen (eds.), *The GlobalArctic Handbook*,
https://doi.org/10.1007/978-3-319-91995-9_16

supportive for the subnational tier of the Arctic cooperation exists in the region? What are the consequences, both domestic and international, of the subnational actors' paradiplomacies?

16.2 Theoretical Framework

Theoretically, this study is based on the concepts of "new regionalism" and "global region". In contrast with traditional integration theories developed by the neoliberal international relations paradigm, these approaches to region-building are multicentric rather than state-centric. As Fredrik Söderbaum puts it "New regionalism—a range of formal/informal mid-level 'triangular' relations among not only states but also non-state actors, notably civil societies and private companies—is a central aspect of the 'new' inter- or transnational relations" (Söderbaum and Shaw 2003: 1). In this context, subnational units are seen as both an important driver of regionalization and a tier of regional governance.

Moreover, I see the Arctic as not only a typical geographic region, but also as a global region where not only territorial proximity, but also supranational/exterritorial interactions are very important.[1] In global regionality, cooperative links can be established by any actors, including subnational units, municipalities, organizations, and individuals. The input of state and non-state actors to the global region-building is equally significant and should be thoroughly studied.

One more theoretical concept is important for this study. The concept of paradiplomacy is often used to describe external activities of subnational and non-state players. This concept denotes parallel international activities of subnational and non-state actors that have limited capabilities, in terms of resources and legal powers, in the foreign policy sphere compared to national governments. According to Soldatos (1990) and Duchacek (1986, 1990), who invented the concept, paradiplomacy is a part of the processes of globalization and regionalization, under which sub- and non-state actors play an increasingly influential role in world politics. Regions, municipalities, companies, NGOs, etc., seek to promote international cooperation, and account for a significant part of contemporary cross-border and trans-border contacts.

The phenomenon of paradiplomacy raises new theoretical questions concerning the role of the state, substate, and non-state actors in international affairs, and it also challenges the existing state system and international law that has provided the grounds for the international political order in the Westphalian era (Hobbs 1994; Hocking 1993).

[1]For a detailed account of the concept of global region, see: Farrell et al. 2005, parts 3 and 4.

16.3 Local/Regional Goes Global

Looking back at the Cold War era, when international relations were based on the Westphalian principles with their praise for state sovereignty, only states were seen as fully legitimate international actors. Subnational entities were not allowed to develop external relations of their own or were strictly controlled by central authorities (as in the case of sister relations or city-twinning between the East and West).

However, with the collapse of the Cold War bipolar system, the situation has changed radically, with world politics becoming less state-centric, and other types of actors have been allowed to go international/global. Most Arctic countries do not object to subnational units' external activities, and in some cases they even encourage these actors to develop cooperation with foreign partners.

Both the European academics and practitioners believe that subnational units' international activities can be a serious resource for both them and national governments and they are useful in many ways. The following benefits from paradiplomacy, as a form of jumping scale, may be identified (Handley 2006: 6–8):

- Bolstering economic and business development
- Improving service delivery and problem solving
- Improving transport infrastructure
- Promoting freedom of movement of people, goods, services, and capital
- Accessing EU and other financial institutions in search of funding
- Promoting community well-being
- Promoting stronger community partnerships
- Increasing global and European awareness
- Yielding more intense local government staff development and training
- Providing resources for developing education and culture
- Promoting tolerance and increasing understanding
- Enhancing youth activities.

Nordic Europe is particularly distinctive in regard to successful experimenting with paradiplomacy. It should be noted that the so-called Nordic model is fundamentally based on principles, such as a common labor market within the Nordic context as a part of the Nordic welfare state system. In turn, such a model presupposes the devolution of power and the active role of non-state and subnational actors. In this region, international cooperation is one of the departures that regions and cities use in aspiring to a distinct, visible, and favorable profile and it is, in this sense, part and parcel of their policies of place-marketing and branding in the context of the increasingly intense and transnational regionalization. Moreover, the subnational tier of cooperation is another way for the Nordic countries to strengthen Nordic solidarity and developing common Nordic identity (Olsson 2016).

The case of Greenland is a special one because, while it remains part of the Kingdom of Denmark, it is quite independent in terms of developing its external relations. As Ackrén (2014) noted, Greenland actively uses paradiplomacy, with special emphasis on cooperation with foreign countries and multinational corporations that may be interested in the development of the country's natural resources.

For Canada's northern provinces and the US state of Alaska, international cooperation is an instrument for both solving problems which are common for many Arctic regions, and attracting federal government attention to the hardships of these subnational units. As Nicol (2016: 110–111) puts it, several important external and internal factors dictate devolution of power in the Canadian northern regions, but also encourage the local actors' international contacts.

As for the US, during the American presidency of the Arctic Council (2015–2017), Alaska has revived its cooperation with Canada's and Russia's northern regions to demonstrate that national governments of the Arctic countries should pay more attention to the problems of subnational players.

In January 2015, before taking over the presidency from Canada (in May of that year), the White House issued an executive order "Enhancing Coordination of National Efforts in the Arctic." The executive order created an Arctic Executive Steering Committee to "enhance coordination of Federal Arctic policies across agencies and offices, and, where applicable, with State, local, and Alaska Native tribal governments" (www.akarctic.com/federalreports/). The committee identified four priorities to address: vulnerable coastal and riverside communities threatened by erosion; energy efficiency through renewable energy and weatherization; oil spill response preparedness; and improving consultation with Alaska Natives.

As far as the Russian northern subnational actors are concerned, the initial thrust for their external activities can be also explained by the harsh realities of the 1990s. In the Yeltsin era, many Russian Arctic territories felt almost abandoned by the federal government; they had to seek for survival strategies of their own. Foreign aid and investment were seen as efficient instruments for keeping the local economies afloat. In fact, given the broad autonomy of the members of the Russian Federation in the Yeltsin period, the northern regions managed to develop rather diverse international contacts.

However, as the socio-economic situation in Russia under the Putin regime improved, subnational entities tended to see international cooperation as an integral part of their strategy of sustainability rather than a strategy of survival. This paradigmatic shift in subnational units' motivation has entailed a radical change in their attitudes to and approach vis-à-vis paradiplomacy (Joenniemi and Sergunin 2014). The Russian subnational actors became more pragmatic and rational in their cooperation with foreign partners. Collaborative projects became less ambitious and more realistic. Overall, they boiled down to the rather practical needs of those engaging in cooperation.

Thus, European and Russian regions and municipalities now tend to coalesce across borders in order to solve concrete and shared problems; this is done for reasons of their own and by employing the competence that they themselves harbor. They aim to add to their strength by transgressing various borders—whether conceptual, identity-related, or spatial—by joining forces in the context of various regional endeavors or, for that matter, through lobbying in various broader contexts. In essence, the driving force, which is spurred by various economic, social, cultural, and environmental concerns, amounts increasingly to that of self-interest. This also implies that the pursuance of paradiplomacy has become less chaotic and more prioritized; in essence, it has been subordinated to the long-term developmental strategies of subnational actors.

As far as other motives of paradiplomacy are concerned, some subnational actors have been interested in partaking in the national government's decision-making in terms of stating their view prior to a final decision being reached or an international treaty being signed. For example, Norway's Sør-Varanger community and Russia's Murmansk Region wanted to be involved in preparing international agreements where their status has been affected (visa regime, delimitation of maritime spaces, establishment of special economic zones and customs regimes, etc.).

Furthermore, previously remote (and, in the case of Russia, closed and barred) spaces of the Arctic—with regions/cities at the edge of this space being unavoidably seen as peripheral—have been opening up as these subnational entities aim to benefit from cross- and trans-border networking. The Arctic subnational actors demonstrate that marginally/peripherally located actors can successfully play with their unique position, both domestically (in relation to the center) and internationally (with similar marginal and/or central actors). Marginal actors can make use of their geographic location by means such as acquiring the role of mediator or "bridge" between countries. They can turn their marginality from a disadvantage to a resource and transform themselves from remote and provincial territories to attractive places that host intense international flows of goods, services, capital, technologies, and people. On a more general plane, paradiplomacy contributes to debordering and desovereignization in a globalizing world.

16.4 Capacity-Building Strategies

These strategies include a wide array of methods, discussed below.

One crucial area consists of *attracting investments*. Both domestic and foreign investments are searched for; this applies both to the existing sectors (such as mining industries or transport infrastructure) as well as potential projects (such as new railways connections and industrial parks).

Branding/PR strategies: In order to attract foreign investors and gain national support, the Arctic regions and municipalities have launched active PR campaigns. For example, they arrange exhibitions, hold "cooperation days" and arrange festivals in towns and regions with which they are related, take part in international fairs, and advertise themselves in the partner cities' and regions' mass media. In addition, the regional and municipal leaders undertake regular trips abroad for public relations purposes. Many Nordic and Russian northern communities run bilingual (and even multilingual) periodicals and websites oriented towards audiences in the regional neighborhood. The main message of this campaign is to present partners as creative and innovative platforms rather than remote and depressive areas.

Increasing familiarity: Paradiplomatic actors differ in terms of their history and current status. This difference can affect the outcomes of interaction either by intensifying the relationship or by problematizing the encounter. Differences may create curiosity, fascination, and nostalgia that promotes cooperation between them, but if they are perceived as too great, they can lead to aversion, resentment, and avoidance. These differences can promote a sense of familiarity or unfamiliarity. The success or failure of cooperative projects depends largely on the interplay of familiarity and unfamiliarity among actors (Joenniemi and Sergunin 2014; Scott 2013; Spierings and Velde 2008, 2013a, b).

Familiarity relies on the utilization of a shared cultural heritage and experiences of cooperation, combined with a downplaying of negative historical memories related to conflicts, although familiarity may include an intensified recall of past negative experiences. Unfamiliarity, in contrast, is associated with a view of cooperation as something entirely new and previously unexplored. If it manifests itself as fear of the unknown, unfamiliarity may impede the construction of a cross-border sense of community, but it can also evoke fascination through the seductive attraction of the relatively unknown. The most important question here is how the past is interpreted, including which view of any given encounter prevails on both sides of a border that is significantly changing its meaning.

The issue of familiarity is especially important for relations between the Russian northern subnational units and Norwegian Arctic regions.

Although Murmansk enjoyed rather intense international contact, even in the Soviet period, and the era of openness imposed no particular frustrations on its inhabitants, the rest of the Russian North remained behind the Iron Curtain during the Cold War and needed time to familiarize itself with its neighborhood. In some cases, the familiarization occurred so quickly and intensely that it led to the emergence of transnational spaces. For example, the town of Kirkenes (in northern Norway), 7000 inhabitants and growing, has been a major location for Russian–Norwegian contacts since the 1990s. It is multicultural in the sense that, in addition to the Norwegian majority, its region has a Saami population, a considerable number of Finnish speakers, and an increasing number of Russians (Russians account for about 10% of the city's population) (Rogova 2008: 29).

As Rogova (2009) has noted, a considerable number of Russians living in the Murmansk Region now view the Norwegian–Russian border as a shared borderland. The border has become far less divisive in terms of culture and identity, as well as politics and administration. Rogova (2009: 31) claimed that a borderland has emerged "which is neither Russia nor Norway to the full extent." Russians visiting Kirkenes do not have the feeling of being abroad; this is reflected in the fact that Kirkenes is known as "Kirsanovka" or "Kirik," which in the language used in the Murmansk region carries connotations of a small, local, nearby entity or village. People often visit the town to shop or to fly out of the Kirkenes airport.

As mentioned above, the familiarity strategy largely depends on shared historical memories. In one respect, the Norwegian–Russian cross-border cooperation and familiarity strategy can draw on the legacy of Pomor trade. These coastal trading contacts, which lasted for nearly three centuries until the Russian Revolution of

1917, were quite important for the development of the northern areas. The legacy is frequently invoked, with today's cooperation and border crossing presented as a return to traditional customs.

Another memory relates the brutal behavior of the German troops stationed in Finnmark until the Red Army freed the area in 1944. The Cold War period, with its emphasis on enmity, influenced Norwegian people's views of Russians. The negative views have, however, gradually faded in the 1990s. For instance, joint anniversary celebrations of the liberation of the Murmansk Region and East Finnmark from Nazi occupation in October 1944 have become traditional.

Accommodating foreign consular offices and trade missions: To maintain sustainable relations with neighboring foreign countries and facilitate travel for its citizens, some Arctic regions and municipalities have favored the establishment of foreign consulates and representative offices. For example, the Russian cities of Arkhangelsk and Murmansk host Norwegian consulates, while Petrozavodsk accommodates a Finnish consulate.

Cooperation with international organizations: In the hope of getting international assistance and recognition, many northern European municipalities try to develop relations with regional and sub-regional organizations, such as the Barents/Euro-Arctic Council (BEAC), Arctic Council, Council of Europe, European Congress of Municipal and Regional Governments, and the European Regions Assembly. For example, the BEAC has both its international headquarters and Norwegian national offices in Kirkenes and its office in Murmansk. These offices help to coordinate cooperative projects at the regional/municipal level, including the Kirkenes-Nikel twinning project, launched in 2008. The Sør-Varanger community, being a part of the Finnmark county, cooperates with the Norwegian Association of Local and Regional Authorities, which, in turn, is a member of the Council of European Municipalities and Regions (2013: 11). Both municipalities have contacts with the City Twins Association (CTA) and may even think about joining it in due time. Cooperation with international organizations is important for regions and municipalities, not only in terms of survival or getting additional resources for their developmental programs, but also in terms of opening them up for the world-wide processes of globalization and regionalization.

The use of the "treaty-making power": Most of the Arctic subnational units have the right to conclude direct agreements with the same-type international partners. There is understanding between the center and local/regional actors that such agreements do not have the status of fully fledged international treaties (this is still considered as a national center's prerogative), and should be concluded with the partners located at the same level rather than with foreign governments. Moreover, they should be prepared in consultation with Foreign Ministries.

Despite occasional collisions with national capitals, many Arctic regions and municipalities continue to see involvement in quasi-"treaty-making" strategy as an effective instrument for building their capacities and enhancing domestic and international prestige.

16.5 Paradiplomacies: Problem-Solving/Functional Cooperation

Cooperation between Arctic subnational units aim to help these actors solve the numerous problems they face and promote their further sustainable development. Their cooperation covers the following areas.

Industrial and agricultural cooperation: For example, Alaska actively cooperates with neighboring Canadian regions: Yukon, Northwest Territories and British Columbia. For instance, since 1990 Alaska has participated in the US Western States/British Columbia Oil Spill Tax Force (http://www.encyclopedia.com/places/united-states-and-canada/us-political-geography/alaska). Their agricultural cooperation is especially impressive. For example, in 2012 Alaska exported agricultural goods worth $800,000 to Canada and imported $7 million of agricultural goods from Canada (Canadian Ministry of Agriculture and Agri-Food 2013).

Developing transport infrastructure: As far as the subnational actors from Northern Europe are concerned, the Swedish and Norwegian northern regions have a long-standing cooperation on shipping iron ore from the LKAB mine in Kiruna to sea ports. In particular, the company LKAB Malmtrafik pursues logistics and ore transport on the ore railway from the mining areas to the harbors in Luleå (Sweden) and Narvik (Norway). Narvik is LKAB's biggest harbor and has a capacity of almost 20 million tons per year (http://www.lkab.com/en/About-us/Overview/Operations-Areas/Narvik/). Narvik's harbor is ice-free all year round and is sufficiently deep for ocean-going ships.

Norwegian and Russian regional authorities are modernizing the E-105 highway from Murmansk to Kirkenes (via the Borisoglebsk-Storskog border-crossing) and building several shortcuts that bypass Pechenga and Nikel making thus the distance shorter. The Norwegian side has also modernized its part of the E-105 as well. Both projects should be completed by 2014, although the Russian part of the highway is still under construction.

The North American partners take care of the Alaska Highway, which extends 1523 mi (2451 km) from Dawson Creek, British Columbia, to Fairbanks, Alaska, and it is the only total road link with the rest of the United States (http://www.encyclopedia.com/places/united-states-and-canada/us-political-geography/alaska).

The Alaska Marine Highway System provides a year-round ferry service, not only to over 30 communities throughout southeast and southwest Alaska, but also extends from Bellingham, Washington (US) to Prince Rupert, British Columbia (Canada). This ferry system extends over 3500 route mi (532 km) and connects communities with each other, with regional centers, and with the continental road system (http://www.encyclopedia.com/places/united-states-and-canada/us-political-geography/alaska).

Creating a common labor market: One aspect of cooperation between the Arctic subnational units consists of relations in the sphere of work. For example, the northern provinces of Finland, Sweden and Norway have a common labor market, although two former countries belong to the EU while Norway remains outside the EU.

To give another example, there was a considerable shortage of skilled labor force in Sør-Varanger, and Finnmark more generally, in the late 2000s; this has become more acute with the re-opening of iron mines in the vicinity of Kirkenes in 2009. Efforts have been made to improve the flow of labor in the Barents region between Norway and Russia; for example, by providing training for jobs in the offshore sector and by changing experience in the promotion of an inclusive labor market. There were plans to partially staff the Kirkenes mines with workers from Nikel who could commute daily or work on the shift basis. However, these plans were postponed because the introduction of new visa/work permit regulations was delayed and the Norwegian trade unions opposed these plans.

Promoting visa facilitation regime: The intensified contacts between Finnmark and the Murmansk regions have entailed the need for liberalization of the visa regime for the border residents. In that context, agreement was reached on a local border traffic zone and the introduction of a border resident ID card on November 2, 2010. People who live within the 30 km border area on the Norwegian and Russian sides are eligible to receive a three-year ID card and be able to cross the border without a visa and stay on the other side up to 15 days each time (Soglashenie mezhdu Pravitel'stvom Korolevstva Norvegija i Pravitel'stvom Rosssiyskoy Federatsii 2010). These arrangements covered the whole Sør-Varanger community (with exception of the Saami village of Neiden), as well as the Russian towns of Nikel, Zapolyarny, Pechenga and Korzunovo.

The agreement was ratified by both the Norwegian and Russian sides in early 2011. However, it entered into force only on May 29, 2012, one and a half years after it had been signed. The numerous technical difficulties, which ranged from providing border residents with reliable ID cards to renovation of the Borisoglebsk-Storskog border-crossing, have repeatedly delayed its implementation. It was estimated that there was a need for 9000 cards on the Norwegian and 35,000 IDs on the Russian side (Lebed 2012). At the Kiruna BEAC summit on June 4, 2013, the Russian and Norwegian prime ministers signed a protocol to extend the 2010 agreement to Neiden, thus covering the whole Sør-Varanger community by a visa-free regime (Pogoretskaya 2013a).

As the Norwegian Foreign Minister Jonas Gahr Støre underlined on one occasion, Norway was one of the first Schengen countries to undertake such an experiment on the liberalization of the visa regime (Lebed 2012).

In July 2015, the US and Russia introduced a visa-free regime for the indigenous peoples of Alaska and Chukotka, who may visit each other on a visa-free basis for up to 90 days. The special border-crossings were established in Nome and Gambell (Alaska) and Anadyr', Providence, Lavrentiya and Welen in Chukotka (http://www.arctic-info.ru/news/23-07-2015/korennie-jiteli-cykotki-i-alaski-mogyt-gostit_-dryg-y-dryga-bez-viz/).

Environment protection: The ecological situation in the High North is an area of concern for many Arctic subnational actors. For example, ongoing climate change throughout Alaska and Northwest Canada has the potential to affect terrestrial ecosystems and the services that they provide to the people living in the region. These services include the provisioning of food and fiber by Alaskan ecosystems; the importance of ecosystems to recreation, cultural, and spiritual activities of people in Alaska and Northwest Canada; and the role Alaskan ecosystems play in regulating the climate system. To cope with these challenges, these North American regions launched a research project on the basis of the University of Alaska that aims to develop, test, and apply the Integrated Ecosystem Model for Alaska and Northwest Canada to forecast how landscape structure and function might change in response to how climate change influences interactions among disturbance regimes, permafrost integrity, hydrology, vegetation succession, and vegetation migration (McGuire and Rupp 2013).

Yet another example is the plan to create a US-Russian natural park for the protection of biodiversity in the Bering Strait region, with the provisional name of *Beringia*. This project is crucial for the local economy, which is heavily dependent on the fishery. The planned park could be based on the experiences of the existing ethno-natural park, established in 1993, with the same name on the Russian side of the Bering Strait (see the *Beringia* park's web-site: http://beringiapark.ru/).

The ecological situation in the Russian Arctic is especially difficult. As a result of intensive industrial and military activities in the region, many Arctic areas are heavily polluted and pose serious health hazards. Russian scientists identified 27 so-called impact zones where pollution has led to environmental degradation and increased morbidity among the local population (see Fig. 16.1). The main impact zones include the Murmansk Region (10% of total pollutants in the 27 impact zones), Norilsk urban agglomeration (more than 30%), West Siberian oil and gas fields (more than 30%) and the Arkhangelsk Region (around 5%) (Dushkova and Evseev 2011; Ekologicheskoe Sostoyanie Impactnykh Raionov 2012). In sum, approximately 15% of the AZRF territory is polluted or contaminated (Kochemasov et al. 2009). In the case of the Murmansk Region, the problem has a trans-boundary nature because some of the biggest pollutants, such as nickel combines, are located closely to the Norwegian border.

Although it should be national governments who should be in charge with environmental problems of such magnitude the Russian and Norwegian regions and communities also try to contribute to solving these problems. For example, Sør-Varanger community (Finnmark) and the Pechenga district (Murmansk Region) have a cooperative program that includes monitoring of the ecological situation on the Norwegian-Russian border.

Education: Networking Arctic regions pay a great attention to cooperation between local universities. For example, the Alaska State University has cooperative agreements with the North-Eastern Federal University (Yakutsk) and Northern-Arctic Federal University (Arkhangelsk) in Russia. The University of Lapland (Finland) cooperates with the Petrozavodsk State University (Karelia), Northern-Arctic Federal University and Arctic University of Norway (University of Tromsø, Norway).

In December 2012, an ambitious project to establish an international Master's program in Border Studies (Bordology) was launched by the University of Nordland and Murmansk State Institute for Humanities. The main idea of the project was to train specialists in border management for the Norwegian and Russian border regions. The group of 20 students were recruited from candidates with bachelor degrees and

Fig. 16.1 The map of impact zones in the Russian Arctic *Source* Dushkova and Evseev (2011, 2)

composed of 10 Russian representatives (five slots are reserved for students from Nikel and Pechenga) and 10 Norwegian representatives who are being trained in Nikel, Murmansk, and Kirkenes for 3.5 years. The program is a combination of class-room lectures and seminars with web-based distant-learning technologies. The courses are taught in English by teachers from Norwegian and Russian universities and practitioners from various governmental and municipal bodies. The program's curriculum is built on the Bologna Process principles and should result in students receiving a MA degree from the University of Nordland (Bulygin 2012; Grimmer 2013; Pogoretskaya 2013b). For the time-being it is difficult to say whether the project will be a success story or not, but the demand for specialists in border management is very high in both twinning communities.

Another example of "educational paradiplomacy" is the Calotte Academy. The international travelling symposium, Calotte Academy, organized since 1991, is an annual doctoral summer school for Ph.D. candidates from the Arctic states, as well as from the observer countries of the Arctic Council. It is designed to promote interdisciplinary discourse as well as academic and policy-oriented dialogue between senior researchers, early-career scientists and advanced graduate students, and other northern stakeholders, such as policymakers, civil servants, and community leaders and planners. It is a "school of dialogue" and participatory in nature—the principle is to share knowledge and experiences with communities. (http://arcticpolitics.com/index.php/calotte-academy).

All northern universities are united in a network-type University of the Arctic, which was launched on June 12, 2001 and endorsed by the Arctic Council (see UArctic website: http://www.uarctic.org/).

Cultural strategies: The Arctic subnational actors view cooperation in the cultural sphere as an important component of their paradiplomacies. In the case of the North American and Nordic players, it can be explained first and foremost by the existence of the related indigenous peoples. For example, the Innuits from Alaska, North Canada, and Greenland, as well as the Saami from Finland, Sweden and Norway have well-established cultural cooperation for many years. Since the 1990s they have also tried to involve the Russian Innuits and Saami into this kind of cooperation.

Another example is the well-established and multifaceted cultural cooperative ties between the Sør-Varanger community and the Pechenga district.

The core of twinning cultural activities is the *Barents Spektakel*, a festival that has been held in Kirkenes annually since 2004. As the organizers describe it, "[t]he festival is a cultural-political cocktail with contemporary art, performances, literature, theatre, film, seminars and concerts as ingredients, spiced with the current issues related to the Barents Region and the High North in general" (Barentsspektakel website). The main aim of the festival is to promote cultural contacts between different countries and peoples of the Barents region, with the goal of developing a common Barents/Arctic culture. Along with cultural, the festival has a clear political connotation: to encourage the states and people of the region to solve existing disputes over territories, borders, natural resources, environmental problems, and so on in a peaceful and non-violent way.

In addition to the *Barents Spektakel*, the two twinning communities have established numerous direct contacts between local writers, poets, artists, actors, dancers, libraries, and museums that formed a dense network of cultural cooperative ties in the region.

16.6 The Institutional Framework

The pursuance of paradiplomacy obviously calls for a favorable institutional setting. A proper and supportive institutional framework allows various subnational units to be both active and successful in their paradiplomatic initiatives.

As Fig. 16.2 indicates, the Arctic institutional network includes several layers.

On the top, ***supranational***, level, are institutions set up by the **EU**, the largest regional actor. For example, the **European Territorial Cooperation** (ETC), previously known as INTERREG Community Initiatives, has been part of EU policy since 1990, providing a framework for the implementation of joint actions and policy exchanges between national, regional, and local actors from different member states and neighboring countries. The ETC has grown from a relatively small INTERREG program to a fully fledged strand of the EU regional policy with its separate regulatory framework envisaged for the period 2014–2020.

In 2007–2013 the Kolarctic program was run by the CBC (cross-border cooperation) program of the European Neighborhood Partnership Instrument (ENPI). The Kolarctic program area includes the Norwegian provinces of Nordland, Trøms, and Finnmark, the Swedish Norrbotten, the Finnish Lapland, and three Russian subna-

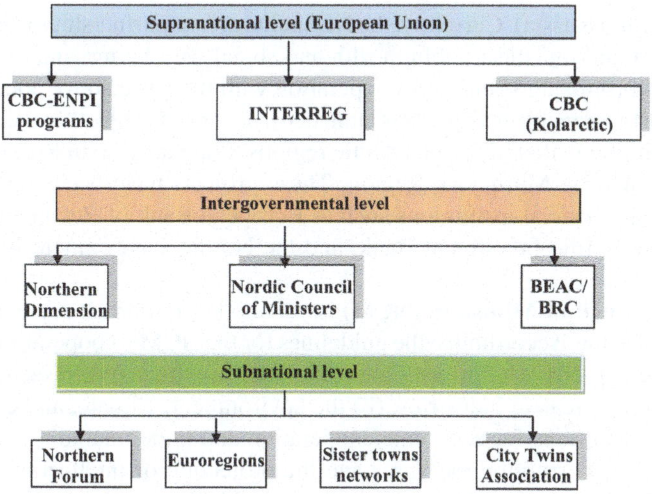

Fig. 16.2 Institutional framework for subnational cooperation in the Arctic region

tional units—the Arkhangelsk and Murmansk regions and the Nenets Autonomous District. The Republic of Karelia and Leningrad region have been eligible for some Kolarctic-related projects as well. The Finnish province of Lapland was responsible for the administration of the program. About 50 projects related to the development of economic and transport infrastructures, logistics, small and medium-size business, innovative entrepreneurship, preservation of the indigenous peoples' economies and cultures, research, and education were supported and implemented by the Kolarctic program in northern Russia (http://www.kolarcticenpi.info/ru).

As to the future of the ETC, three strands (cross-border, transnational, and inter-regional) will be maintained in the financial period of 2014–2020. This plurality will no doubt facilitate its implementation and the use of the already gained experience.

To avoid unnecessary inter-institutional duplication, it is important that the ETC places stronger emphasis on the thematic concentration and strengthened links to other EU programs. However, it should be guaranteed that the themes that the European Commission presents as priority ones are sufficient to cover the differing needs of cross-border and trans-border cooperation. A delicate balance between a greater regional flexibility and the need to achieve results with scarce resources at hand has to be found. However, balance can only be achieved if all the parties to the negotiations treat each other as partners.

However, it should be noted that because of the Ukrainian crisis, the funds for the Russian part of the ETC remain frozen and none of the projects have been implemented.

The ***intergovernmental*** level is represented by several institutions. The **Northern Dimension** (ND), which was initially a EU project oriented towards cooperation with the Baltic Sea and Northern Europe regions, has been transformed to a system of equally funded partnerships between the EU and three neighboring countries

(Iceland, Norway, and Russia). Currently, the ND includes four partnerships (on environment; transport and logistics; public health and social well-being; and culture), which are seen as promising venues for cooperation with the three above countries. Since 2007 (when the transformed ND was launched), dozens of projects in the above areas have been implemented in various Arctic regions. For example, in Russia they included Karelia and the Murmansk Region. These projects have been supported by the international financial institutions such as European Bank of Reconstruction and Development, Nordic Investment Bank, and the Nordic Environment Finance Corporation.

The **Nordic Council of Ministers** (NCM) is yet another important regional and intergovernmental actor. According to the guidelines for the NCM's cooperation with northern Russia 2009–2013, its priority areas include: (a) education, research and innovation, including creative industries; (b) the environment, climate, and energy; (c) promotion of conditions for economic cooperation and trade, including legislative cooperation, anti-corruption measures and the protection of intellectual rights and patents; (d) the ND's partnerships, especially for public health and environment; and (e) promotion of democracy and civic society through cooperation on local government and good governance, cooperation between parliamentarians, co-operation between the media and journalists, and co-operation between NGOs (Nordic Council of Ministers Nordic Council of Ministers 2009: 2–3). The NCM has several information offices in northern Russia.

The problem with the ND partnerships and the NCM is that they have a multi-focused agenda because their activities cover not only the Barents and Arctic regions, but also the Baltic Sea area. It seems that both institutions should avoid duplication and there is clearly a need to establish an improved division of labor between them. This is especially important in view of the scarcity of resources available to the regional actors.

In institutional terms, the North European "flank" is covered by the Barents Euro-Arctic cooperation. Along with the inter-ministerial **BEAC** (Barents-Euro-Arctic Council), there is the **BRC** (Barents Regional Council), which includes 13 counties from Finland, Norway, Sweden, and Russia (five of which belong to the Russian North). For example, at its Kirkenes meeting in June 2013, the BRC adopted a new *Barents Program 2014–2018* with the aims of promoting creative businesses and fast-growing enterprises in the region; increasing cross-border cooperation to achieve economies of scale and quality of life; supporting joint management and preservation of natural resources; implementing a joint climate change adaptation; enhancing innovation and research cooperation by increasing critical mass; focusing on missing cross-border links in the transport infrastructure; fostering mobility across the borders for workers, enterprises, tourists and students; and focusing on cultural cooperation in order to develop mutual understanding and regional development (The Barents Euro-Arctic Council 2013).

Given the numerous overlaps with the "sister" institutions involved in cooperation at the subnational level (ND, NCM, Arctic Council), BEAC and BRC are seeking synergy with them. These two councils have managed to establish cooperation on the project level with the above bodies in areas such as climate change research and elimination of the Barents environmental hot spots.

In addition to supranational and intergovernmental levels, there is also a purely ***subnational*** layer represented by the Northern Forum, City Twins Association, sister towns networks, and Euroregions. These organizations are important in terms of encouraging paradiplomacy in the Arctic region as they operate at the subregional and municipal levels. The problem with the upper institutional levels is that they are run by the EU bodies and/or national governments, not by subnational units themselves, and therefore aim mostly at the macro- rather than mezo- and micro-regional levels, neglecting cooperation between the Arctic substate units. In contrast with the governmentally sponsored institutions, the above fora were created by subnational units themselves in a bottom-up manner.

The Northern Forum is a non-profit, international body composed of 18 subnational units from eight countries. Along with the regions and municipalities from the Arctic states (Canadian Yukon Territory, Finnish Lapland, Icelandic Akureyri and Russian Chukotka, Khanty-Mansiysk Autonomous Area, Krasnoyarsk Province, Murmansk Region, Yakutiya-Sakha, and Yamal-Nenets Autonomous Area) it includes the South Korean Ganwon Province and the Japanese Hokkaido Prefecture, which are interested in developing cooperation with the High North partners (http://www.northernforum.org/en/the-northern-forum/northern-forum). The Forum was established in 1991 with the aim of developing horizontal ties between the Arctic subnational actors, and initially had 11 members from nine countries. However, for various reasons the Forum was unable to prove its efficiency and lost some of its members. The NF Secretariat moved from Fairbanks to Yakutsk (Russia) because Alaska quit the Forum.

Euroregions are another potential venue for cooperation between the North European subnational units. These are, in essence, administrative-territorial units designed to promote cross-border cooperation between neighboring local or regional authorities in countries that share land or maritime borders. In fact, they constitute well-known mechanisms of cooperation between regions and quite popular in the EU countries. However, there was only one Euroregion in the sub-Arctic region: Karelia has participated in the *Euroregio Karelen* together with three Finnish regional councils—Kainuu, Northern Karelia and Ostrobothnia (http://euregio.karelia.ru/article/52?lang=eng). However, this does not preclude other North European and Russian regions and municipalities from launching new Euroregions.

To coordinate and institutionalize twinning between municipalities, the **City Twins Association** (CTA) was established in December 2006. Altogether, the CTA includes 14 cities, including four pairs in Northern Europe: Valka–Valga (Latvia–Estonia), Imatra–Svetogorsk (Finland–Russia), Narva–Ivangorod (Estonia–Russia), and Tornio–Haparanda (Finland–Sweden) (City Twins Association 2010). Only one pair (Tornio–Haparanda) is located in the sub-Arctic area. However, the CTA should still be taken into account because the Kirkenes–Nikel pair has a plan to join the CTA at some point.

A proper division of labor between all these actors is called for. For example, the BRC and ETC could be especially useful in developing and implementing joint projects in areas such as environment protection, energy, development of local transportation, cross-border infrastructure, public–private partnerships, and fund-raising for specific projects. In some spheres, such as regional transport systems, public health and quality of life, science, education, and culture, the ND and NCM could take the lead. The CTA is helpful in sharing best practices in urban development as well as solving common municipal problems.

To sum up, almost all of the involved actors recognize that their task is to ensure the rightful architectural and financial demands for further cooperation in the Arctic.

16.7 In Lieu of Conclusion: Implications of Paradiplomacy

The role of subnational actors in the Arctic region has changed substantially over the last two or three decades. These actors have been gradually transformed from passive policy-takers to policy-shapers. Subnational units have become active players not only in their realms, but also on all levels of Arctic governance.

The substate units do not hesitate to use paradiplomacy to protect their interests and establish horizontal, network-type links with foreign partners.

There has been a clear paradigmatic shift in subnational units' motivation with regard to paradiplomacy. Instead of romanticism in the case of the West's Arctic substate actors (such as demonstrations of northern solidarity) or a survival instrument in the case of the Russian Arctic regions and municipalities (in the 1990s), paradiplomacy is now seen by these actors as an important component of their sustainable development strategy. Paradiplomatic activities became better organized and coordinated with the center's policies, as well as subordinated to pragmatic/realistic objectives.

The subnational actors managed to develop an arsenal of specific methods for paradiplomacy that fall into two categories: capacity-building strategies and problem-solving/functional cooperation. The former includes attracting investments, branding/PR strategies, increasing familiarity, accommodation of foreign consular offices and trade missions, cooperation with international organizations, and the use of the "treaty-making power". Functional cooperation covers such spheres as: industries and agriculture, development of the regional transport infrastructure, creation of a common labor market, promotion of a visa facilitation regime, environmental protection, education, and development of cultural contacts.

The Arctic substate units have managed—with the help of national governments and, on some occasions, without it—to exploit the institutional network that has been shaped by supranational (EU), intergovernmental (ND, NCM, BEAC/BRC), and subnational actors (NF, Euroregions, CTA, sistership arrangements) and is now available in the Arctic region. However, this rather dense network clearly needs better coordination, organization, and division of labor in order to eliminate bottlenecks, bureaucratic procedures, parallelisms, and duplications.

It should be noted that the aspirations of subnational actors and the center often overlap. Their interests are compatible in matters such as the promotion of cross- and trans-border trade, attracting foreign investment and know-how, development of cross- and trans-border transport infrastructures, facilitation of a visa regime for the residents of border regions, ecological projects, indigenous peoples' cooperation, tourism, youth and sport cooperation, and cultural and academic exchanges. Since some Arctic subnational units were able to solve these problems themselves, the national governments did not need to intervene in these spheres and spend funds for these purposes.

In the North American and Nordic countries, the subnational tier of international cooperation has become an integral part of national foreign policy. In most cases, paradiplomacy complements rather than contradicts the center's international course.

In the case of Russia, the center-periphery relations remain uneasy. Moscow still believes that, under adverse conditions (such as severe economic crisis), paradiplomacy may amount to a further disintegration of Russia's single economic, financial, administrative, and cultural space. Furthermore, it may be conducive to the rise of some rather parochial interest groups as well as the emergence of self-willing and outward-oriented local elites, and the outcome may amount to partial regionalization and privatization of security and military structures. The negative record can also include inconsistencies in the application of international strategies caused by the regional elites intervening in the decision-making process, and could even—at least theoretically—contribute to the rise of separatism and secessionism, which could result in disintegration of the country.

On the other hand, the Kremlin cannot ignore the fact that the gradually growing international activities of subnational actors also bring a number of positive changes. First and foremost, paradiplomacy encourages further democratization of Russia's rather centralized administrative system, including managing the external relations of regions and municipalities. It has also, as part of the devolution process, helped to discredit the "top-down" model of the Russian federalism and encouraged its replacement with a "bottom-up" process that has lively grass-roots. Moreover, international cooperation has allowed many regions, particularly some remote and border-located regions, not only to survive the transition period, but also to turn their marginality into an advantage and a feature.

Paradiplomacy will undoubtedly continue to play an important transformative role in Russia's future. Rather than contributing to disintegration, as has been sometimes feared, it can actually serve as a catalyst for the pursuance of successful reforms within the country and the creation of the Arctic governance system.

References

Ackrén M (2014) Greenlandic paradiplomatic relations. In: Heininen L (ed) Security and sovereignty in the North Atlantic. Palgrave, Basingstoke, pp 42–61

Bulygin A (2012) Speztialist po graniztevedeniyu – real'nost [Specialist in border studies is a reality]. Pechenga, 8 Sept. http://pechenga-gazeta.ru/index.jsp?view=article&id=6009. Accessed 1 Aug 2013 (in Russian)

Canadian Ministry of Agriculture and Agri-Food (2013) Alaska-Canada: agricultural trade = $8 million in 2012. http://publications.gc.ca/collections/collection_2015/aac-aafc/A77-4-2012-eng.pdf. Accessed 5 Nov 2016
City Twins Association (2010) History of the city twins association. http://www.citytwins.org/en/page/13/. Accessed 1 Aug 2013
Council of European Municipalities and Regions (2013) National associations of regional and local authorities. List of CEMR members. http://www.ccre.org/docs/list_of_members_en.pdf. Accessed 1 Aug 2013
Duchacek I (1986) The territorial dimension of politics: within, among, and across nations. Westview, Boulder
Duchacek I (1990) Perforated sovereignties: towards a typology of new actors in international relations. In: Michelmann H, Soldatos P (eds) Federalism and international relations: the role of subnational units. Clarendon Press, Oxford, pp 1–33
Dushkova D, Evseev A (2011) Analiz Techogennogo Vozdeistviyana Geosistemy Evropeiskogo Severa Rossii [Analisys of technogenic impact on geosystems of the European Russian North]. Arktika i Sever (4):1–34. http://narfu.ru/upload/iblock/673/16.pdf. Accessed 23 Nov 2015 (in Russian)
Ekologicheskoe Sostoyanie Impactnykh Raionov Sushi Arkticheskoi Zony Rossiyskoi Federatsii [The environmental situation in the impact zones of the terrestrial parts of the Arctic zone of the Russian Federation] (2012). http://www.arcticonline.ru/ekologiya/ekologicheskoe-sostoyanie-impaktnyhrajonov-sushi-arkticheskoj-zony-rossijskoj-federacii. Accessed 23 Nov 2015 (in Russian)
Farrell M, Hettne B, van Langenhove L (eds) (2005) Global politics of regionalism: theory and practice. Pluto Press, London
Grimmer T (2013) Politicheskaya filosopfiya – osnofa uspekha [Political philosophy is a precondition for success]. Pechenga, 6 Feb. http://pechenga-gazeta.ru/index.jsp?view=article&id=6449. Accessed 1 Aug 2013 (in Russian)
Handley S (2006) Take your partners. The local authority handbook on international partnerships. London: Local Government International Bureau. http://www.lga.gov.uk/lga/aio/190428. Accessed 27 May 2011
Hobbs H (1994) City hall goes abroad: the foreign policy of local politics. Sage, Thousand Oaks
Hocking B (1993) Localizing foreign policy: non-central governments and multilayered diplomacy. Palgrave Macmillan, New York
Joenniemi P, Sergunin A (2014) Paradiplomacy as a capacity-building strategy: the case of Russia's Northwestern subnational actors. Prob Post-Commun 61(6):18–33
Kochemasov YV, Morgunov BA, Solomatin VI (2009) Ekologo-ekonomicheskaya Otsenka Perspectivy Razvitiya Arktiki [Ecological-economic assessment of perspectives of the arctic's development]. http://www.perspektivy.info/rus/ekob/arktika_perspektivy_razvitija_2009–04–24.htm. Accessed 23 Nov 2015 (in Russian)
Lebed Y (2012) Bol'shie vozmozhnosti dlya zhitelei raiona [Greater opportunities for the district's residents]. Pechenga, 2 June. http://pechenga-gazeta.ru/index.jsp?view=article&id=5619. Accessed 1 Aug 2013 (in Russian)
McGuire D, Rupp S et al (2013) Annual 2012 progress report on the integrated ecosystem model for Alaska and Canada Project. Fairbanks: University of Alaska. https://csc.alaska.edu/sites/default/files/2012-IEM_Annual_Report.pdf. Accessed 6 Nov 2016
Nicol H (2016) Ripple effects: devolution, development and state sovereignty in the Canadian North. In: Heininen L (ed) Future security of the global Arctic: state policy, economic security and climate. Palgrave, Basingstoke, pp 99–120
Nordic Council of Ministers (2009) Guidelines for the Nordic council of ministers' cooperation with North-West Russia 2009–2013. www.norden.org/en/nordic-councilof-ministers/ministers-for-co-operation-mr-sam/russia/documents/guidelines-for-the-nordic-council-of-ministers-co-operation-with-northwest-russia. Accessed 12 Sept 2012
Olsson M (ed) (2016) Encyclopedia of the Barents Region. Pax Forlag, Oslo

Pogoretskaya O (2013a) Prioritet sotrudnichestva – prostye grazhdane [Ordinary citizens is a priority for cooperation]. Pechenga, 12 June. http://pechenga-gazeta.ru/index.jsp?view=article&id=6896. Accessed 1 Aug 2013 (in Russian)
Pogoretskaya O (2013b). Uchis', student! [Learn, student!]. Pechenga, 1 June. http://pechenga-gazeta.ru/index.jsp?view=article&id=6855. Accessed 1 Aug 2013 (in Russian)
Rogova, A. (2008). From rejection to re-embracement. Language and identity of the Russian speaking minority in Kirkenes, Norway. Kirkenes: Barents Institute
Rogova A (2009) Chicken is not a bird—Kirkenes is not abroad. Borders and territories in perception of the population in a Russia-Norwegian borderland. J North Stud 1:31–42
Scott J (2013) Construction of familiarity in finnish-Russian Karelia: shifting uses of history and the re-interpretation of regions. Eur Plan Stud 21(1):75–92
Söderbaum F, Shaw T (eds) (2003) Theories of new regionalism. Palgrave, Basingstoke
Soglashenie mezhdu Pravitel'stvom Korolevstva Norvegija i Pravitel'stvom Rosssiyskoy Federatsii ob uproshchenii porjadka vzaimnyx poezdok zhiteley prigranichnyx territoriy Korolevstva Norvegija i Rossiyskoy Federatsii, Oslo, 2 November 2010 [Agreement between the government of the Kingdom of Norway and the government of the Russian Federation on facilitation of mutual trips of the residents of the border areas of the Kingdom of Norway and Russian Federation]. http://www.regjeringen.no/upload/JD/Vedlegg/Grenseboeravtala_RUS.pdf. Accessed 1 Aug 2013 (in Russian)
Soldatos P (1990) An explanatory framework for the study of federal states as foreign-policy actors. In: Michelmann H, Soldatos P (eds) Federalism and international relations: the role of subnational units, Clarendon Press, Oxford, 34–53
Spierings B, Velde M (2008) Shopping, border, and unfamiliarity: consumer mobility in Europe. Tijdschrift Voor Economische en Sociale Geografie 99(4):497–505
Spierings B, Velde M (2013a) Cross-border mobility, unfamiliarity, and development policy in Europe. Eur Plan Stud 21(1):1–4
Spierings B, Velde M (2013b) Cross-border differences and unfamiliarity: shopping mobility in the Dutch-German Rhine–Waal Euroregion. Eur Plan Stud 21(1):5–23
The Barents Euro-Arctic Council (2013) The Barents Cooperation. www.barentsinfo.fi/beac/docs/Barents_Cooperation_information_English_October_2013.pdf. Accessed 19 Jan 2014

Chapter 17
Sámi People at Different Levels of Decision-Making Processes in the Global Arctic

Laura Olsén

17.1 Introduction

This chapter discusses the Sámi people's position at different levels of decision-making processes, from national to international levels (see also Olsén 2016). The Sámi are represented politically in Norway, Sweden, and Finland through three Sámi parliaments, one in each country. The Sámi people in Russia do not have their own publically recognized Sámi parliament but there are nongovernmental organizations (NGOs) such as the Kola Saami Association and the Saami Association of the Murmansk Region (International Work Group for Indigenous Affairs 2017, 60). The Sámi people also have their own pan-Sámi organizations, such as the Saami Council, in which Sámi people from all four countries are represented (Saami Council n.d.; Baer 2017, 3; Heininen 2002), and the Sámi Parliamentary council, which is the cooperational body of three Sámi parliaments and has two permanent representatives of the Russian Sámi (Sámi Parliamentary Council n.d.). The Sami Council has relatively strong status as an actor in international politics especially in the Arctic region and is represented at the Arctic Council as a permanent participant, and in the Arctic Council Indigenous Peoples Secretariat. It also has consultant status in the United Nations Economic and Social Council and observer status in the Barents Euro-Arctic Council Working Group of Indigenous People (Representation n.d.).

From legal and institutional perspectives, the position of the Sámi people could theoretically be argued as relatively good, especially in the Nordic countries. The position of the Sámi people has remarkably improved through the Saami Council also at the international level during the last several decades, so the Sámi play a role in, and are connected to, global politics (see also Heininen 2002). As Lindroth and Sinevaara-Niskanen (2018) argue the position of indigenous peoples has remarkably improved in international relations, but they have also had to adapt themselves to the

L. Olsén (✉)
University of Lapland, Rovaniemi, Finland
e-mail: laura.olsen@ulapland.fi

M. Finger and L. Heininen (eds.), *The GlobalArctic Handbook*,
https://doi.org/10.1007/978-3-319-91995-9_17

system of international politics to be heard in a system largely shaped by Western ideology. This does not mean that they would have a policy-making role, but rather a policy-shaping one in international relations (Lindroth and Sinevaara-Niskanen 2018).

Despite the relatively good position of the Sámi parliaments in three Nordic countries, it has recently been demonstrated that the real abilities of the Sámi Parliaments to affect national decision-making processes are rather limited. One of the most recent examples in Finland is the negotiation process of the agreement concerning fishing rights in the River Teno,[1] in which the Sámi Parliament was not properly consulted. Despite strong Sámi opposition, the Finnish and Norwegian states ratified the agreement. According to the statement of the Sámi Parliament in Finland (12.8.2016) the agreement violated the Sámi people's rights to cultural and property protection, and self-determination (Sámi Parliament in Finland 2016, 1). This process also gained much international attention, and finally led a Sámi activist group to establish a moratorium to an island located in the middle of the river (YLE 2017). The Deputy Chancellor of Justice in Finland made a decision, with which the Finnish Ministry of Agriculture and Forestry neglected to comply to a certain extent its obligation to negotiate with the Sámi Parliament, as set down in the Act on the Sámi Parliament (974/1995) in Section Nine (Deputy Chancellor of Justice 2017, 3).

This chapter briefly discusses important developments in the political history of the Sámi people from the late 19th century, analyzes the changes in their position in decision-making processes, and views the impacts of the emergence of international discourse concerning indigenous peoples' rights on the Sámi position in Nordic societies and internationally. After the historical overview, the current developments are discussed. The situation of the Sámi in national, regional, and international decision-making structures will be evaluated in general. This chapter will build a general picture of the situation, and the analysis is based mostly on previous studies of the issue.

Globalization had a significant role in shaping modern Arctic indigenous communities in many ways. At the same time, Arctic indigenous peoples have played an important role in the global political indigenous movement (Tennberg 2010, 264). The development of indigenous peoples' own administrative systems and structures has given them better standing in decision-making structures than they previously had, such as at the beginning of the 20th century. On the other hand, this development has also forced indigenous peoples to follow Western ideology in these processes and accept conditions that they do not always set themselves (e.g., Lindroth and Sinevaara-Niskanen 2018). Finally, the question arises: *Which kind of position do indigenous peoples, in this case the Sámi, really have in decision-making processes in the globalizing Arctic.*

[1]Teno is a river in Sápmi with great cultural and traditional livelihood value for the Sámi. It also marks a border between Finland and Norway.

17.2 Basis for the Analysis

When analyzing indigenous peoples' position in decision-making, their right to self-determination and participatory rights will be viewed from the perspective of human rights and international law. These two concepts are strongly linked to indigenous peoples' position in decision- making processes. The idea of self-determination has its roots in history and has played different roles in development of societies, from political and ideological principles to the universal human rights (Henriksen 2009, 7). The right to self-determination as a human right is recognized in many instruments of international law, such as in the International Covenant on Civil and Political Rights (1966, Article 1) and the International Covenant on Economic, Social and Cultural Rights (1966, Article 1).

For many years, states did not really recognize the idea of indigenous peoples' self-determination. Actors, such as the Human Rights Committee, only started to apply self-determination, as recognized in Article 1 of the International Covenant on Civil and Political Rights to indigenous people starting in 1999 in concluding observations related to country reports. (Heinämäki 2017, 26; see e.g., Concluding Observations of the Human Rights Committee on Canada 1999; Concluding Observations of the Human Rights Committee on Norway 1999). States gradually first started to recognize indigenous peoples' right to self-determination mostly when referring to internal self-governance and later by adopting the UN Declaration on the Rights of Indigenous Peoples (UNDRIP 2007), at least in principle (according to Article 3), as giving indigenous peoples the unqualified right to self-determination (Heinämäki 2017, 26; Xanthaki 2009; Heinämäki and Kirchner 2017, 229). However, Article 4 of UNDRIP stresses: "*Indigenous peoples, in exercising their right to self-determination, have the right to autonomy or self-government in matters relating to their internal and local affairs, as well as ways and means for financing their autonomous functions*" (UNDRIP 2007). Article 46, Paragraph 1 also mentions territorial integrity (Xanthaki 2009; Heinämäki and Kirchner 2017, 229). Xanthaki (2009) notes that, in the context of UNDRIP, the right to self-determination is not limited to just self-government of indigenous peoples, even though it focuses on that. Furthermore, UNDRIP does not extend to secession (Xanthaki 2009; see also Heinämäki and Kirchner 2017, 229). The debate remains open as to what *unqualified right to self-determination* exactly means in practice in the context of each state (Heinämäki and Kirchner 2017, 229).

UNDRIP stresses indigenous peoples' right to participate in different contexts, such as decision-making in matters that would affect their rights (UNDRIP 2007, Articles 18, 27 & 41; Henriksen 2009, 17–18; Xanthaki 2009). Heinämäki and Kirchner (2017) closely connect meaningful, effective participation to indigenous peoples' right to self-determination of economic, social, and cultural development, as stated in the annex of UNDRIP (Heinämäki and Kirchner 2017, 230). According to Heinämäki and Kirchner, this kind of participation means "*the right to consultation or even FPIC with respect to land and resource use and other important matters, such as true and meaningful participation in the relevant national and international decision-making*"

(Heinämäki and Kirchner 2017, 230). Therefore, indigenous peoples' right to free, prior, and informed consent (FPIC) also has an important role in implementing their right to self-determination (Heinämäki and Kirchner 2017, 230).

In the final report of the *Expert Mechanism on the Rights of Indigenous Peoples* (2011, 17), concerning indigenous peoples' right to participate in decision-making, the principle of FPIC is highlighted as one of the most important principles in protecting indigenous peoples' right to participation. It functions in both national and international laws. At the national level, it is part of the decision-making process in issues that concern and affect indigenous communities (Heinämäki and Kirchner 2017, 234). At the international level, FPIC is integral in implementing indigenous peoples' rights to self-determination and participation (Rombouts 2014, 11; Heinämäki and Kirchner 2017, 230, 234).

Another important instrument for indigenous peoples' rights is the ILO Convention No. 169, (1989), which also emphasizes the participation rights of indigenous peoples and states' responsibilities to consult indigenous peoples on "*legislative and administrative measures, which may affect them directly*" (Article 6, para. 1a; see also Articles 2, 7 & 15; Ward 2011, 60; Heinämäki and Kirchner 2017, 238–242). Unlike the ILO Convention No. 169, UNDRIP is not legally binding, but in many ways, it reflects international law and UN human rights treaties recognizing indigenous peoples' rights (Xanthaki 2009; Åhren 2016, 103–105; Heinämäki and Kirchner 2017, 228–233).

As this chapter will demonstrate, the development of indigenous peoples' rights and legal status at the international level has also had remarkable effects at national and local levels, and mutually actions taken by indigenous communities and activists at the local level have directly and indirectly affected their rights at the international level.

17.3 Losing Their Rights to Participate on Decision-Making Processes?

The region that the Sámi have traditionally inhabited for thousands of years has been a stage for states' power games for centuries. Especially during the last few centuries, different conflicts and political ambitions have shaped the region's borders and their regulations. In 1751, the border between Norway and Sweden was established, and the Sámi officially became citizens of one state. To avoid separating different Sámi groups, and to guarantee their rights to traditional usage of renewable natural resources in both countries, the states implemented an addendum to the border treaty (Lantto 2010, 544–546; Pedersen 2012, 61–68; Allard and Funderud Skogvang 2015, 4). This was called the *Lapp Codicil*.

The states' politics concerning the position and the rights of the Sámi started to shift in the 19th century. In 1809, Russian occupied Finland, and the border between Norway and Finland was closed for reindeer husbandry in 1852, and between Sweden and Finland in 1889 (Lantto 2010, 547–548). As a result, the Sámi lost their right to herd reindeer on both sides of the borders (Lantto 2010, 547–548; Allard and Funderud Skogvang 2015, 4). The 19th century also saw a period of Norwegianization politics in Norway. The aim of this was to assimilate the Sámi to Norwegian society by, for example, trying to forbid the use of Sámi languages at schools. Changes were also implemented to other structures of Norwegian society and were reflected in national legislation (e.g., Minde 2005).

According to Mörkenstam (2017), the Swedish state approached the Sámi with "divide and rule" politics, which originated in the late 19th century. Mörkenstam refers to the development of Swedish national legislation concerning the rights of the Sámi, in which those who were not herding reindeer were not taken into consideration regarding Sámi rights. This was a result of the first reindeer-grazing act enacted in 1886 (Mörkenstam 2017, 363). This led to a policy in which those Sámi who were occupied in livelihoods other than reindeer herding were assimilated to Swedish society (Mörkenstam 1999, 145–150, cited in Mörkenstam 2017, 363). Furthermore, reindeer-herding Sámi were segregated from the rest of society (Mörkenstam 1999, 145–150, cited in Mörkenstam 2017, 363; Lantto and Mörkenstam 2008). After becoming independent in 1917, Finland followed the example of its neighboring countries to a certain extent in its stance on the Sámi (e.g., Lehtola 2015a, 74). However, the official approach was different, since the emphasis was on the common needs of all citizens of Finland (Valkonen 2017, 178; Lehtola 2012, 16–17). Under this guise, the state could ignore Sámi issues (Valkonen 2017, 178; Lehtola 2012, 16–17).

The situation of the Sámi in Russia is essentially different from the Nordic countries, in which the Sámi are the only people who have been officially recognized as indigenous. There are more than 40 different groups of indigenous people in Russia (e.g., Rohr 2014, 8). Berg-Nordlie (2015) suggests that the political history of the Russian Sámi should be examined and explained in the context of Russia's general indigenous policy. The Russian Sámi have gone through dramatic changes in their political history during different eras, from Russia's imperial indigenous policy during the 19th and early 20th centuries, to "the Soviet spring of the indigenous minorities," and to the harsh period of enforced collectivization and removals by Stalinists (Berg-Nordlie 2015).

The Sámi had their own traditional ways to manage land use through the Siida system and customary law, and their own culture and traditional way of living. The Sámi had to gradually adapt to the arrival of settlers and the conditions set by external authorities. The legal conception of the states, especially concerning rights about land ownership and use, affected the dynamics between people living in the Sámi homeland region (Lehtola 2015b; Pääkkönen 2008, 196–202). Christianity also played a role in the colonial process, and its effects on the Sámi culture are still seen today (see e.g., Lehtola 2015a, 61–63, b, 25).

By viewing the development of indigenous peoples' position in international politics during this period, we see similarities at global and regional levels in three Nordic countries. Moretti (2012) argued that the legal status of nomadic people changed between the 16th and 19th centuries. They changed from being the subjects of international law to being objects to the domestic law of the states that conquered their lands (see also Heinämäki and Kirchner 2017, 225). Similar developments took place in the Nordic countries in which the Sámi live.

17.4 Fighting for Their Rights—From Local to Global, and Back

The emergence of the first Sámi organizations and strong personalities in Sámi movements, such as Elsa Laula in the early 20th century (e.g., Lantto and Mörkenstam 2008, 30), and the beginning of Nordic Sámi cooperation in the 1950s (e.g., Rantala 2004, 2–4; Valkonen 2017, 180), gradually started to affect the Sámi's position in the political decision-making processes. The development of the position of the Sámi has been different in each country (e.g., Minde 2003; Lantto and Mörkenstam 2008; Valkonen 2017; Berg-Nordlie 2015), but it has generally ascended in the same direction. The Sámi have started rightfully to demand their rights as indigenous people, and the situation regarding their position in Nordic societies has progressed.

In the 1960s and 1970s increasing numbers of Sámi went to study at Universities in Sweden, Norway and Finland. The ideas of equality and right to self-determination, based on declarations of human rights and different conventions, became central in the Sámi politics at the time (Minde 2003, 79–80; Lehtola 2005, 18–19, 25). At approximately the same time, the Sámi started to connect their situation to the wider international indigenous discourse (Minde 2003, 79–80) and became active actors in international indigenous movements (e.g., Seurujärvi-Kari 2012, 51–57; Minde 2003, 80; 2008; Valkonen 2017, 180). *Indigenous* as term referring to group of people who have a right to self-determination is fairly new and one of its earliest appearances in a legal context was in the documents of the International Labour Organization in the 1950s (Niezen 2009, 27; see also Lindroth and Sinevaara-Niskanen 2018, 9–11). However, after the 1950s, it provided a common platform for indigenous peoples to fight for their rights and improve their position at national and international levels concerning their rights and in decision-making structures (e.g., Niezen 2009, 27; Lindroth and Sinevaara-Niskanen 2018, 9–11; Smith 1999, 7). Tennberg (2010) claims that the work that Arctic indigenous peoples have done since the 1960s in organizing their own representation and promoting their concerns at the domestic and international level, and the success that has been achieved in the Arctic region can provide models for other indigenous peoples' organizations globally (p. 264).

Despite internationally stronger discourse concerning the rights of indigenous peoples and the Sámi being actively part in its formulation, the Nordic states did not at first consider the Sámi as people to whom conventions of international law would apply. According to Minde (2003), the states' perspective was: "*The Sámi were well integrated in the general life of the community*" (p. 87). None of these three Nordic countries nor the Russian Federation ratified ILO Convention 107.

Nevertheless, some positive development has taken place in Sweden, Finland and Norway. The basis for the Sámi people to establish their own political systems and position in national decision-making structures were created. In Sweden, the Sámi became recognized as minority or ethnic group in the 1960s, and there were attempts to create a more inclusive view of Samihood in the Sámi politics to also take into consideration those Sámi who were not engaged in reindeer herding (see e.g., Lantto and Mörkenstam 2008, 35–36). In Norway, the Norwegian Sami Council was established in 1964, based on the Sami Committee's proposal. The Sámi gradually managed to establish their own political sector (Minde 2003, 78–79). In Finland, the Sámi Delegation, which was a popularly elected institution representing the Sámi and the precursor of the Sámi Parliaments, was established in 1973 and later provided an example for the development of the Sámi parliamentary system also in Sweden and Norway (Lehtola 2005; Mörkenstam et al. 2016, 10–11; Valkonen 2017, 183). It represented the Sámi living in Finland and centrally promoted the interests of the Sámi in Finnish society (Lehtola 2005; Valkonen 2017, 185–186).

Finally, at the end of the 1980s and the beginning of the 1990s, political and legal situations changed and remarkable improvements concerning the Sámi people's right to self-determination and participation happened in three Nordic countries. This was a consequence of a long and complicated course of events, which had its own nuances in each country. First, the Sámi people rising to claim their rights and an active role in international indigenous movements played a crucial part (see e.g., Valkonen 2017, 180; Minde 2003, 2008; Lantto and Mörkenstam 2008, 37). Second, the emergence of international discourse of indigenous peoples' rights gave strength and motivation to Sámi activists and supported their claims at the national level (Minde 2003, 76–87).[2] In parallel, general societal consciousness of these issues increased, which also supported the situation of the Sámi (Minde 2003, 76–87; see also Lantto and Mörkenstam 2008, 37). For example, the position of the Sámi in Finland has improved through cooperation between the state and the Sámi (Minde 2003, 80; Valkonen 2017, 178). The Sámi have actively participated in political activities and constructed an understanding at the state level of their needs and important issues (Minde 2003, 80; Valkonen 2017, 178). Third, the establishment of pan-Sámi organizations gave the Sámi better footing in international forums (Minde 2003, 76–87). Finally, some remarkable events, such as the Alta Dam dispute between the Norwegian state and the Sámi took place in the late 1970s and 1980s, forcing the state to react to keep face as a defender of indigenous peoples' rights (see e.g., Minde 2003, 2008, 67–68). In Sweden, the Taxed Mountain case also pushed general debate at the national level toward the Sámi rights perspective (Lantto and Mörkenstam 2008, 37). These events do not represent a comprehensive list of significant events and developments, but point out general features of the course of events. In 1989, the first Sámi parliament

[2]At the international level, the famous report of the UN Special Raportteur José Martinez Cobo *Study of the Problem of Discrimination against Indigenous Peoples: Final Report* has influenced development of issues concerning indigenous people. The report also emphasizes the importance of indigenous peoples self-governance (Cobo 1981–1983; Heinämäki and Kirchner 2017, 226; Åhren 2016, 85–86).

was opened in Norway, followed by the Sámi parliament in Sweden 1993 and in Finland in 1996.

17.5 Establishing the Sámi Position in Decision-Making

Approaching the end of the 20th century, the Sámi had established their position in Norwegian, Swedish and Finnish societies as recognized indigenous people and had their own political bodies representing them. In Norway, the Sámi people were officially recognized as indigenous people in 1990 after Norway ratified the ILO Convention No. 169 (Allard 2017a, 318). In Sweden, the Sámi were officially declared an indigenous population by the Parliament in 1977 (Government Bill 1976; Report of the Committee for Education and Culture 1976), but are still often treated as an ethnic minority, especially in legislation (Allard 2017b, 343–44). In the Instrument of Government, which is part of the Swedish constitution, the Sámi are referred as Sámi *people* (Ch. 1, S. 2, para. 6), which, according to the Government Bill (2009, 189), indicates that they have a "[…] *special status as a recognized indigenous people in the country*" (Translation by Allard 2017b, 343–344; Government Bill 2009, 189). In Finland, the Sámi are recognized as indigenous people by the Constitution of Finland (1999, Ch. 2, S. 17, para. 3; Heinämäki 2017, 22). In Russia, they are recognized as indigenous, small-numbered people of the North, Siberia, and the Far East of the Russian Federation in the legal context (Rohr 2014, 9). As political bodies, the Sámi Parliaments are important actors in realizing the rights of the Sámi for self-determination in Norway, Sweden and Finland and could provide examples to the development of global models for indigenous self-governance and channels to participate in decision-making (e.g., UN 2011, Article 37; see also Mörkenstam et al. 2016, 10; Josefsen et al. 2015). They can be considered progressive in many ways. However, more resources should be provided by the states for the Sámi Parliaments to support their work in realizing the rights of the Sámi as indigenous people.

However, it is important to note that the positions and mandates of the Sámi parliaments differ in each country. Many factors, from each country's history, to current political and legal systems and situations affect the position of the Sámi parliaments (Mörkenstam et al. 2016; see also Josefsen et al. 2015). In addition, the electoral and representation systems of the Sámi Parliaments vary in each country. In Norway, elections are carried out in 7 different constituencies (Sámi Act in Norway 1987, Ch. 2, S. 4). In Finland and Sweden, elections are held in only one constituency that comprises the entire country (Mörkenstam et al. 2016, 18; Act of the Sámi Parliament in Finland, Ch. 4, S. 20; Act of the Sámi Parliament in Sweden 1992, Ch. 3, S. 3). Also the representation of people differs in each country. In Norway, the Sámi are represented by Sámi Parliament candidates who belong to the political parties (Mörkenstam et al. 2016, 18). In Finland, the Sámi are represented by private candidates, and in Sweden, they are represented "*through a party system where personal votes have a considerable impact*" (Mörkenstam et al. 2016, 18).

The people who are registered in the electoral roll in their country can vote in the Sámi parliamentary elections. In Norway and Sweden, criteria defining the right to register to the electoral roll are similar and listed in Chap. 1, Sect. 2 in the Sámi Parliament Act in Sweden, and in Chap. 2, Sect. 6 in the Sámi Act in Norway. First, the person must identify as Sámi and have (or in Sweden, also have had) Sámi as a domestic language; or have a parent or grandparent (or in Norway, also a great-grandparent) whose domestic language is or was Sámi; or their parent is or has been registered in the Sámi electoral register (Act of the Sámi Parliament in Sweden, Ch. 1, S. 2; Sámi Act in Norway, Ch. 2, S. 6; see also Mörkenstam et al. 2016, 19).

The criteria differ in Finland, since in addition to someone's self-identification as a Sámi, they or at least one of their parents or grandparents, must have learned Sámi as a first language; or a person is a descendent of a person who was entered in a land, taxation, or population register as a mountain, forest, or fishing Lapp; or at least one of their parents has, or could have been, registered as an elector for an election to the Sámi Delegation of the Sámi Parliament (Act on the Sámi Parliament in Finland, Ch. 1, S. 3). In Finland, unlike in Norway and Sweden, the definition of the Sámi, and who is eligible to register to the electoral register has provoked much discussion. However, because of the complicated, multidimensional and conflicting nature of this question, it will not be further analyzed in this chapter. Nevertheless, the situation led to a point in 2015 in which the Finnish Supreme Administrative Court decided, against the will of the Sámi Parliament of Finland, that 93 people were eligible to be registered to the electoral roll of the Sámi Parliament (Heinämäki 2017, 86–175; see also Mörkenstam et al. 2016, 33; see also Valkonen et al. 2017).

At the national level in the Nordic countries, the Sámi Parliaments are the principle bodies representing the Sámi in political decision-making, but their mandates differ in each country, according to issues defined in the Acts on the Sámi Parliaments (see Mörkenstam et al. 2016, 14). In Sweden, the Sámi Parliament has a legal position as a government agency "*with primary task of monitoring questions related to Sámi culture in Sweden*" (Act on Sámi Parliament in Sweden, Ch. 1, S. 1; see also Mörkenstam et al. 2016, 14). In Finland: "*The Sámi, as an indigenous people, have linguistic and cultural autonomy in the Sámi homeland as provided in this Act and in other legislation. For the tasks relating to cultural autonomy the Sámi shall elect from among themselves a Sámi Parliament*" (Act on Sámi Parliament in Finland, Ch. 1, S. 1). Authorities should also "*negotiate with the Sámi Parliament in all far reaching and important measures which may directly and in a specific way affect the status of the Sámi as an indigenous people and which concern the following matters in the Sámi homeland*" (Act on Sámi Parliament in Finland, Ch. 2, S. 5; see also Mörkenstam et al. 2016, 14–15). In Norway, the Sámi Parliament should be given an opportunity by other public bodies, to express an opinion before decisions are made on "*any matter that in the view of the parliament particularly affects the Sami people*" (Sámi Act in Norway, Ch. 2, S. 1 & S. 2; see also Mörkenstam et al. 2016, 14–15). In addition, the Skolt Sámi have their own administrative system representing them in Finland, and their rights are protected also by the Skolt Act (1995).

In addition to the Acts on the Sámi Parliament, other legislation affects the Sámi people's position in each country. Only Norway has so far ratified the ILO Convention

No. 169, which sets different premises for the actualization of Sámi people's rights. The Norwegian government has also enacted the Consultation Agreement with Sámi Parliament (Den nye sameretten, NOU 2007), which improves Sámi people's position and possibilities in influencing the drafting process of legislation affecting them (Allard 2015, 49–55). In addition, Norway enacted the Finnmark Act (2005, S. 29) in the year 2005 to recognize and identify Sámi people's traditional lands, as required in the ILO Convention No. 169, Article 14 (Ravna 2011, 423). This identification process is carried out by the Finnmark Commission, which was established for this purpose (Ravna 2011, 423).

In Sweden, the Reindeer Herding Act (Reindeer-Herding Act 1971, 437) is the primary instrument regulating the Sámi people's legal position, but they are also protected to some degree by the Instrument of Government, which is one of the fundamental laws in the Swedish Constitution (Bengtsson 2015, 65; Allard 2015, 50; see also Lantto and Mörkenstam 2008). Compared to Norway and Finland, the protection provided by the constitution to the Sámi in Sweden is thought to be the weakest (Allard 2015, 50–51). In Finland, the Sámi people have their rights protected by the Finnish constitution (1999, Ch 2, S. 17, para. 3; e.g., Heinämäki 2017, 22), and the Sámi homeland region is defined in the Act on the Sámi Parliament (1995, Ch. 1, S. 4). Many other laws, especially concerning the states' duty to consult Sámi on matters such as land-use issues, have a significant impact on the Sámi people's position on decision-making and actualizing their rights as indigenous people (see Heinämäki 2017; Allard 2017a, b). The legislation affecting the Sámi people's rights to continue their traditional livelihoods is important and closely connected to their culture (see Heinämäki 2010).

Recently states have also undertaken some legislative measures, which have been argued to hinder the Sámi people's possibilities to conduct their traditional livelihoods, which has provoked much discussion both in the Sámi and national media (see e.g., Tynkkynen 2017; Alajärvi 2017; Hirvasvuopio 2016; Länsman 2015; Wesslin 2016; see also Sámi Parliament in Finland 2016; Heinämäki et al. 2017). Many of these legislative measures have other cumulative affects on Sámi culture and society in general. Therefore, Sámi people's participation in negotiation processes and consultation on these matters should be ensured (Heinämäki et al. 2017). New Sámi activist groups have started to push for national implementation of their internationally recognized rights (e.g., Ellos Deatnu, n.d.; Suohpanterror 2018).

Although the Sámi people's position in national decision-making processes has improved during the last decades, certain issues remain, in which the voices of the Sámi are not sufficiently heard. Since indigenous peoples rights to participate are a crucial element in actualizing their right to self-determination (Heinämäki 2017, 28), the situations in Finland, Sweden, and Norway could be viewed critically. However, it is important to remember that the situation varies in each country and is not evaluated in this chapter.

17.6 Establishing a Position in International Decision-Making in the Arctic

The Nordic Sámi cooperation started in the 1950s, and the Nordic Sámi Conference was organized in Sweden in 1953. The theme of the conference was an activation of the Sámi people (Rantala 2004, 2–4). To be able to adapt to the new culture and society, the Sámi felt that they should be active actors (Rantala 2004, 2–4). In the context of Sámi politics the discussion on being active and on their need to adapt were thus visible already in the 1950's. Adaptability and active agency can be seen as key elements of being also resilient (Lindroth and Sinevaara-Nskanen 2018, 87), and Sámi have in many ways demonstrated their resilience throughout the history. However, the idea that indigenous peoples are often also in the international politics nowadays expected to be adaptive and resilient can be seen also as a modern way to govern indigenous lives (e.g. Lindroth and Sinevaara-Niskanen 2018). For example, when it comes to the living conditions of indigenous peoples, there is constantly a demand for indigenous peoples to adapt to changes in their living environment in the name of development (Lindroth and Sinevaara-Niskanen 2018).

In 1956, the Nordic Sámi Council was established (Rantala 2004, 2–4; see also Valkonen 2017, 180), followed by other Sámi organizations after that (e.g., Minde 2003, 80). The Nordic Sámi Council started to organize conferences every three years, in which issues concerning Sámi political matters were discussed (Minde 2003, 80; Rantala 2004, 2–4; Valkonen 2017, 180). Conferences were also a forum to launch statements and strengthen the idea of the Sámi as one people in the Nordic countries (Minde 2003, 81). In the 1970s, the Sámi people's international cooperation with other indigenous peoples around the world started to increase. In 1973, the Sámi organizations were invited to the Conference for the Arctic Peoples, organized in Copenhagen, and the representative of the Nordic Sámi Council was later invited to the preliminary meetings for the World Council of Indigenous Peoples (WCIP) (Minde 2003, 81–87).

Participation in the international indigenous peoples' cooperation had positive effects on the work of the Sámi organizations at the national and regional levels (Minde 2003, 81–87). Since the 1970s, when indigenous peoples' movements started to use the term *indigenous people*, it provided them means to raise and express their common voice and use it in their struggle for self-determination at the global and local levels (Smith 1999, 7). This is clearly reflected in Sámi politics in the Nordic countries.

Nowadays, the Nordic Sámi Council is known as the Saami Council, which is a voluntary NGO that has Sámi member organizations from all four countries in which the Sámi live. The main objectives of the Saami Council are to: Safeguard the interests of the Sámi as people; strengthen the solidarity among the Sámi as one people and indigenous people; work to ensure that the rights of the Sámi as indigenous people are recognized in the national legislations of the states; and promote internationally the rights of the Sámi and other indigenous people (Saami Council Charter 2008, S. 2, Goals).

The Saami Council is represented in international and regional institutions, and organizations such as the Barents Euro-Arctic Council Working Group of Indigenous Peoples (WGIP), the United Nations Economic and Social Council, Arctic Council Indigenous Peoples' Secretariat (IPS) and the Arctic Council.[3] The role of the Saami Council varies in each institution. In the UN Economic and Social Council, it has consultant status, which means that it has a right to participate in meetings in the UN system and have an impact on different issues in this council (Representation n.d.; Baer 2017, 3). Indigenous people currently have only NGOs participating in the UN's work with consultant status. Self-governmental institutions, such as the Sámi Parliaments' do not have this possibility, but they participate to the work of the UN through the Permanent Forum on Indigenous Issues (UNPFII) and the Expert Mechanism on the Rights of Indigenous Peoples (EMRIP) (Baer 2017; Heinämäki 2017, 28–29). The establishment of UNPFII should be celebrated as a great achievement on the path in improving indigenous peoples' status in international relations and decision-making, but its mandate to produce recommendations without having tools to implement or monitor them has also been criticized (Lindroth and Sinevaara-Niskanen 2018, 9).

The Saami Council has observer status in the WGIP, which has advisory status to the Barents Euro-Arctic Council and the Barents Regional Council. The goals for the indigenous peoples' cooperation in the region are to secure "*rights, foundation for trade, society, culture and language through implementation of the Action Plan of Indigenous Peoples of the BEAR*" (Action Plan for Indigenous Peoples in the Barents Euro-Arctic Region 2016–2018, 2017, 7). Therefore, the Saami Council also participates in policy-shaping in the regional cooperation in the Barents region.

Since this chapter emphasizes the Sámi people's position in decision-making in the Arctic, one of the most important organizations in this context is the Arctic Council, and the Saami Council's position in it. The Saami Council has been involved in the Arctic Council's work since it was established in 1996 by the Ottawa Declaration, in which the Saami Council, the Inuit Circumpolar Conference, and the Association of Indigenous Minorities of the North, Siberia and the Far East of the Russian Federation were given the status of permanent participants in the Arctic Council (Ottawa Declaration 1996, S. 2; see also Heinämäki 2010, 67). The Saami Council and two other aforementioned organizations were already involved in environmental cooperation in the Arctic before the Arctic Council was established. In the Declaration on the Protection of Arctic Environment and the Arctic Environmental Protection Strategy (AEPS), they were invited as observers to the Meetings on the Arctic Environment of the eight Arctic countries (AEPS 1991, 42; see also Heinämäki 2010, 67).

Although Arctic indigenous people were already involved with regional cooperation concerning the Arctic environment (e.g., Heinämäki 2010, 67), their position improved remarkably when they achieved permanent participant status in the Arctic Council (see also Koivurova and Heinämäki 2006, 104; see also e.g., Shadian 2017). As stated in the Ottawa Declaration: "*The category of Permanent Participation is*

[3]In addition, the Saami Council is also represented in the Lássságámmi Foundation and the Saami Parliamentary Council (Representation n.d.).

created to provide for active participation and full consultation with Arctic indigenous representatives within the Arctic Council" (Ottawa Declaration 1996, S. 2). In practice, this means that Arctic indigenous people negotiate and attend the same meetings with the eights Arctic states and can also submit statements and proposal concerning decisions to be made (Koivurova and Heinämäki 2006, 104; Arctic Council 1998, Part 2, S. 5; Heinämäki 2010, 67). The indigenous people should be fully consulted on decisions made in the Arctic Council, but final decisions are still made by the states in consensus (Koivurova and Heinämäki 2006, 104. See also Lindroth and Sinevaara-Niskanen 2018).

Despite the fact that indigenous representatives are in the Arctic Council in a policy-shaping role, their position is still relatively strong (e.g., Koivurova andand Heinämäki 2006). They can shape and have a real impact on the work of Arctic Council, which has a significant role in the Arctic region. Thus, the Arctic Council could provide an example for other regional and international forums on including indigenous peoples in decision-making processes concerning social, economic, and especially environmental issues, which are important for indigenous peoples and directly affect them in many ways (e.g., Heinämäki 2010; Koivurova and Heinämäki 2006; Lindroth and Sinevaara-Niskanen 2018, 8). The work of the Arctic Council is progressive in many ways, especially when it comes to including indigenous knowledge (Heinämäki 2010, 68; Lindroth and Sinevaara-Niskanen 2018, 9; Arctic Council 2017). Nevertheless, it is also important to recognize that the Arctic Council only makes non-binding decisions, which are seen as soft law instruments. The soft law has its value in developing international norms, but it has been also questioned if indigenous peoples' position in the Arctic Council would be as strong if the Council had the power to make legally binding decisions (Lindroth and Sinevaara-Niskanen 2018, 45–46; Koivurova and Heinämäki 2006).

However, indigenous peoples' status as permanent participants in the Arctic Council could also be argued to have an impact on their position in the Arctic cooperation in general from a normative perspective. As an example, five coastal Arctic states (A5) organized the Ilulissat Ministerial Meeting in 2008, concerning Arctic Ocean governance to ensure stability and strengthen the position of A5-states in the region (Heininen 2016, 25–26). Noncoastal states and representatives of the Arctic indigenous peoples in the Arctic Council were excluded from the meeting, which naturally provoked opposition amongst them (Keil and Knecht 2017, 9). Since the establishment of the Arctic Council, Arctic indigenous peoples had participated in decision-making processes concerning Arctic issues through the Council. The Inuit Circumpolar Council reacted to the Ilulissat Ministerial Meeting by issuing a Circumpolar Inuit Declaration on Sovereignty in the Arctic (2009), in which it appealed also to its status as a permanent participant in the Arctic Council.

17.7 Conclusion

In this chapter, I have outlined the general development of the position of the Sámi people in Norway, Sweden, and Finland. The Russian Sámi are included in the analysis in only a few places because the societal structures and legislation in the Russian Federation differ in many ways from the Nordic countries. In this context, comparison between the Sámi in the Nordic countries and in the Russian Federation would not serve my purpose. There are similarities in the outlines of the development of the position of indigenous peoples in many of the Arctic countries, but such development is unique in all of them. This chapter concentrates on the Sámi people and their representative bodies and organizations because they have been involved in international indigenous movements since their early stage and have established their position in national, regional, and international decision-making processes.

The position of the Sámi people in decision-making processes has changed from being the people who originally inhabited the northern parts of Norway, Sweden, Finland and Russia with their traditional Siida system and customary law, to the people who were divided by the establishment of state borders and had to adapt to the new social order of settlers (Lehtola 2015b; see also Pääkkönen 2008, 196–202). After starting to claim their rights as indigenous people, supported by international indigenous movements and actively participating in building their movement (e.g., Tennberg 2010), the Sámi have built their own political bodies representing them at the national level. They have also been involved in important international process in strengthening indigenous peoples' position in international politics and developing new forms to include their voice in decision-making processes also at the international level.

In Sámi political activism, starting at the latter half of the 20th century, local is very much interconnected with global. The development of the international discourse of indigenous peoples' rights reflects in many ways the development of the Sámi people's position in Nordic societies. However, there are still issues that have serious effects on the living conditions and culture of the Sámi, but their opinions are not taken into consideration. These issues are often connected to their rights to affect decisions-making processes concerning, for example, land use in their homeland region. These issues are not discussed comprehensively in this chapter, but they are still significant obstacles in implementing the rights of indigenous people at the national level (see in general Heinämäki et al. 2017). The new wave of Sámi activism has risen, and it seems to be again, in many ways, interconnected to strengthening indigenous peoples activism in other parts of the world (e.g., Standing with the Rock 2018). Whereas during previous waves of Sámi activism, they rose to claim the rights that belonged to them as indigenous people, now it is time for them to claim implementation of these rights at national levels. The means used in modern activism have changed and, thanks to the worldwide web, the message of indigenous people changes from local to global within an unprecedented amount of time.

References

About the Saami Council. http://www.saamicouncil.net/en/about-saami-council/

Act on the Sámi Parliament (1995) (974) (Fin.). (Unofficial translation, legally binding only in Finnish and Swedish. Ministry of Justice, Finland)

Act of the Sámi Parliament in Sweden (1992) (1433) (Swe.)

Action Plan for Indigenous Peoples in the Barents Euro-Arctic Region 2016–2018 (2017) Working Group of Indigenous Peoples in the Barents Euro-Arctic Region (WGIP). https://www.barentsinfo.fi/beac/docs/ActionPlan_2017-2018_ENG.pdf

AEPS (1991) Declaration on the Protection of Arctic Environment and the Arctic Environmental Protection Strategy

Åhren M (2016) Indigenous Peoples' status in the international legal system. Oxford University Press, Oxford

Alajärvi M (2017) Miksi Saamelaiset Vastustavat Uutta Tenon Sopimusta? Utsjoella Konsertoidaan Isosti Vastustuksen Tueksi. YLE. https://yle.fi/uutiset/3-9738772

Allard C (2015) Some characteristic features of Scandinavian laws and their influence on Sami matters. In: Allard C, Funderud Skogvang S (eds) Indigenous rights in Scandinavia: autonomous Sami Law, Ashgate Publishing Limited, Surrey, 49–64)

Allard C (2017a) Norway. In: Heinämäki L, Allard C, Kirchner S, et al (eds) Saamelaisten Oikeuksien Toteutuminen: Kansainvälinen Oikeusvertaileva Tutkimus. Publications of the government's analysis, assessment and research activities 4/2017. Prime Minister's Office, Rovaniemi, 311–340

Allard C (2017b) Sweden. In: Heinämäki L, Allard C, Kirchner S et al (eds) Saamelaisten Oikeuksien Toteutuminen: Kansainvälinen Oikeusvertaileva Tutkimus. Publications of the government's analysis, assessment and research activities 4/2017. Prime Minister's Office, Rovaniemi, 341–360

Allard C, Funderud Skogvang S (2015) Introduction. In: Allard C, Funderud Skogvang S (eds) Indigenous Rights in Scandinavia: Autonomous Sami Law, Ashgate Publishing Limited, Surrey

Arctic Council (1998) Arctic Council Rules of Procedures. SAO REPORT (Iqaluit): Annex 1

Arctic Council (2017) Agreement on enhancing international Arctic scientific cooperation

Baer L-A (2017) International Expert Group Meeting on the Theme Implementation of the United Nations Declaration on the Rights of Indigenous Peoples: The Role of the Permanent Forum on Indigenous Issues and other indigenous specific mechanisms (Article 42) (25–27 January 2017, New York) (PFII/2017/EGM). http://www.un.org/esa/socdev/unpfii/documents/2017/egm/LarsAndersBaer.pdf

Bengtsson B (2015) Reforming Swedish Sami legislation: A survey of the arguments. In: Allard C, Funderud Skogvang S (eds) Indigenous rights in Scandinavia: autonomous Sami law. Ashgate Publishing Limited, Surrey, 49–64

Berg-Nordlie M (2015) Two centuries of Russian Sámi policy: arrangements for autonomy and participation seen in light of imperial, Soviet and federal indigenous minority policy 1822–2014. Acta Borealia 32(1):40–67. https://doi.org/10.1080/08003831.2015.1030849

Circumpolar Inuit Declaration on Sovereignty in the Arctic (2009) Inuit Circumpolar Council

Cobo JM (1981–1983) Study of the problem of discrimination against indigenous populations: final report submitted by the special Rapporteur, Mr. José Martínez Cobo. (30 July 1981E/CN.4/Sub.2/476-5 August 1983E/CN.4/Sub.2/1983/21). https://www.un.org/development/desa/indigenouspeoples/publications/2014/09/martinez-cobo-study/

Concluding Observations of the Human Rights Committee on Canada (1999) (UN Doc. CCPR/C/79/Add. 105)

Concluding observations of the human rights committee on Norway (1999) (UN Doc. CCPR/C/79/Add. 112)

Deputy Chancellor of Justice (2017) Päätös: Tenojoen kalastussopimuksen valmistelu (Dnro. OKV/12/21/2016). https://www.okv.fi/media/filer_public/a0/56/a056167f-2a5a-43d7-a7c8-9883a8c18c40/okv_12_21_2016.pdf

Den Nye Sameretten (2007)

Ellos Deatnu (n.d.) https://ellosdeatnu.com/

Expert Mechanism on the Rights of Indigenous Peoples (2011) Final study on indigenous peoples and the right to participate in decision-making. Report of the Expert Mechanism (A/HRC/EMRIP/2011/2)

Finnmark Act (2005) (85)

Government Bill (1976) Om Insatser för Samerna (Prop. 1976/77:80) (Swe)

Government Bill (2009) En Reformerad Grundlag (Prop. 2009/10:80) (Swe)

Heininen L (2002) The Saami as a pan-national actor. In: Karppi K, Eriksson J (eds) Conflict and cooperation in the north. Kulturgräns norr, Kulturens Frontlinjer. Umeå, pp 223–238

Heininen L (2016) Security of the global Arctic in transformation—changes in problem definition of security. In: Heininen L (ed) Future security of the global Arctic. State policy, economic security and climate. Palgrave Macmillan, Palgrave Pivot, Basingstoke, 12–34

Heinämäki L (2010) The right to be part of nature: indigenous peoples and the environment. Acta Universitatis Lapponiensis 180. Lapland University Press, Rovaniemi

Heinämäki L (2017) Saamelaisten oikeudet Suomessa. In: Heinämäki L, Allard C, Kirchner S et al (eds) Saamelaisten Oikeuksien Toteutuminen: Kansainvälinen Oikeusvertaileva Tutkimus. Publications of the government's analysis, assessment and research activities 4/2017. Prime Minister's Office, Rovaniemi, 22–85

Heinämäki L, Allard C, Kirchner S et al (2017) Saamelaisten Oikeuksien Toteutuminen: Kansainvälinen Oikeusvertaileva Tutkimus. Publications of the government's analysis, assessment and research activities 4/2017. Prime Minister's Office, Rovaniemi, 341–360

Heinämäki L, Kirchner S (2017) Assessment on recent developments regarding indigenous peoples' legal status and rights in international law: With special focus on free, prior and informed consent. In: Heinämäki L, Allard C, Kirchner S et al (eds) Saamelaisten Oikeuksien Toteutuminen: Kansainvälinen Oikeusvertaileva Tutkimus. Publications of the government's analysis, assessment and research activities 4/2017. Rovaniemi, Prime Minister's Office, 224–282

Henriksen JB (2009) Sami self-determination: autonomy and self-government: education and culture. Gáldu Čála, J Ind Rights, (2). http://galdu.custompublish.com/getfile.php/3131423.2388.xbarcaypvc/2_2009_sami_selfdetermination_eng_.pdf

Hirvasvuopio S (2016) Saamelaisneuvoston Presidentti Metsähallituslaista: Suomessa Trendinä, ettei Saamelaisten Oikeuksia Oteta Huomioon. YLE. https://yle.fi/uutiset/3-8740167

ILO Convention 169 (1989) Convention Concerning Indigenous and Tribal Peoples in Independent Countries

Instrument of Government with Amendments (2010) (Swe)

International Covenant on Civil and Political Rights (1966)

International Covenant on Economic, Social and Cultural Rights (1966)

International Work Group for Indigenous Affairs (2017) The Indigenous World 2017. https://www.iwgia.org/images/documents/indigenous-world/indigenous-world-2017.pdf

Josefsen E, Mörkenstam U, Saglie J (2015) Different institutions within similar states: The Norwegian and Swedish Sámediggis. Ethnopolitics 14(1):32–51. https://doi.org/10.1080/17449057.2014.926611

Keil K, Knecht S (2017) Introduction: The Arctic as a globally embedded space. In: Keil K, Knecht S (eds) Governing Arctic Change: Global Perspectives. Palgrave MacMillan, 1–18 https://doi.org/10.1057/978-1-137-50884-3

Koivurova T, Heinämäki L (2006) The participation of indigenous peoples in international norm-making in the Arctic. Polar Record 42(221):101–109. https://doi.org/10.1017/S0032247406005080

Lantto P (2010) Borders, citizenship and change: The case of the Sami people, 1751–2008. Citizsh Stud 14(5):543–556. https://doi.org/10.1080/13621025.2010.506709

Lantto P, Mörkenstam U (2008) Sami rights and Sami challenges. Scandinavian J History 33(1):26–51. https://doi.org/10.1080/03468750701431222

Lehtola V-P (2005) Saamelaisten Parlamentti. Saamelaiskäräjät 2005. Jyväskylä: Gummerus Kirjapaino Oy

Lehtola V-P (2012) Saamelaiset Suomalaiset: Kohtaamisia 1896–1953. Suomalaisen Kirjallisuuden Seura, Helsinki

Lehtola V-P (2015a) Saamelaiset: Historia, yhteiskunta, taide. Puntsi, Inari

Lehtola V-P (2015b) Sámi histories, colonialism, and Finland. Arctic Anthropology 52(2):22–36. https://doi.org/10.3368/aa.52.2.22

Lindroth M, Sinevaara-Niskanen H (2018) Global politics and its violent care for indigeneity: Sequels to colonialism. Palgrave Macmillan, Cham, Switzerland

Länsman K (2015) Kiistelty Metsähallituslaki Etenee – "Ei Tässä Saamelaisten Asemaa Heikennetä." YLE. https://yle.fi/uutiset/3-8538502

Minde H (2003) The Challenges of Indigenism: The Struggle for Sami Land Rights and Self-Government in Norway 1960–1990. In: Jentoft S, Minde H, Nilsen R (eds) Indigenous peoples: resource management and global rights. Eburon Academic Publishers, Delft, The Netherlands, pp 75–106

Minde H (2005) Assimilation of the Sami—implementation and consequences. Gáldu Čála—J Ind Peoples Rights (3)

Minde H (2008) The destination and the journey: indigenous peoples and the United Nations from the 1960s through 1985. In: Minde H (ed) Indigenous peoples: self-determination, knowledge, identity. Delft, Eburon Academis Publishers, The Netherlands, 49–86

Moretti M (2012) International law and nomadic people. Author House, Central Milton Keynes, UK

Mörkenstam U (1999) Om "Lapparnens Privilegier". Föreställningar om Samiskhet i Offentlig Svensk Samepolitk 1883–1997. Stockholm Studies in Politics, Stockholm, 67

Mörkenstam U, Josefsen E, Nilsson R (2016) The Nordic Sámidiggis and the limits of indigenous self-determination. Gáldu Čálá, J Ind Peoples Rights, (1)

Mörkenstam U (2017) Some remarks on the criteria to register in the Sámi electoral roll in Sweden. In: Heinämäki L, Allard C, Kirchner S et al (eds) Saamelaisten Oikeuksien Toteutuminen: Kansainvälinen Oikeusvertaileva Tutkimus. Publications of the government's analysis, assessment and research activities 4/2017. Prime Minister's Office, Rovaniemi, 361–366

Niezen R (2009) The rediscovered self: indigenous identity and cultural justice. McGill-Queen's University Press, Montreal

Olsén L (2016) Sámi people in decision-making processes: Analyzing Sámi people's possibilities to participate and influence on decision-making processes. Presentation at the Arctic Frontiers. http://www.arcticfrontiers.com/wp-content/uploads/downloads/2016/Arctic%20Frontiers%20Science/Presentations%2028%20January%202016/Part%20II%20Arctic%20stewardship/13_Laura_Olsen.pdf

Ottawa Declaration (1996) Declaration on the Establishment of the Arctic Council: Joint Communique of the Governments of the Arctic Countries on the Establishment of the Arctic Council

Pedersen S (2012) The coastal Sámi of Norway and their rights to traditional marine livelihood. Arctic Rev Law Polit 3(1):51–80. https://arcticreview.no/index.php/arctic/article/view/29

Pääkkönen E (2008) Saamelainen etnisyys ja pohjoinen paikallisuus- Saamelaisten etninen mobilisaatio ja paikallisperustainen vastaliike. Acta Universitatis Lapponiensis 153. Lapin yliopistokustannus, Rovaniemi

Rantala L (2004) Historiikki: Saamelaisneuvosto 50 vuotta. 18. Saamelaiskonferenssi 7.-9.10.2004, Honningsvåg

Ravna Ø (2011) The process of identifying land rights in parts of Northern Norway: Does the Finnmark Act prescribe an adequate procedure within the national law? In: Gudmundur A, Koivurova T, Hossain K (eds) The Yearbook of Polar Law vol 3, 423–453, Brill. https://doi.org/10.1163/22116427-91000066

Reindeer-Herding Act (1971) (437) (Swe.)

Report of the Committee for Education and Culture (1976) Med Anledning av Propositionen 1976/77:80 om Insatser för Samerna Jämte Motioner (Kru 1976/77:43) (Swe)

Representation (n.d.) http://www.saamicouncil.net/en/about-saami-council/representations/

Rohr J (2014) Indigenous Peoples in the Russian Federation. IWIGA Report 18. http://www.iwgia.org/iwgia_files_publications_files/0695_HumanRights_report_18_Russia.pdf

Rombouts SJ (2014) Having a say—indigenous peoples, international law and free, prior and informed consent. Wolf, Oisterwijk

Saami Council Charter (2008) *Saami Council*

Sámi Act in Norway (1987) (56)

Sámi Parliament in Finland (2016) Lausunto: Suomen ja Norjan välisen Kalastuksesta Tenojoen Vesistössä Tehdyn Sopimuksen Allekirjoittaminen (Dnro 379/D.a.2/2016). http://www2.samediggi.fi/index2.php?option=com_docman&task=doc_view&gid=2900&Itemid=10

Sámi Parliamentary Council (n.d.) http://www.samediggi.fi/sami-parliamentary-council/?lang=en

Seurujärvi-Kari I (2012) Ale Jaskkot Eatnigiella: Alkuperäiskansaliikkeen ja Saamen Kielen Merkitys Saamelaisten Identiteetille. Helsingin yliopisto, Helsinki

Shadian JM (2017) Remaining political space: the limits of Arctic indigenous self-determination in international governance. In: Keil K, Knecht S (eds) governing Arctic change: global perspectives. Palgrave MacMillan, 43–58 https://doi.org/10.1057/978-1-137-50884-3

Skolt Act (1995) (253) (Fin.)

Smith LT (1999) Decolonizing methodologies: research and indigenous peoples. Zed Books Ltd, London

Suohpanterror (2018) https://suohpanterror.com/

Standing with the Rock (2018) http://standwithstandingrock.net/

Tennberg M (2010) Indigenous peoples as international political actors: a summary. Polar Record 46(238):264–270. https://doi.org/10.1017/S0032247409990398

Tynkkynen J (2017) Perustuslakivaliokunta: Saamelaisten Oikeudet Turvattava Tenolla Paremmin. YLE. https://yle.fi/uutiset/3-9476277

UN (2011) Report of the Special Rapporteur on the situation of human rights and fundamental freedoms of indigenous people, James Anaya: The situation of the Sami in the Sápmi region of Norway, Sweden and Finland (A/HRC/18/XX/Add.Y)

UNDRIP (2007) United nations declaration on the rights of indigenous peoples (A/RES/61/295)

Valkonen S (2017) Kuvaus saamelaisten itsehallinnon kehittymisestä sekä saamelaisuuden määrittelyn ja saamelais- ja alkuperäiskansakeskustelun taustoista ja tapahtumista Suomessa. In: Heinämäki L, Allard C, Kirchner S et al (eds) Saamelaisten Oikeuksien Toteutuminen: Kansainvälinen Oikeusvertaileva Tutkimus. Publications of the government's analysis, assessment and research activities 4/2017. Prime Minister's Office, Rovaniemi, 176–223

Ward T (2011) The right of free, prior, and informed consent: Indigenous Peoples' Participation Rights within international law. Northwest J Int Human Rights 10(2):54–84. https://scholarlycommons.law.northwestern.edu/njihr/vol10/iss2/2

Wesslin S (2016) Uusi Kalastuslaki Herättää Tunteita Ylä-Lapissa: Lohestus on Osa Paikallista Kulttuuria. YLE. https://yle.fi/uutiset/osasto/sapmi/uusi_kalastuslaki_herattaa_tunteita_yla-lapissa_lohestus_on_osa_paikallista_kulttuuria/8934425

Xanthaki A (2009) Indigenous rights in international law over the last 10 years and future developments. Melb J Int Law 10(1):27–37

YLE (2017) Sámi activists occupy island in protest at Tenojoki fishing rules. https://yle.fi/uutiset/osasto/news/sami_activists_occupy_island_in_protest_at_tenojoki_fishing_rules/9717663

The manufacturer's authorised representative in the EU is Springer Nature Customer Service Centre GmbH, Europaplatz 3, 69115 Heidelberg, Germany. If you have any concerns regarding our products, please contact ProductSafety@springernature.com

Printed and bound by CPI Group (UK) Ltd, Croydon, CR0 4YY
30/09/2026
02225658-0009